Welcome to

McGraw-Hill Education SAT Subject Test: Physics

How to Use This Book

The SAT Physics test covers a very large amount of material, and your preparation time may be short. That is why it is important to use your study time wisely. This book provides a comprehensive review of everything you need to know for the test, and it has been organized to make your study program practical and efficient. It will help you:

- Identify the physics topics that you most need to focus on
- Familiarize yourself with the test format and test question types
- Review all the basic physics you need to know for the test
- Check your progress with questions at the end of each review chapter
- Practice your test-taking skills using sample tests

The following four-step study program has been designed to help you make the best use of this book.

1 Take the Diagnostic Test

Once you have read through this chapter, start your preparation program by taking the Diagnostic Test. This test is carefully modeled on the actual SAT Physics test in terms of format, types of questions, and topics tested. Take the Diagnostic Test under test conditions and pay careful attention to the one-hour time limit. When you complete the test, score yourself using the scoring information at the end of the test. Then read through the explanations to see which test topics gave you the most trouble. Look for patterns. Did you miss questions in one or two specific subject areas?

Did specific question formats give you trouble? When did you need to guess at the answer? Use your results to identify the topics and question types that were most difficult for you. Once you know your physics strengths and weaknesses, you'll know which subjects you need to focus on as you prepare for the exam.

2 Review the Test Topics

This book provides a full-scale review of all the topics tested on the SAT Physics test. Once you have identified the topics that give you the most trouble, review the relevant chapters. You do not need to work through the review chapters in the order in which they appear. Skip around if you like, but remember to focus on the topics that gave you the most trouble on the Diagnostic Test.

Each review chapter ends with practice problems that you can use to see how well you have mastered the material. If you get a problem wrong, go back into the chapter and reread the section that covers that particular topic.

Make a study schedule. If you have the time, plan to spend at least two weeks working your way through the review chapters. Be sure to set aside enough time at the end of your schedule to take the practice tests at the end of the book. However, if you do not have much time before the test, you will need to shorten your review time and focus instead entirely on the practice tests.

3 Build Your Test-Taking Skills

As you work through the examples and review questions in each review chapter, you'll become familiar with the kinds of questions that appear on the SAT Physics test. You'll also practice the test-taking skills essential for top scores. These include:

- The ability to recall and comprehend major concepts in physics and to apply them to solve problems
- The ability to interpret information gained from observations and experiments
- The ability to make inferences from experimental data, including data presented in graphs and tables

4 Take the Practice Tests

Once you have completed your review of all the SAT Physics topics, get ready for the real exam by taking the two practice tests at the back of this book. When you take each test, try to simulate actual test conditions. Sit in a quiet room, time yourself, and work through as much of the test as time allows. The tests are ideal for practice because they have been constructed to reflect the actual exam as closely as possible. The directions and practice questions are very much like those on the real test. You'll gain experience with the test format, and you'll learn to pace yourself so that you can earn the maximum number of points in the time allowed.

Each test will also serve as a review of the topics tested because complete explanations are provided for every question. The explanations can be found at the end of each test. If you get a question wrong, you'll want to review the explanation carefully. You may also want to go back to the chapter in this book that covers the question topic.

At the end of each test you'll also find scoring information. Calculate your raw score, then use the table provided to find your approximate scaled score. The scaling on the real test may be slightly different, but you'll get a good idea of how you might score on the actual test.

Top 15 Things You Need to Know About Physics

This feature provides a list of 15 topics that are essential to understanding the physics tested on the SAT Subject Test in Physics. These topics may be tested directly or indirectly as the basis for more specific questions.

1 Newton's First Law of Motion

Three fundamental principles, called Newton's First, Second, and Third Laws, form the basis of classical, or Newtonian, mechanics and have proved valid for all mechanical problems not involving speeds comparable with the speed of light and not involving atomic or subatomic particles. Newton's First Law states that an object not subjected to external forces remains at rest or moves with constant speed in a straight line. This is also known as the Law of Inertia. See Chapter 4.

2 Newton's Second Law of Motion

The acceleration of the object is directly proportional to the resultant external force acting on the object and is inversely proportional to the mass of the object. See Chapter 4.

3 Newton's Third Law of Motion

If two objects interact, the force exerted by the first object on the second object (called the *action force*) is equal in magnitude and opposite in direction to the force exerted by the second object on the first object (called the *reaction force*). See Chapter 4.

4 Newton's Law of Universal Gravitation

Any two bodies in the universe attract each other with a force that is directly proportional to the product of their masses and inversely proportional to the square of the distance between them. This law has been superseded by Einstein's Theory of General Relativity, but it continues to be used as an approximation of the effects of gravity in most applications. See Chapter 9.

5 Einstein's Special Theory of Relativity

Einstein's theory about the relationship between space and time is based on two postulates: (1) that the laws of physics are unchanging in all inertial systems and (2) that the speed of light in a vacuum is the same for all observers, regardless of the motion of the light source. See Chapter 19.

6 Principles of Kinematics

Know how to calculate the speed, velocity, and acceleration of an object. See Chapter 3.

7 First Law of Thermodynamics

The First Law of Thermodynamics (the Law of Conservation of Energy) states that energy cannot be created or destroyed; therefore the change in the internal energy of a system is equal to the sum of the heat added to the system and the work done on the system minus the heat loss of the system and the work done by the system. See Chapter 11.

8 Second Law of Thermodynamics

The Second Law of Thermodynamics states that entropy cannot decrease and heat cannot be transferred from a colder to a hotter body within a system without net changes occurring in other bodies within that system. See Chapter 11.

9 Properties of Waves

Know how to calculate the amplitude, wavelength, and frequency of waves, including electromagnetic waves. See Chapters 15 and 16.

10 Law of Conservation of Energy

This law describes the principle that energy cannot be created nor destroyed, although it can be changed from one form to another. After Einstein announced the $E = mc^2$ equation, this law was developed to combine the Law of Conservation of Mass formulated by Antoine Lavoisier in 1785 with the Second Law of Thermodynamics (the Law of Conservation of Energy). See Chapter 5.

11 Law of Conservation of Momentum

The Law of Conservation of Momentum states that the momentum will remain constant no matter what until and unless any external force comes into action. For a collision occurring between object X and object Y in an isolated system, the momentum lost by object X is equal to the momentum gained by object Y. See Chapter 7.

12 Coulomb's Law

Coulomb's Law describes the electrostatic interaction between electrically charged particles. The law states that the force between two point charges acts in the direction of the line between them and is directly proportional to the product of their electric charges divided by the square of the distance between them. The law was essential to the development of the theory of electromagnetism and is analogous to Isaac Newton's inverse-square Law of Universal Gravitation. Coulomb's Law can be used to derive Gauss's Law and vice versa. See Chapter 12.

13 Principles of Fluid Mechanics

Pascal's Principle states that pressure applied to a confined fluid is transferred equally throughout the fluid. Archimedes' Principle states that the magnitude of a buoyant force is equal to the weight of the fluid displaced. Bernoulli's Principle states that the pressure of a fluid decreases as the velocity of the fluid increases. See Chapter 6.

14 Ohm's Law

Ohm's Law defines the relationships between power, voltage, current, and resistance. The law states that the current through a conductor between two points is directly proportional to the potential difference across the two points. See Chapter 13.

15 Quantum Mechanics

Quantum mechanics is the body of scientific principles that explains the behavior of matter and its interactions with energy on the scale of atoms and subatomic particles. It attempts to describe and account for the properties of molecules and atoms and their constituents. These properties include the interactions of the particles with one another and with electromagnetic radiation. See Chapter 18.

General Test-Taking Strategies

Relax. If you studied and practiced, you will know what to expect on test day. If you start to get stressed, allow yourself a moment to relax. Put your pencil down. Close your eyes. Take a couple of deep breaths. Then return to the test with a calm mind.

Read Carefully. Be sure that you know what the question is asking. Some answer choices may be partial answers or values of variables other than the one you are asked about.

Keep Moving. You cannot let one question eat up a huge amount of your time. Mark the question and return to it at the end of the test if you have time remaining. Don't think about the pesky question while you are working the next question either. Move on, and let it go!

Guessing. The SAT Subject Tests do have a small penalty for wrong answers, so random guessing may not help you. You get one point for each correct answer and lose a fraction of a point for each incorrect answer. Leaving a question blank neither gains nor costs you anything. Educated guessing, however, can work in your favor.

Use Process of Elimination. If you know that one answer is wrong, cross it off. The more you can do this, the better will be your odds of choosing the correct answer.

Estimate. When you are making an educated guess on a calculation, be sure that you eliminate any choices that are too big or too small to be the value you are looking for.

Use Your Test Booklet. Your test booklet is your scratch paper. Use it as much as you need. Write out your calculations, cross off answer choices you know are wrong, circle questions you want to revisit, underline important points, and make notes when needed.

Manage Your Answer Sheet. Choose only one answer for each question, and make sure that your response completely fills the appropriate bubble on the answer sheet. You may want to work one or two pages in your test booklet and then bubble in the answers for those pages all at once. Be sure that you bubble in your answers on the correct question numbers. If you skip a question, it is easy to get your numbering off on the answer sheet, so double-check before marking an answer. Do not make any stray marks on the answer sheet. These can invalidate your response to a question. Completely erase any answers that you change.

Keep Track of Time. Wear a watch (without an alarm) to the test so that you can track your time. Your test is only one hour long, and it will go by quickly. When you are down to the last five minutes, make sure that your answer sheet is in order and that any questions you want to answer are answered on it.

Eight-Week Study Plan*

Week 1 (8 weeks before your test date):

- Register for your test.
- Review the front pages of this book, including this study plan.
- Read *Part I*, up to the Diagnostic Exam.
- Take the *Diagnostic Exam.*
- Schedule your study time on your calendar so that you do not let other things push it aside.
- Read over *Appendix A: Mathematics Review* to be sure that you are up to speed on the math you will need for physics questions.
- Read *Chapter 1: Measurements and Data Displays.*

Week 2 (7 weeks before your test date):

- Read and work through *Chapter 2: Vectors.*
- Read and work through *Chapter 3: Kinematics.*
- Read and work through *Chapter 4: Dynamics.*
- Read over the terms listed in the *Glossary* at the back of this book. Make flash cards for any terms with which you are not completely familiar.

Week 3 (6 weeks before your test date):

- Read and work through *Chapter 5: Work, Energy, and Power.*
- Read and work through *Chapter 6: Fluid Mechanics.*
- Read and work through *Chapter 7: Linear Momentum.*
- Make flash cards for the formulas in *Appendix B*, and make sure that you know what each formula means.

Week 4 (5 weeks before your test date):

- Read and work through *Chapter 8: Circular and Rotational Motion.*
- Read and work through *Chapter 9: Gravity.*
- Read and work through *Chapter 10: Heat and Temperature.*
- Read and work through *Chapter 11: Thermodynamics.*

Week 5 (4 weeks before your test date):

- Read and work through *Chapter 12: Static Electricity.*
- Read and work through *Chapter 13: Current and Circuits.*
- Read and work through *Chapter 14: Magnetism.*
- Review all your flash cards.

Week 6 (3 weeks before your test date):

- Read and work through *Chapter 15: Waves.*
- Read and work through *Chapter 16: Light.*
- Read and work through *Chapter 17: Optics.*
- Review all your flash cards, and mark any that you do not know well.

Week 7 (2 weeks before your test date):

- Read and work through *Chapter 18: Atomic Physics.*
- Read and work through *Chapter 19: Special Relativity.*
- Take *Practice Test 1*, and review it thoroughly.
- Review any of the Topic Review chapters that cover topics on which you missed questions on the practice test.

Week 8 (the week leading up to your test):

- Take *Practice Test 2*, and review it thoroughly. Do some extra work on any topic areas with which you still struggle.
- Find your testing center, and be sure that you know how to get there and how long it will take you on the morning of the test.

The day before your test:

- Organize your ID and supplies. Gather your Admission Ticket, your photo ID, your pencils (two No. 2 pencils with erasers), a watch (without any beeping alarm), a snack for any breaks you have between multiple subject tests, and a sweater or light jacket.
- Do *not* try to study the night before the test. Cramming does not work well. You have studied already, and your brain needs a break. Do something relaxing, such as watching a movie or playing a game.
- Get plenty of sleep.

Test day:

- Get up early.
- Flip through the book to warm up your brain. Do not redo practice questions; just read a bit about each topic.
- Eat breakfast, but nothing too heavy that will make you sluggish.
- Double-check your supplies.
- Arrive early at the testing center.

**If you have fewer than eight weeks to study, you can modify this plan to speed things up.*

McGraw-Hill Education

SAT* SUBJECT TEST PHYSICS

Third Edition

CHRISTINE CAPUTO

New York Chicago San Francisco Athens London Madrid
Mexico City Milan New Delhi Singapore Sydney Toronto

1 2 3 4 5 6 7 8 9 LHS 23 22 21 20 19 18

ISBN 978-1-260-13538-1
MHID 1-260-13538-1

e-ISBN 978-1-260-13539-8
e-MHID 1-260-13539-X

Special Contributor: Wendy Hanks

Contents

PART I **Introduction**

All About the SAT Subject Test in Physics 3

Strategies for Top Scores 9

PART II **Diagnostic Test**

Diagnostic Test 15

Answer Sheet 15

Answer Key 36

Answers and Explanations 37

PART III **Physics Topic Review**

CHAPTER 1 **Measurements and Data Displays** 45

CHAPTER 2 **Vectors** 50

Vectors and Scalars 50

Review Questions 55

Answers and Explanations 58

CHAPTER 3 **Kinematics** 59

Review Questions 67

Answers and Explanations 71

CHAPTER 4 **Dynamics** 73

Review Questions 79

Answers and Explanations 83

CHAPTER 5 **Work, Energy, and Power** 84

What Is Work? 84

Calculating Work 84

Review Questions 89
Answers and Explanations 93

CHAPTER 6 **Fluid Mechanics** 94
Review Questions 100
Answers and Explanations 104

CHAPTER 7 **Linear Momentum** 105
Review Questions 109
Answers and Explanations 112

CHAPTER 8 **Circular and Rotational Motion** 114
Review Questions 119
Answers and Explanations 123

CHAPTER 9 **Gravity** 124
Review Questions 128
Answers and Explanations 131

CHAPTER 10 **Heat and Temperature** 133
Review Questions 138
Answers and Explanations 141

CHAPTER 11 **Thermodynamics** 142
Review Questions 147
Answers and Explanations 149

CHAPTER 12 **Static Electricity** 150
Review Questions 154
Answers and Explanations 157

CHAPTER 13 **Current and Circuits** 159
Review Questions 165
Answers and Explanations 168

CHAPTER 14 **Magnetism** 169
Review Questions 179
Answers and Explanations 182

CHAPTER 15 **Waves** 184
Review Questions 191
Answers and Explanations 194

CHAPTER 16 **Light** 195
Review Questions 198
Answers and Explanations 201

CHAPTER 17 **Optics** 202
Review Questions 210
Answers and Explanations 213

CHAPTER 18 **Atomic Physics** 215
Review Questions 224
Answers and Explanations 227

CHAPTER 19 **Special Relativity** 228
Review Questions 231
Answers and Explanations 233

PART IV **Practice Tests**

Practice Test 1 237
Answer Sheet 237
Answer Key 251
Answers and Explanations 252

Practice Test 2 257
Answer Sheet 257
Answer Key 271
Answers and Explanations 272

Appendixes

APPENDIX A **Mathematics Review** 279

APPENDIX B **Summary of Important Formulas** 282

APPENDIX C **Values of Trigonometric Functions** 285

APPENDIX D **International Atomic Masses** 287

APPENDIX E **Score Sheet** 290

Glossary 291

PART I

Introduction

All About the SAT Subject Test in Physics

The SAT Subject Tests

What Are the SAT Subject Tests?

The SAT Subject Tests (formerly called the SAT II tests and the Achievement Tests) are a series of college entrance tests that cover specific academic subject areas. Like the better-known SAT test, which measures general verbal and math skills, the SAT Subject Tests are given by the College Entrance Examination Board. Colleges and universities often require applicants to take one or more SAT Subject Tests along with the SAT. SAT Subject Tests are generally not as difficult as Advanced Placement tests, but they may cover more than is taught in basic high school courses. Students usually take an SAT Subject Test after completing an Advanced Placement course or an Honors course in the subject area.

How Do I Know If I Need to Take SAT Subject Tests?

Review the admissions requirements of the colleges to which you plan to apply. Each college will have its own requirements. Many colleges require that you take a minimum number of SAT Subject Tests—usually one or two. Some require that you take tests in specific subjects. Some may not require SAT Subject Tests at all.

When Are SAT Subject Tests Given, and How Do I Register for Them?

SAT Subject Tests are usually given on six weekend dates spread throughout the academic year. These dates are usually the same ones on which the SAT is given. To find out the test dates, visit the College Board Web site at www.collegeboard.org. You can also register for a test at the Web site. Click on the tabs marked "students" and follow the directions you are given. You will need to use a credit card if you register online. As an alternative, you can register for SAT Subject Tests by mail using the registration form in the *Student Registration Guide for the SAT and SAT Subject Tests*, which should be available from your high school guidance counselor.

How Many SAT Subject Tests Should I Take?

You can take as many SAT Subject Tests as you wish. According to the College Board, more than one-half of all SAT Subject Test takers take three tests, and about one-quarter take four or more tests. Keep in mind, though, that you can take only three tests on a single day. If you want to take more than three tests, you'll need to take the others on a different testing date. When deciding how many SAT Subject Tests to take, base your decision on the requirements of the colleges to which you plan to apply. It is probably not a good idea to take many more SAT Subject Tests than you need. You will probably do better by focusing only on the ones that your preferred colleges require.

Which SAT Subject Tests Should I Take?

If a college to which you are applying requires one or more specific SAT Subject Tests, then of course you must take those particular tests. If the college simply requires that you take a minimum number of SAT Subject Tests, then choose the test or tests for which you think you are best prepared and likely to get the best score. If you have taken an Advanced Placement course or an Honors course in a particular subject and done well in that course, then you should probably consider taking an SAT Subject Test in that subject.

When Should I Take SAT Subject Tests?

Timing is important. It is a good idea to take an SAT Subject Test as soon as possible after completing a course in the test subject, while the course material is still fresh in your mind. If you plan to take an SAT Subject Test in a subject that you have not studied recently, make sure to leave yourself enough time to review the course material before taking the test.

What Do I Need on the Day of the Test?

To take an SAT Subject Test, you will need an admission ticket to enter the exam room and acceptable forms of photo identification. You will also need two number 2 pencils. Be sure that the erasers work well at erasing without leaving smudge marks. The tests are scored by machine, and scoring can be inaccurate if there are smudges or other stray marks on the answer sheet. Any devices that can make noise, such as cell phones or wristwatch alarms, should be turned off during the test. Study aids such as dictionaries and review books, as well as food and beverages, are barred from the test room.

The SAT Physics Test

The SAT Physics test is a one-hour exam consisting of 75 multiple-choice questions. According to the College Board, the test measures the following knowledge and skills:

- Ability to recall and understand important physics concepts and to apply those concepts to solve physics problems
- Knowledge of simple algebraic, trigonometric, and graphical relationships and principles of ratio and proportion, and ability to apply those principles to solve physics problems
- Knowledge of the metric system of units

According to the College Board, the questions on the test are distributed by topic in approximately the following percentages:

SAT Physics Test Topics

TOPIC	APPROXIMATE PERCENTAGE OF TEST
Mechanics	36-42
Electricity and magnetism	18-24
Waves	15-19
Heat, kinetic theory, and thermodynamics	6-11
Modern physics	6-11
Miscellaneous (measurement, math skills, laboratory skills, history of physics, etc.)	4-9

The College Board advises that because high school physics courses can vary, you are likely to encounter questions on topics that are unfamiliar.

About one-quarter to one-third of the test questions will require you to recall and understand concepts and information. About one-half the questions will require you to apply a single physics concept. The remaining one-quarter of the questions will require you to recall and relate more than one physics concept.

What School Background Do I Need for the SAT Physics Test?

The College Board recommends that you have the following before taking the SAT Physics test:

- A one-year college prep course in physics
- Algebra and trigonometry courses
- Physics laboratory experience

How Is the SAT Physics Test Scored?

On the SAT Physics test, your "raw score" is calculated as follows: you receive one point for each question you answer correctly, but you lose one-quarter of a point for each question you answer incorrectly. You do not gain or lose any points for questions that you do not answer at all. Your raw score is then converted into a scaled score by a statistical method that takes into account how well you did compared to others who took the same test. Scaled scores range from 200 to 800 points. Your scaled score will be reported to you, your high school, and to the colleges and universities you designate to receive it. Scoring scales differ slightly from one version of the test to the next. The scoring scale provided in Appendix E in this book is only a sample that will show you your approximate scaled score.

When Will I Receive My Score?

Scores are mailed to students approximately three to four weeks after the test. If you want to find out your score a week or so earlier, you can do so for free by accessing the College Board Web site at www.collegeboard.org.

How Do I Submit My Score to Colleges and Universities?

When you register to take the SAT or SAT Subject Tests, your fee includes free reporting of your scores to up to four colleges and universities. To have your scores reported to additional schools, visit the College Board Web site or call (866)756-7346. You will need to pay an additional fee.

SAT Physics Question Types

Part A of the SAT Physics Test Consists of Classification Questions.

Each set of classification questions includes five lettered choices that are used to answer all questions in the set. The choices may consist of words, equations, graphs, sentences, diagrams, or data that are generally related to the same topic. Each question in the set must be evaluated individually. Any choice may be the correct answer to more than one question in the set.

Example:

Directions: Each set of lettered choices refers to the numbered questions of statements immediately following it. Select the one lettered choice that best answers each question or best fits each statement, and then fill in the corresponding oval on the answer sheet. A choice may be used once, more than once, or not at all in each set.

Questions 9 and 10 relate to the following.

(A) period
(B) wavelength
(C) kinetic energy
(D) frequency
(E) amplitude

9. Which quantity is maximized when the displacement of a mass on a pendulum from its equilibrium position is zero?

10. Which quantity is measured in hertz?

To answer question 9, you need to know about the motion of a pendulum. The displacement of the mass from the equilibrium position is zero when the mass is at the bottom, or center, point of the swing. At this point, the speed of the mass is greatest and the kinetic energy is maximized. The correct answer is C.

To answer question 10, you must be familiar with this unit of measure. You may recall that 1 hertz (Hz) equals 1 cycle per second. The quantity that measures cycles per second is frequency, so the correct answer is D. Another way to approach this question is to identify the units of each quantity listed. For example, period measures an amount of time, so its unit may be seconds. Wavelength and amplitude measure distance, so their units may be centimeters or meters. Kinetic energy is measured in joules or other units of energy.

Part B of the SAT Physics Test Consists of Five-Choice Multiple-Choice Questions.

Each five-choice multiple-choice question can be written as either an incomplete statement or as a question. You are to select the choice that best completes the statement or answers the question.

Example:

Directions: Each of the questions or incomplete statements below is followed by five suggested answers or completions. Select the one that is best in each case and then fill in the corresponding oval on the answer sheet.

$${}^{15}_{7}\text{N} + {}^{1}_{1}\text{H} \rightarrow {}^{12}_{6}\text{C} + X$$

54. A physicist is studying the nuclear reaction represented above. Particle X is which of the following?

(A) ${}^{1}_{1}\text{H}$
(B) ${}^{2}_{1}\text{H}$
(C) ${}^{-1}_{0}e$
(D) ${}^{1}_{0}\text{H}$
(E) ${}^{4}_{2}\text{He}$

Question 54 tests your understanding of nuclear reactions and equations. First, you must recognize the information provided by the symbols. In the symbol ${}^{A}_{Z}\text{X}$, where X is the chemical symbol for the element, A is the atomic mass number, and Z is the atomic number. Second, you must recall that matter is conserved in all natural processes. Therefore, the equation must balance to represent this fact. $15 + 1 = 12 + A$, which yields $A = 4$. To solve for Z, use $7 + 1 = 6 + Z$. Therefore, $Z = 2$. The missing particle is, therefore, described by ${}^{4}_{2}\text{He}$, which is choice E.

Some Five-Choice Completion Questions May Have More than One Correct Answer or Solution.

A special type of five-choice completion question contains several statements labeled by Roman numerals. One or more of these statements may correctly answer the question. The statements are followed by five lettered choices, with each choice consisting of some combination of the Roman numerals that label the statements. You must select from among the five lettered choices the one that gives the combination of statements that best answer the question. Questions of this type are spread throughout the more standard five-choice completion questions.

Example:

29. In which of the following examples is the net force acting on the object equal to zero?

I. A soccer ball rolls to a stop.
II. A person holds an elevator door open.
III. A child rides on a carousel horse at a carnival.

(A) I only
(B) II only
(C) III only
(D) I and II only
(E) I and III only

To answer this question, you must recall that according to Newton's first law, a net force of zero must be acting on an object if the object maintains a constant velocity. Though no one is kicking the soccer ball any longer, a force must be acting on it because it is slowing to a stop. The force acting on the ball is friction. I is incorrect.

The person pushing on the elevator door is exerting a force on the door. However, neither the person nor the door is moving. Because the door is not moving, its velocity is constant at zero. This means that the net force acting on the door must also be zero. II is, therefore, correct.

The child riding on the carousel is moving at a constant speed. However, because the direction is constantly changing, the velocity is also changing. This means that the net force acting on the child is not zero. III is incorrect.

The net force is zero only in statement II, so choice B is the correct answer.

Some Five-Choice Completion Questions Relate to Common Material.

In some cases, a set of five-choice completion questions relate to common material that precedes the set. That material may be a description of a situation, a diagram, or a graph. Although the questions are related, you do not have to know the answer to one question in a set to answer a subsequent question correctly. Each question in the set can be answered directly from the material given for the entire set of questions.

Example:

Questions 36 and 37: A crane is lifting an object with a mass of 500 kilograms at a constant velocity to a height of 20 meters over a period of 5 seconds. The crane then holds the object in place for 30 seconds.

36. How much power does the crane expend in lifting the object?

(A) 25 W
(B) 1.3×10^2 W
(C) 5.0×10^3 W
(D) 1.0×10^4 W
(E) 2.0×10^4 W

37. How much power does the crane expend to hold the object in place?

(A) 0 W
(B) 3.3×10^2 W
(C) 1.5×10^3 W
(D) 6.0×10^3 W
(E) 2.0×10^4 W

To answer question 36, you need to know that power is a measure of work divided by time. In addition, work is a measure of force multiplied by displacement. The object is lifted with constant velocity. Therefore, the net force acting on it is zero. The force exerted by the crane must be equal and opposite to the weight of the object. The weight of the object is (500 kg)(10 m/s^2) = 5.0×10^3 N. The power is then determined by the following:

$$P = \frac{W}{t} = \frac{5.0 \times 10^3 \text{ N}(20 \text{ m})}{5 \text{ s}} = 2.0 \times 10^4 \text{ W}$$

The correct answer is E.

To answer question 37, you must recognize that even though a force is exerted to hold the object in place, no work is done on the object if it does not move any distance. If no work is done, no power is expended. Therefore the correct answer is A.

Strategies for Top Scores

When you take the SAT Physics test, you'll want to do everything you can to make sure you get your best possible score. That means studying right, building good problem-solving skills, and learning proven test-taking strategies.

Here are some tips to help you do your best.

Study Strategies

- **Get to know the format of the exam.** Use the practice tests in this book to familiarize yourself with the test format, which does not change from year to year. That way, you'll know exactly what to expect when you see the real thing on test day.
- **Get to know the test directions.** If you are familiar with the directions ahead of time, you won't have to waste valuable test time reading them and trying to understand them. The format and directions used in the practice exams in this book are modeled on the ones you'll see on the actual SAT Physics exam.
- **Study hard.** If possible, plan to study for at least an hour a day for two weeks before the test. You should be able to read this entire book and complete all five practice exams during that time period. Be sure to write notes in the margins of the book and paraphrase what you read. Make study cards from a set of index cards. Those cards can "go where you go" during the weeks and days before the test. If you are pressed for time, focus on taking the five practice exams, reading the explanations, and reviewing the particular topics that give you the most trouble.

Problem-Solving Strategies

- **Know what the question is asking.** While this tip may sound obvious, it is crucial that you read the question carefully to identify the information you are seeking. If you jump to the answer choices before completing the question, you may miss a relationship that you need to identify. It is equally important to go back and check the question after completing a calculation. For some questions, you may stop too soon or take the calculation too far. Take time to check that you have answered the question being asked.
- **Solve problems in whatever way is easiest for you.** There are usually several ways to solve any problem in physics and arrive at the correct answer. For example, when converting units some students prefer to use a dimensional analysis whereas others prefer to set up a proportion. Do what is easiest for you. Remember that the SAT exam is all multiple choice. That means that no one is going to be checking your work and judging you by which solution method you chose, so solve the problem any way you like.
- **Make sure you read all relevant information.** There may be additional information that is required to answer the question. Look for descriptive material that may be provided along with a graph or diagram.
- **Know your formulas.** You will not be allowed to bring a calculator to the test. You are also not allowed to bring in any sheets of useful information. Roughly three-quarters of the test requires you to use formulas. If you do not know basic formulas such as how force relates to mass and acceleration, $F = ma$, you are sure to lose easy points.

Many formulas will come easily as you study physics. Others may be difficult for you to remember. If this is the case, look them over just before the test. You may wish to jot down those formulas on the top or back of the question booklet before you begin the test so you don't forget them. Keep in mind that merely memorizing formulas will not be enough. You also need to understand them. Only rarely do questions ask you to simply plug numbers into a formula. More often you need to rearrange or relate various formulas to solve a problem.

- **Pay attention to units of measure.** The test questions predominantly use the metric system. Familiarize yourself with the units of measurement for common physical quantities. Include units in your calculations. If the outcome of a calculation does not yield the proper unit, you may have used information incorrectly.
- **Estimate when possible.** Once you know what a question is asking, it is helpful to get a rough idea of what the answer should look like through estimation. Of course this strategy is helpful only for questions involving calculations. Estimation is a good way to avoid wrong answers when you are making an educated guess.
- **Identify all labels on graphs and diagrams.** About one-quarter of the questions on the test will involve graphs or diagrams. When you encounter such a question, take a moment to review the information provided. For example, identify the quantities plotted on the axes of a graph. Then read the related question and answer choices. Knowing what you are dealing with before you read the question can help you identify the correct answer.
- **Write down any information you need to answer a question.** *Do not hesitate to draw or write on your question booklet.* If a diagram is not provided with a question, draw a rough sketch of the information described. Field lines, velocity vectors, and graphs are just some of the topics that will become much easier to work with once you have drawn them.

  Write down formulas or equations you may need. You may find it helpful to write down formulas related to a topic. If, for example, you are dealing with a question about energy, write down such equations as $KE = \frac{1}{2}mv^2$ and $PE = mgh$. If you are unsure of the answer, it may be helpful to plug in the given values. Some rearranging and rewriting may lead you in the right direction.
- **Pay attention to words in questions such as EXCEPT, NOT, ALWAYS, and NEVER.** Some questions include qualifying words in capital letters. These words change the way you need to approach the question.

Tips for Test Day

- **Don't panic!** Once test day comes, you're as prepared as you're ever going to be, so there is no point in panicking. Use your energy to make sure that you are extra careful in answering questions and marking the answer sheet.

- **Use your test booklet as scratch paper.** Your test booklet is not going to be reused by anyone when you're finished with it, so feel free to mark it up in whatever way is most helpful to you. Circle important words, underline important points, write your calculations in the margins, and cross out wrong answer choices.
- **Be careful when marking your answer sheet.** Remember that the answer sheet is scored by a machine, so mark it carefully. Fill in answer ovals completely, erase thoroughly if you change your mind, and do not make any stray marks anywhere on the sheet. Also, make sure that the answer space you are marking matches the number of the question you are answering. If you skip a question, make sure that you skip the corresponding space on the answer sheet. Every 5 or 10 questions, check the question numbers and make sure that you are marking in the right spot. You may want to mark your answers in groups of 5 or 10 to make sure that you are marking the answer sheet correctly.
- **Watch the time.** Keep track of the time as you work your way through the test. Try to pace yourself so that you can tackle as many of the 75 questions as possible within the 1-hour time limit. Check yourself at 10- or 15-minute intervals using your watch or a timer.
- **Don't panic if time runs out.** If you've paced yourself carefully, you should have time to tackle all or most of the questions. If you do run out of time, don't panic. Make sure that you have marked your answer sheet for all the questions that you have answered so far. Then look ahead at the questions you have not yet read. Can you answer any of them quickly, without taking the time to do lengthy calculations? If you can, mark your answers in the time you have left. Every point counts!
- **Use extra time to check your work.** If you have time left over at the end of the test, go back and check your work. Make sure that you have marked the answer sheet correctly. Check any calculations you may have made to make sure that they are correct. Take another look at any questions you may have skipped. Can you eliminate one or more answer choices and make an educated guess? Resist the urge to second-guess too many of your answers, however, as this may lead you to change an already correct answer to a wrong one.

PART II

Diagnostic Test

Diagnostic Test

Treat this diagnostic test as the actual test, and complete it in one 60-minute sitting. Use the following answer sheet to fill in your multiple-choice answers. Once you have completed the practice test:

1. Check your answers using the Answer Key.
2. Review the Answers and Explanations.
3. Complete the Score Sheet to see how well you did.

Answer Sheet

Tear out this answer sheet and use it to complete the practice test. Determine the BEST answer for each question. Then fill in the appropriate oval.

1. A B C D E	**21.** A B C D E	**41.** A B C D E	**61.** A B C D E
2. A B C D E	**22.** A B C D E	**42.** A B C D E	**62.** A B C D E
3. A B C D E	**23.** A B C D E	**43.** A B C D E	**63.** A B C D E
4. A B C D E	**24.** A B C D E	**44.** A B C D E	**64.** A B C D E
5. A B C D E	**25.** A B C D E	**45.** A B C D E	**65.** A B C D E
6. A B C D E	**26.** A B C D E	**46.** A B C D E	**66.** A B C D E
7. A B C D E	**27.** A B C D E	**47.** A B C D E	**67.** A B C D E
8. A B C D E	**28.** A B C D E	**48.** A B C D E	**68.** A B C D E
9. A B C D E	**29.** A B C D E	**49.** A B C D E	**69.** A B C D E
10. A B C D E	**30.** A B C D E	**50.** A B C D E	**70.** A B C D E
11. A B C D E	**31.** A B C D E	**51.** A B C D E	**71.** A B C D E
12. A B C D E	**32.** A B C D E	**52.** A B C D E	**72.** A B C D E
13. A B C D E	**33.** A B C D E	**53.** A B C D E	**73.** A B C D E
14. A B C D E	**34.** A B C D E	**54.** A B C D E	**74.** A B C D E
15. A B C D E	**35.** A B C D E	**55.** A B C D E	**75.** A B C D E
16. A B C D E	**36.** A B C D E	**56.** A B C D E	
17. A B C D E	**37.** A B C D E	**57.** A B C D E	
18. A B C D E	**38.** A B C D E	**58.** A B C D E	
19. A B C D E	**39.** A B C D E	**59.** A B C D E	
20. A B C D E	**40.** A B C D E	**60.** A B C D E	

PART A

Directions: Each set of lettered choices refers to the numbered questions of statements immediately following it. Select the one lettered choice that best answers each question or best fits each statement, and then fill in the corresponding oval on the answer sheet. A choice may be used once, more than once, or not at all in each set.

Questions 1–5 relate to the following physical principles or topics.

(A) pressure
(B) energy
(C) work
(D) force
(E) centripetal acceleration

Select the quantity that each expression defines.

1. $m\left(\dfrac{v^2 - v_o^2}{2d}\right)d$

2. $\dfrac{4\pi^2 r}{T^2}$

3. $k\dfrac{q_1 q_2}{r}$

4. $\dfrac{nRT}{V}$

5. $\dfrac{m}{t}(v - v_o)$

Questions 6–10

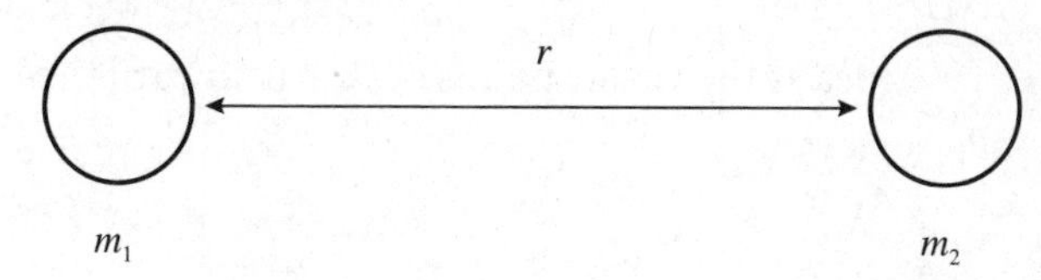

(A) quartered
(B) halved
(C) unchanged
(D) doubled
(E) quadrupled

Two masses, m_1 and m_2, are separated by a distance r, as shown above. Select the effect on the attraction of each of the following changes in mass or distance.

6. m_1 is doubled
7. m_1 and m_2 are doubled
8. r is doubled
9. m_1 is doubled, m_2 is halved
10. m_2 is halved

Questions 11–14 relate to the following directions.

(A) out
(B) right
(C) up
(D) down
(E) zero

Find the direction of the force on a positive charge for each diagram where v is the velocity of the charge, and B is the direction of the magnetic field.

(x) means the vector points inward, into the page.

(•) means the vector points outward, toward the viewer.

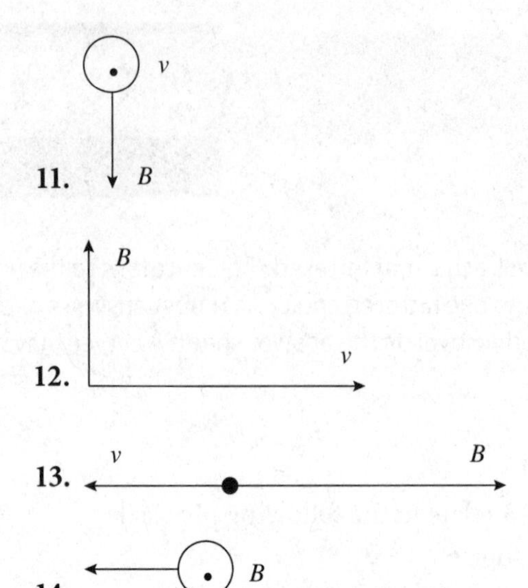

PART B

Directions: Each of the questions or incomplete statements below is followed by five suggested answers or completions. Select the one that is best in each case and then fill in the corresponding oval on the answer sheet.

$$^{27}_{13}\text{Al} + {}^{4}_{2}\text{He} \rightarrow {}^{30}_{15}\text{P} + X + \text{energy}$$

15. A physicist is studying the decay reaction represented above. Particle X is which of the following?

(A) $^{1}_{1}\text{H}$
(B) $^{2}_{1}\text{H}$
(C) $^{-1}_{0}e$
(D) $^{1}_{0}n$
(E) $^{4}_{2}\text{He}$

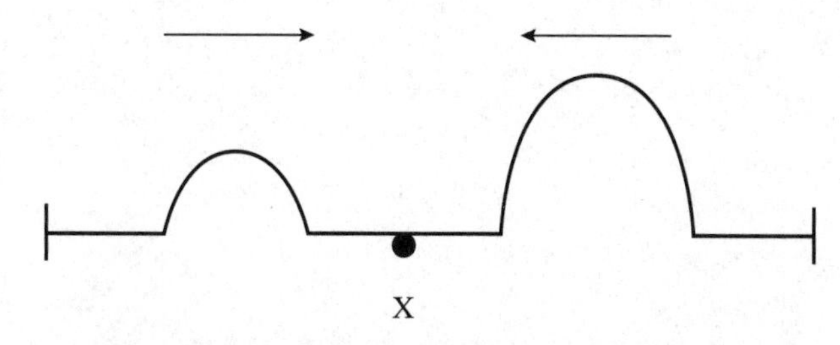

16. The figure above shows two pulses on a string approaching each other. Which of the following diagrams best represents the appearance of the string shortly after the pulses pass point X?

(A)

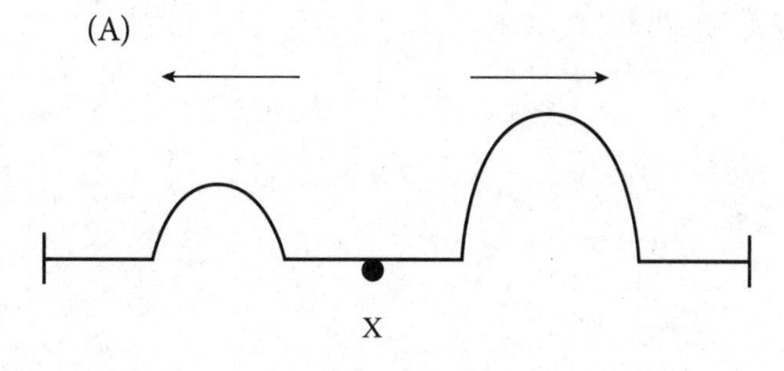

(B)

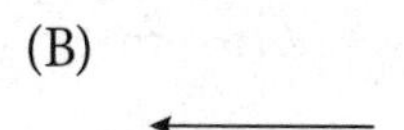

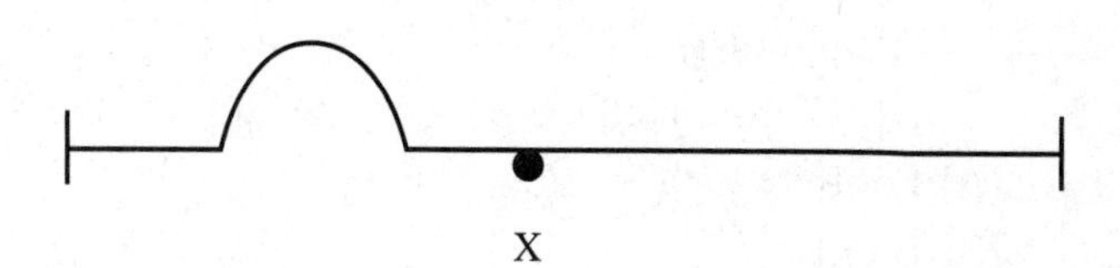

(C)

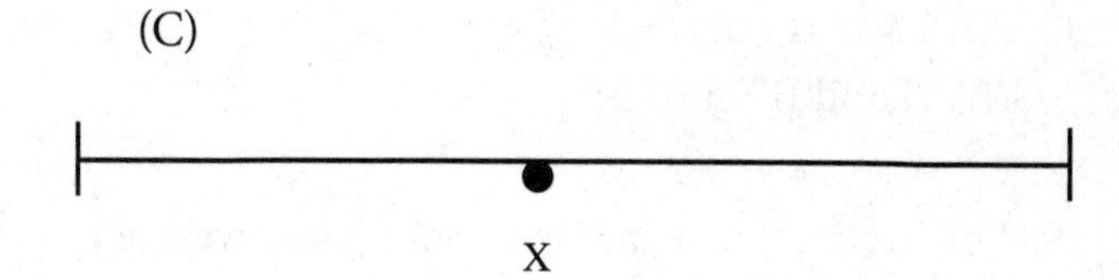

(D)

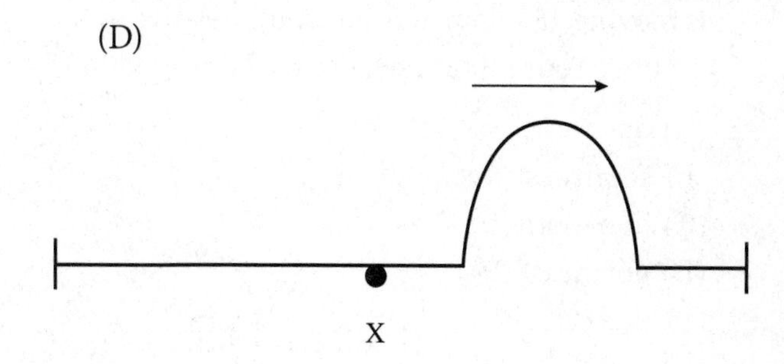

(E)

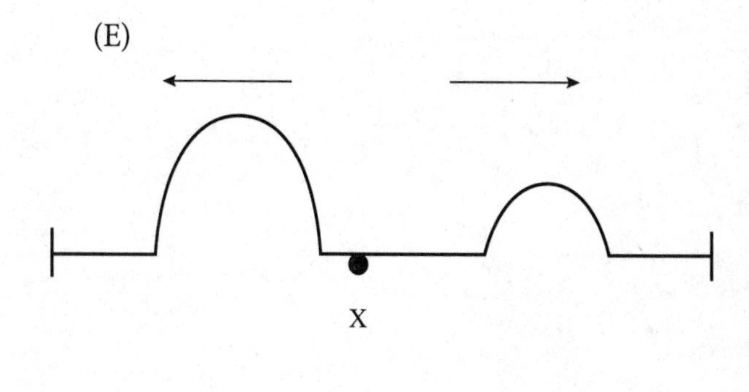

GO ON TO THE NEXT PAGE

17. If the specific heat of steam is 2.01×10^3 Joules per kilogram per °C, how much energy is removed when 10 kilograms of water is cooled from steam at 120°C to liquid at 110°C?

(A) 2.01 J
(B) 2.01×10^5 J
(C) 4.98×10^4 J
(D) 4.98×10^{-2} J
(E) 6.1×10^4 J

18. Which of these phenomena support(s) the particle model of light?

I. polarization
II. photoelectricity
III. Compton effect

(A) I only
(B) II only
(C) III only
(D) I and II only
(E) II and III only

19. You are in a boat that is traveling due west at 45 kilometers per hour relative to a current that is moving 45 kilometers due south relative to the ground. Your motion relative to the ground is

(A) due west
(B) southwest
(C) due south
(D) northeast
(E) due east

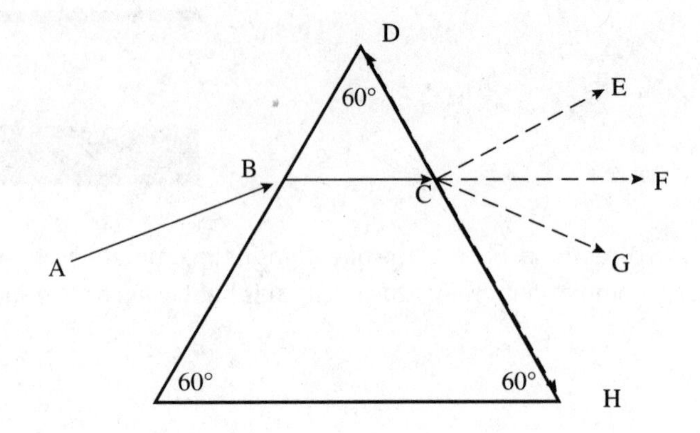

20. In the diagram above, a ray of light strikes a glass prism so that the ray travels through the prism parallel to its base. Which ray correctly shows how the light exits the prism?

(A) $\overrightarrow{CD}$
(B) $\overrightarrow{CE}$
(C) $\overrightarrow{CF}$
(D) $\overrightarrow{CG}$
(E) $\overrightarrow{CH}$

Questions 21 and 22

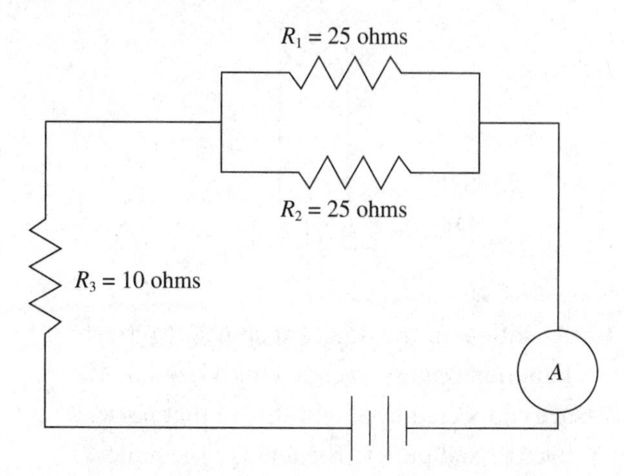

The diagram above shows an electric circuit containing three resistors. The ammeter, *A*, reads 5.0 amperes.

21. What is the voltage across R_3?

(A) 5 V
(B) 10 V
(C) 15 V
(D) 20 V
(E) 50 V

22. What is the current in R_1?

(A) 1.0 A
(B) 2.0 A
(C) 2.5 A
(D) 5.0 A
(E) 10 A

23. An object starts from rest and accelerates at 6.0 meters per second squared. How far will it travel during the first 4.0 seconds?

(A) 10 m
(B) 12 m
(C) 24 m
(D) 36 m
(E) 48 m

24. A heat engine extracts 40 joules of energy from a hot reservoir, does work, then exhausts 30 joules of energy into a cold reservoir. What is the efficiency of the heat engine?

(A) 10%
(B) 25%
(C) 33%
(D) 67%
(E) 75%

25. The magnitude of the charge on an ion depends on which of the following?

I. number of electrons
II. number of protons
III. number of neutrons

(A) I only
(B) III only
(C) I and II only
(D) II and III only
(E) I, II, and III

26. If the mass of a body is doubled while the net force acting on the body remains the same, the acceleration of the body is

(A) halved
(B) doubled
(C) unchanged
(D) quartered
(E) quadrupled

GO ON TO THE NEXT PAGE ➡

Questions 27 and 28

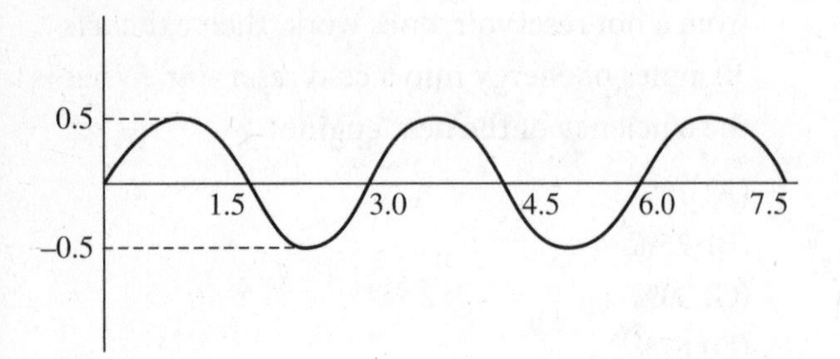

27. What is the wavelength of the wave?

(A) 0.5 m
(B) 1.0 m
(C) 1.5 m
(D) 2.0 m
(E) 3.0 m

28. What is the amplitude of the wave?

(A) 0.5 m
(B) 1.0 m
(C) 1.5 m
(D) 3.0 m
(E) It varies between –0.5 m and +0.5 m.

29. If electricity costs $0.10 per kilowatt-hour, how much does it cost for electricity to operate a 1,200-watt television for 2 hours?

(A) $0.024
(B) $0.24
(C) $2.40
(D) $24.00
(E) $240.00

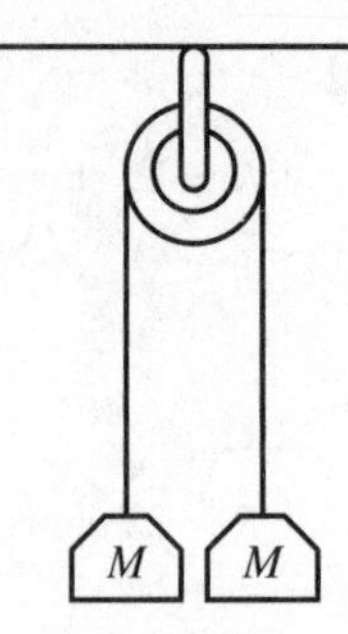

30. As shown in the diagram above, two freely hanging weights, each having a mass of M, are connected by a light thread that passes over a fixed pulley. The mass of the pulley and frictional losses are negligible. If a weight of mass m is added to one of the weights, its downward acceleration in terms of M, m, and g will be

(A) $(2M + m)g/(M + m)$
(B) $(M - m)g/m$
(C) $(M + m)g/m$
(D) $mg/(M + m)$
(E) $mg/(M - m)$

31. According to the kinetic theory of matter, which of these statements is NOT true about the particles that make up a gas?

(A) The particles of a gas obey the laws of classical mechanics and interact only when they collide.
(B) The speeds of particles in a gas are distributed such that the speeds of most molecules are close to the average.
(C) The average translational kinetic energy of particles in a gas is directly proportional to the absolute temperature.
(D) All particles of a gas have the same speed at the same specified temperature.
(E) The separation between particles is, on average, equal to the diameter of each molecule.

32. The magnetic field produced by a long, straight, current-carrying wire

(A) is uniform
(B) is directed radially outward from the wire
(C) forms circles around the wire
(D) increases in strength as the distance from the wire increases
(E) is parallel to the wire

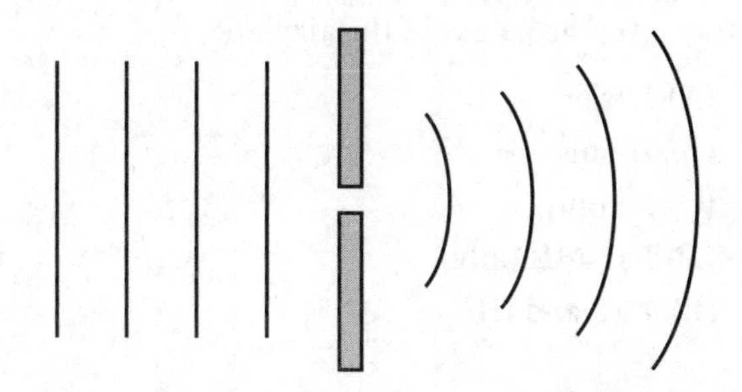

33. In the diagram above, a series of straight wave fronts passes through a small opening in an obstacle. What phenomenon is responsible for this observation?

(A) refraction
(B) photoelectricity
(C) diffraction
(D) dispersion
(E) polarization

34. The resistance of a conductor decreases when it experiences an increase in

I. length
II. temperature
III. cross-sectional area

(A) I only
(B) II only
(C) I and II only
(D) II and III only
(E) III only

35. A cylinder with a piston has a cross-sectional area of 0.010 m^2. How much work can be done by a gas in the cylinder if the gas exerts a constant pressure of 5.0×10^5 Pa on the piston, moving it a distance of 0.020 m?

(A) 0.01 J
(B) 10.0 J
(C) 100 J
(D) 1000 J
(E) 10,000 J

36. An ice cube is placed in a foam cup filled with warm liquid water. Which principle explains why the ice melts and the system eventually reaches a consistent temperature?

(A) First law of thermodynamics
(B) Second law of thermodynamics
(C) Ideal gas law
(D) Conservation of momentum
(E) Archimedes' principle

GO ON TO THE NEXT PAGE ➡

37. The half-life of iodine–131 is approximately 8 days. About what fraction of a sample of iodine–131 will remain after 32 days?

(A) $\frac{1}{2}$

(B) $\frac{1}{4}$

(C) $\frac{1}{8}$

(D) $\frac{1}{16}$

(E) $\frac{1}{32}$

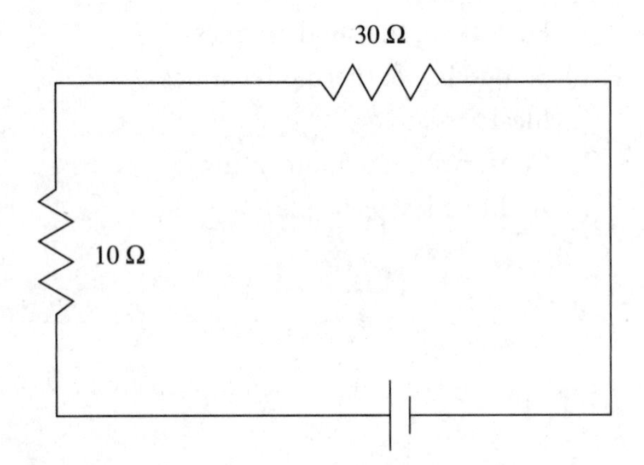

38. In the circuit shown above, the current I_1 in the 10-ohm resistor is related to the current I_2 in the 30-ohm resister by which of the following equations?

(A) $I_1 = \frac{1}{3}I_2$

(B) $I_1 = \frac{2}{3}I_2$

(C) $I_1 = 3I_2$

(D) $I_1 = I_2$

(E) $I_1 = \frac{3}{2}I_2$

39. An airplane is dropping supplies to firefighters battling a blaze in a forest. In addition to knowing the acceleration due to gravity, what factor(s) must the pilot consider in order to determine where to drop the supplies so that they will land beside the firefighters? (Ignore air resistance.)

I. the speed of the airplane
II. the mass of the supplies
III. the height of the airplane

(A) I only
(B) II only
(C) III only
(D) I and III only
(E) I, II, and III

40. In a darkened room, a beam of monochromatic light is shined on an opaque barrier with a single narrow slit. The light that goes through the slit falls on a screen held parallel to the barrier. Which of the following best describes the intensity of the observed pattern plotted against the distance along the screen?

(A)

(B) 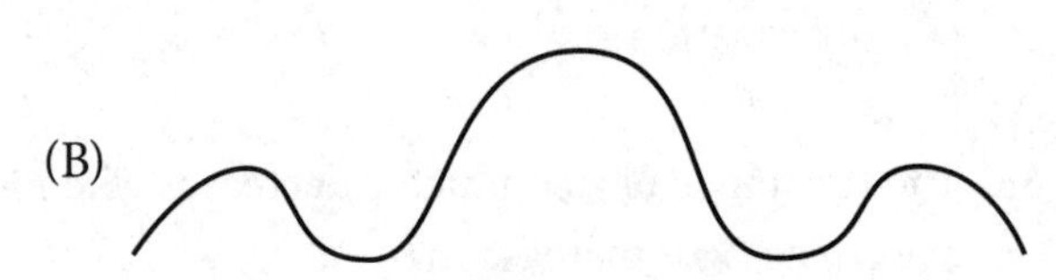

(C) 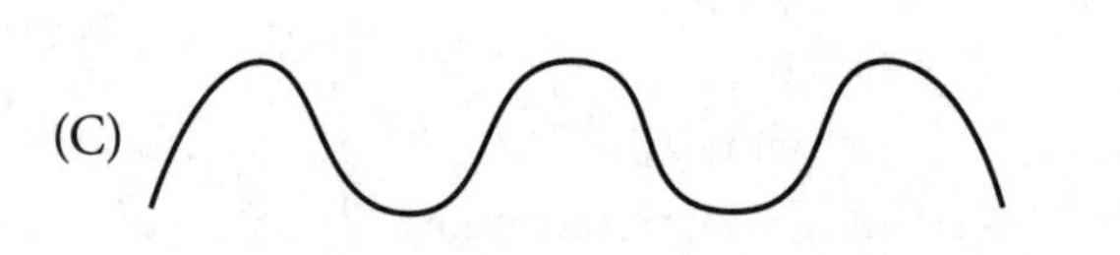

(D) 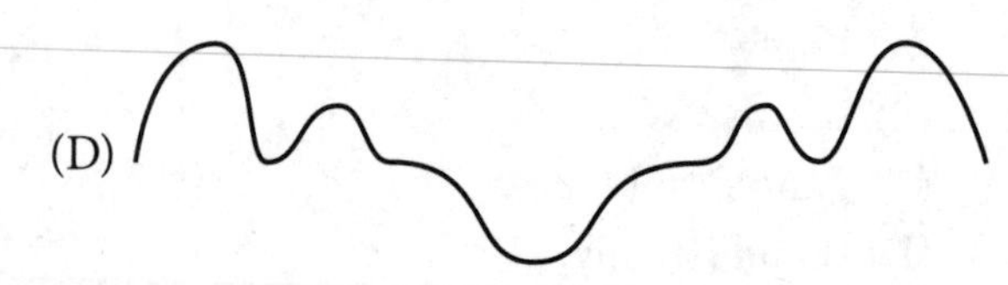

(E) 

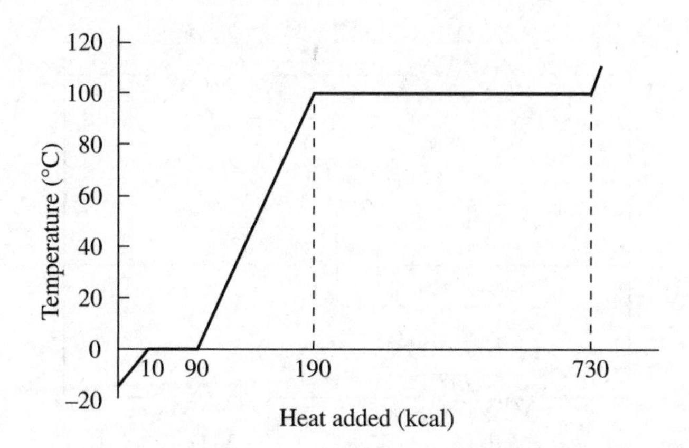

41. The diagram above relates the temperature of a sample of water to the amount of heat added or removed. Based on the diagram, which of these statements is true?

(A) As pressure rises, a greater amount of latent heat is required for a phase change than at lower pressures.
(B) The latent heat of fusion for water is greater than its latent heat of vaporization.
(C) The mass of a sample of water decreases as heat is removed from it during periods of constant temperature.
(D) The amount of latent heat required to change from a solid to a liquid is greater than to change from a liquid to a gas.
(E) A greater amount of latent heat is released when steam changes to liquid water than when the same mass of liquid water changes to ice.

42. A string of length ℓ that is fastened at both ends is plucked in the middle. At the fundamental frequency, the wavelength of the wave in the string is equal to

(A) $\frac{1}{3}\ell$
(B) $\frac{1}{2}\ell$
(C) ℓ
(D) 2ℓ
(E) 4ℓ

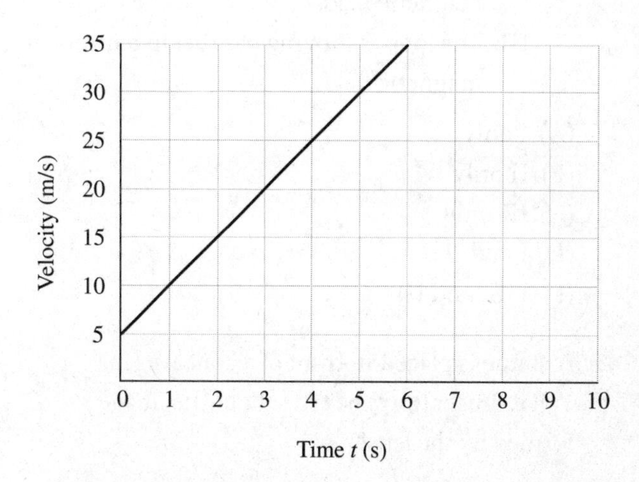

43. The acceleration of a test vehicle is plotted on the graph above. Based on the graph, what is the acceleration of the vehicle?

(A) 0.2 m/s^2
(B) 5 m/s^2
(C) 6 m/s^2
(D) 15 m/s^2
(E) 36 m/s^2

X X X X B
X X X X
X X X X
X X X X

44. A conducting wire loop is located inside a magnetic field. There will be induced current in the wire loop if

I. The magnetic field is increasing.
II. The loop is rotating in a uniform magnetic field.
III. The loop is moving parallel to a uniform magnetic field.

(A) I only
(B) II only
(C) III only
(D) I and II only
(E) I, II, and III

45. A statue is placed in front of a concave lens. What must always be true about the image formed by the lens?

	Type	Size	Orientation
(A)	virtual	smaller	erect
(B)	virtual	smaller	inverted
(C)	virtual	larger	inverted
(D)	real	smaller	erect
(E)	real	larger	inverted

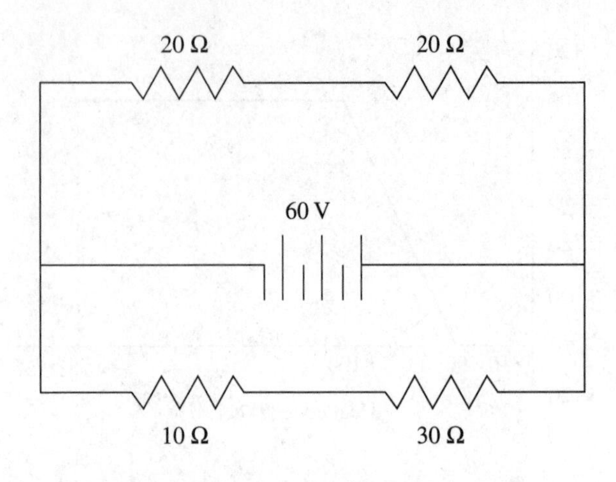

46. What is the current through the 30-ohm resistor in the circuit shown above?

(A) 0.5 A
(B) 1 A
(C) 1.5 A
(D) 2.0 A
(E) 3.0 A

GO ON TO THE NEXT PAGE ➡

47. The graph below plots the velocity, v, of an object over a given period of time, t.

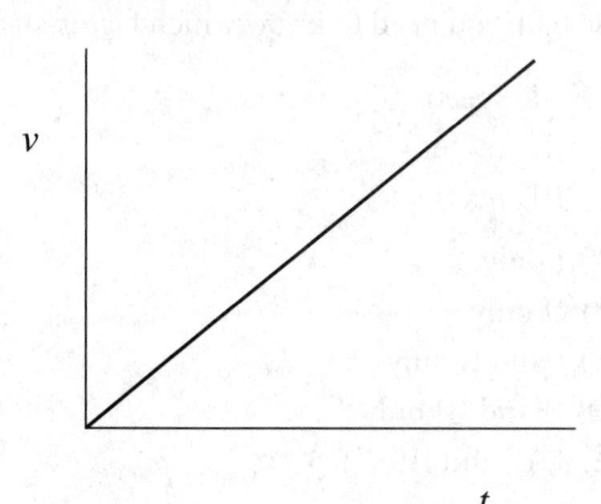

Which graph represents the acceleration, a, of the object during the same period of time?

(A)

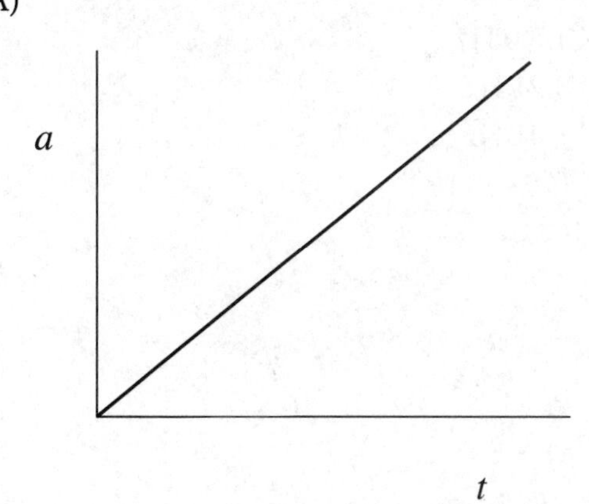

(B)

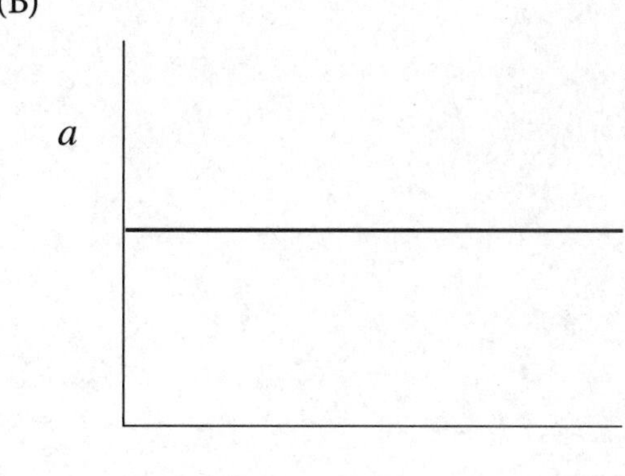

(C)

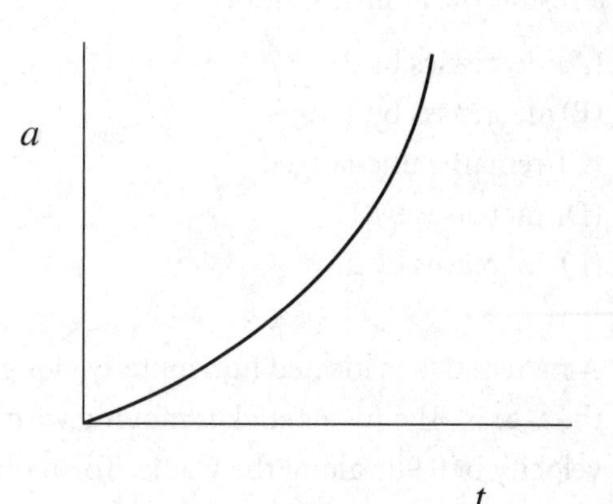

(D)

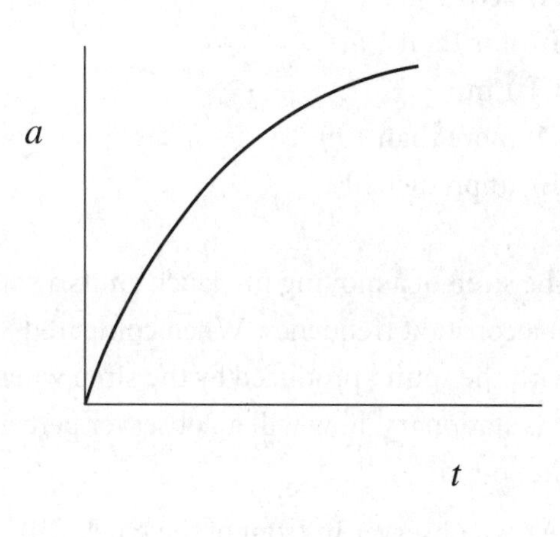

(E)

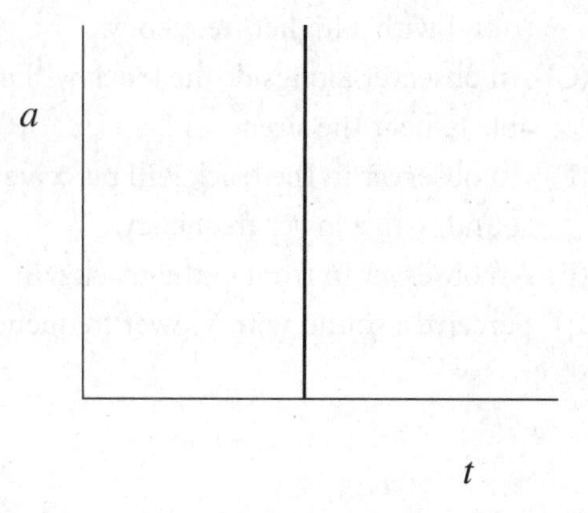

48. The nucleus of an atom emits a beta particle. As a result, the atomic number

(A) decreases by 2
(B) decreases by 1
(C) remains unchanged
(D) increases by 1
(E) increases by 2

49. A meter stick is located horizontally along the *x*-axis. The meter stick is moving with a velocity of 0.90c along the *y*-axis. To an observer standing on the *x*-axis, the length of the stick will be

(A) zero
(B) less than 1 m
(C) 1 m
(D) more than 1 m
(E) unpredictable

50. The siren of a moving fire truck emits a sound at a constant frequency. When compared with the sound produced by the siren when it is stationary, how will an observer perceive the sound?

(A) An observer in front of the truck will perceive a sound with a higher frequency.
(B) An observer behind the truck will perceive a sound with a higher frequency.
(C) An observer alongside the truck will not be able to hear the siren.
(D) An observer in the truck will perceive a sound with a lower frequency.
(E) An observer in front of the truck will perceive a sound with a lower frequency.

51. A ball at the end of a string is swinging in a horizontal circle. To calculate the acceleration of the ball, you need to know which factor(s)?

I. speed
II. radius
III. mass

(A) I only
(B) II only
(C) I and II only
(D) II and III only
(E) I, II, and III

52. The bob of a pendulum completes 21 cycles in 30 seconds. What is its frequency?

(A) 0.7 Hz
(B) 1.4 Hz
(C) 3.0 Hz
(D) 21 Hz
(E) 41 Hz

GO ON TO THE NEXT PAGE ➡

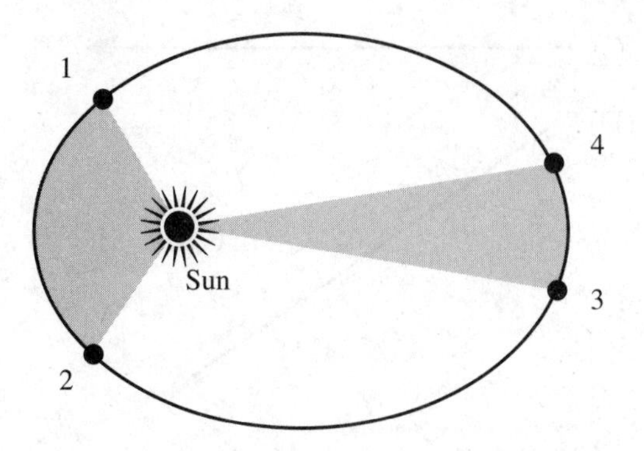

53. The diagram above represents the motion of a planet around the sun as described by Kepler's second law. Based on the diagram, which statement about planetary motion is true?

(A) Planets closer to the sun travel at a greater speed than planets farther from the sun.
(B) Planets move fastest in the part of their orbits where they are closest to the sun.
(C) All planets have the same period of revolution regardless of their mean distance from the sun.
(D) Some planets follow circular paths, whereas others follow elliptical paths.
(E) The period of a planet is determined by its mass, with more massive planets having shorter periods.

54. When the distance between two point charges is doubled, the force between them is

(A) quadrupled
(B) doubled
(C) unchanged
(D) halved
(E) quartered

55. A transformer changes 12 volts to 24,000 volts. There are 10,000 turns in the secondary coil. How many turns are in the primary? (Assume 100% efficiency.)

(A) 2
(B) 4
(C) 5
(D) 10
(E) 12

Questions 56 and 57 refer to the graph below, which represents the speed of an object moving along a straight line. The time is represented by ***t***.

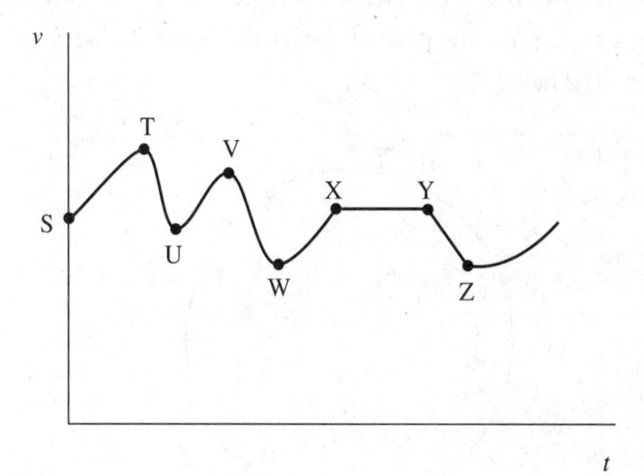

56. During which interval is the object's speed increasing?

(A) VW
(B) TU
(C) UV
(D) XY
(E) YZ

57. During which interval is the acceleration constant, but not zero?

(A) ST
(B) TU
(C) UV
(D) XY
(E) YZ

GO ON TO THE NEXT PAGE ➡

58. A source emits a sound with a frequency of 2.6×10^4 hertz. If the speed of the sound is 3.9×10^3 meters per second, what is the wavelength of the sound?

(A) 6.8×10^{-4} m
(B) 3.0×10^{-2} m
(C) 1.5×10^{-1} m
(D) 1.5×10^{2} m
(E) 2.2×10^{4} m

59. A negatively charged particle is moving to the right in a plane perpendicular to a uniform magnetic field when it enters the field. If the magnetic field is into the page, which of the following drawings represents the path of the particle once it enters the field?

(A)

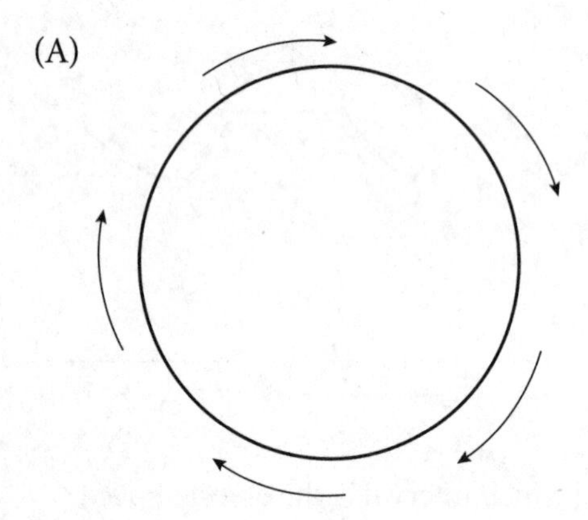

(B)

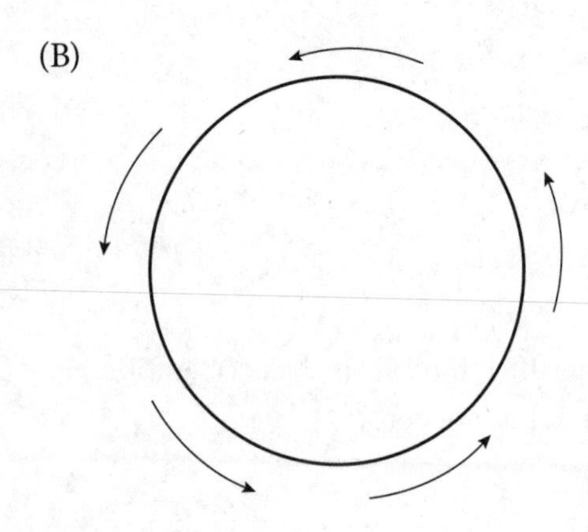

(C)

(D)

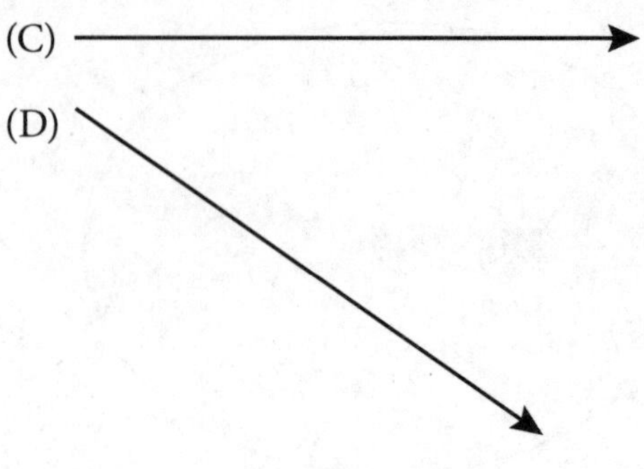

(E)

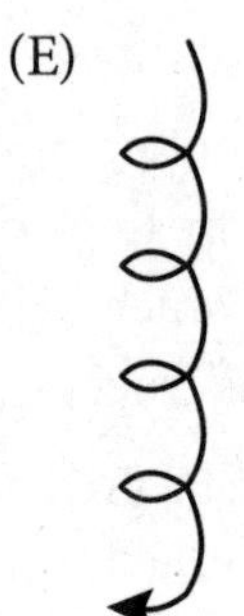

60. A compact disc is spinning in a CD player. The disc has a small blue marking at one point on its edge. Which of the following is true of the acceleration of the marking while the disc is in motion?

(A) It is zero at all times.
(B) Its magnitude and direction are constant.
(C) It gradually decreases over time.
(D) It is constant in magnitude, but not direction.
(E) It causes the CD to spin faster.

61. The energy conversion that takes place in a generator is

(A) electrical to mechanical
(B) mechanical to electrical
(C) electrical to heat
(D) chemical to electrical
(E) heat to chemical

GO ON TO THE NEXT PAGE ➡

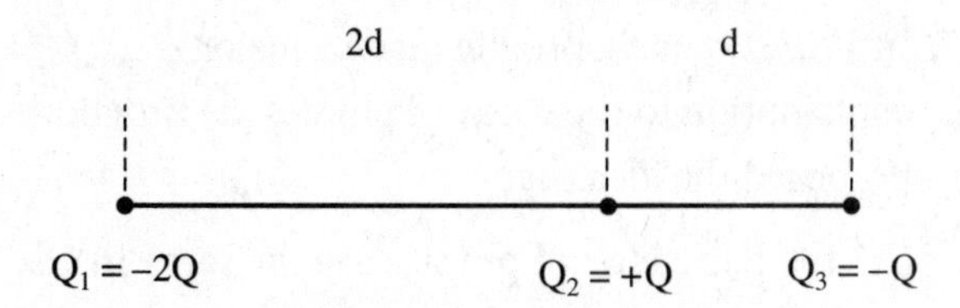

62. What is the magnitude and direction of the net electric force on particle Q_3 in the figure above, due to the other two charges?

(A) The net force is zero.

(B) The net force is $\frac{7}{9}\frac{kQ^2}{d^2}$ to the right.

(C) The net force is $\frac{7}{9}\frac{kQ^2}{d^2}$ to the left.

(D) The net force is $\frac{1}{3}\frac{kQ^2}{d^2}$ to the right.

(E) The net force is $\frac{4}{3}\frac{kQ^2}{d^2}$ to the left.

Questions 63 and 64 relate to the graph below, which shows the net force **F** in newtons exerted on a 2-kilogram block as a function of time t in seconds. Assume the block is at rest at $t = 0$ and that **F** acts in a fixed direction.

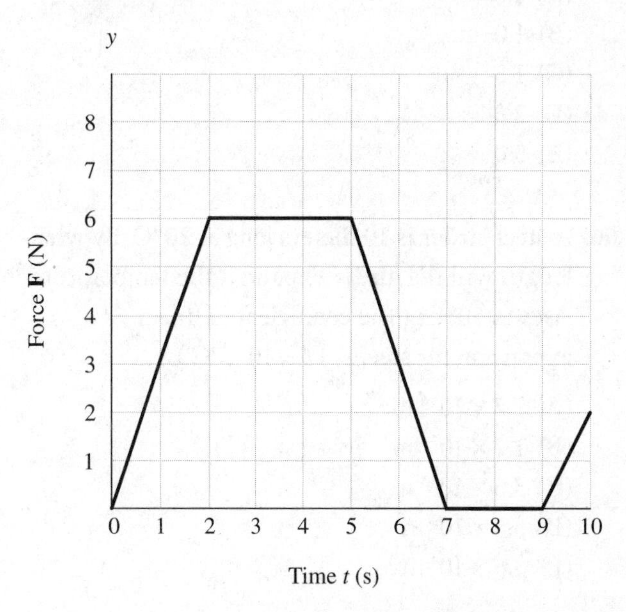

63. Which statement is true about the motion of the block during the time interval from 2 to 5 seconds?

(A) The block is not moving.
(B) The acceleration is constant.
(C) The mass of the block is changing.
(D) The speed of the block is constant.
(E) The force on the block is decreasing.

64. The acceleration of the block at $t = 3$ seconds is

(A) $\frac{1}{3}$ m/s²
(B) 2 m/s²
(C) 3 m/s²
(D) 6 m/s²
(E) 18 m/s²

GO ON TO THE NEXT PAGE ➡

65. An object is placed 30 centimeters in front of a converging lens of focal length 10 centimeters. If the image is 15 centimeters from the lens, what is the magnification?

(A) 0.5
(B) 1.0
(C) 1.5
(D) 2.0
(E) 5.0

66. A steel girder is 10 meters long at 20°C. By what length will the girder expand if the temperature rises to 50°C? (The coefficient of linear expansion for steel is 12×10^{-6}/°C.)

(A) 1.2×10^{-6} m
(B) 1.2×10^{-4} m
(C) 3.6×10^{-3} m
(D) 3.6×10^{3} m
(E) 3.6×10^{4} m

67. In 1923, Louis de Broglie made a major contribution to the study of physics. de Broglie proposed the idea that

(A) the basic laws of physics are the same in all inertial reference frames
(B) light consists of fluctuating electric and magnetic fields
(C) electrons circle the nucleus of an atom in stationary states
(D) light can be described in terms of discrete units called photons
(E) material particles have wavelengths related to their momentum

68. A football player kicks a football from ground level at an angle of 45° above the horizontal. The player then kicks the ball again with the same speed, but at an angle of 30° above the horizontal. Which of the following diagrams correctly shows the trajectories of the balls?

(A)

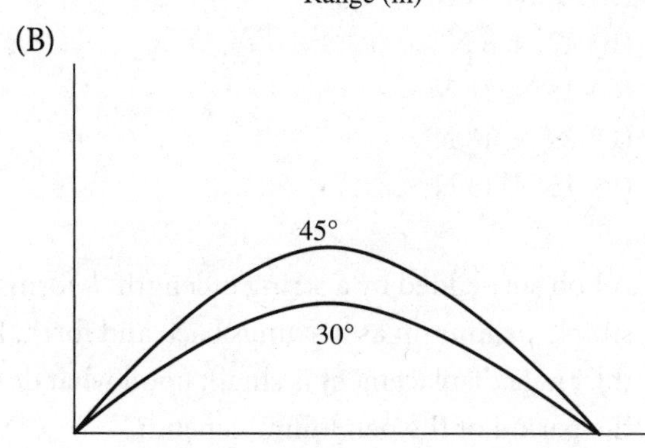

(B)

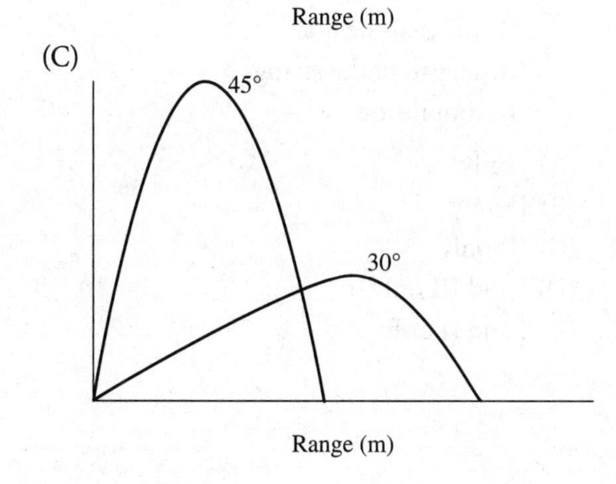

(C)

45°
30°
Range (m)

(D)

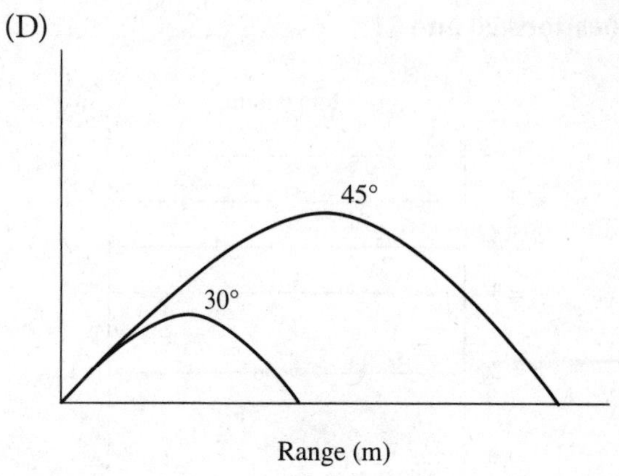

(E)

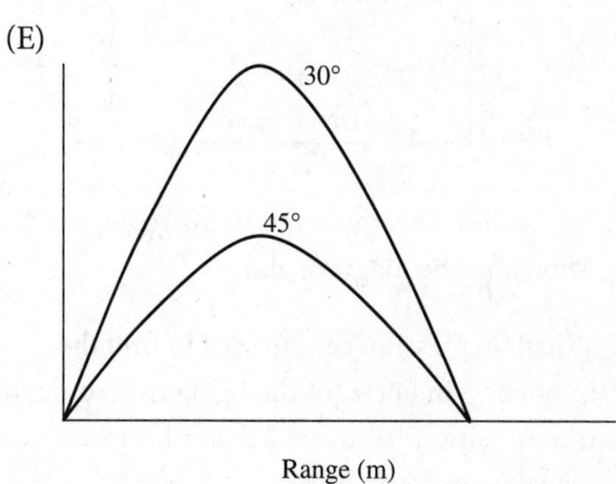

69. Which of these measurements has three significant digits?

I. 39.2 g
II. 0.05103 cm
III. 1860 L

(A) I only
(B) II only
(C) I and II only
(D) II and III only
(E) I and III only

Diagnostic Test

Questions 70 and 71

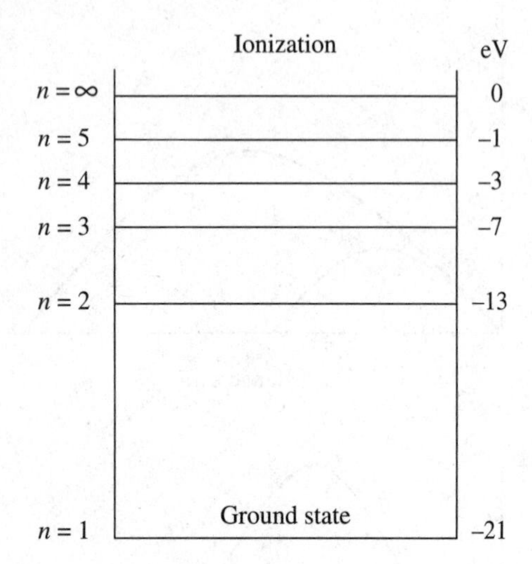

Some of the energy levels of hydrogen are shown in the diagram above.

70. Which expression can be used to find the frequency, in hertz, of the light emitted during the transition from $n = 2$ to $n = 1$, in terms of Planck's constant, h?

(A) 7 eV/h
(B) 7 eV $\times h$
(C) 13 eV/h
(D) 13 eV $\times h$
(E) 21 eV/h

71. Which of these transitions will result in the emission of a photon with the least amount of energy?

(A) $n = 4$ to $n = 3$
(B) $n = 3$ to $n = 2$
(C) $n = 5$ to $n = 3$
(D) $n = 4$ to $n = 2$
(E) $n = 3$ to $n = 1$

72. A flower is placed into a clear vase of water. To an observer on the side of the vase, the stem appears to be bent where it enters the water. What phenomenon causes the stem to appear bent?

(A) polarization
(B) photoelectricity
(C) diffraction
(D) reflection
(E) refraction

73. Two forces are being exerted on an object in the horizontal direction. If these are the only forces acting on the object, which pair of forces can result in a net force of 18 newtons?

(A) 18 N, 18 N
(B) 12 N, 8 N
(C) 13 N, 31 N
(D) 22 N, 18 N
(E) 15 N, 13 N

74. A bob suspended by a string of length, l, forms a simple pendulum as it swings back and forth. If the angle displacement is small, upon what does the period of the pendulum depend?

I. mass of the bob
II. length of the string
III. amplitude

(A) I only
(B) II only
(C) III only
(D) I and III only
(E) I and II only

75. In 1911, Ernest Rutherford and his associates conducted an experiment in which they bombarded a thin, gold foil with fast-moving alpha particles. What surprising observation did they make during this experiment?

(A) Most particles experienced wide-angle deflections.
(B) A small number of particles were redirected toward their source.
(C) All particles passed directly through the foil without deflection.
(D) Some particles were absorbed by the foil and disappeared.
(E) Many particles experienced a change in charge as they hit the foil.

STOP

If you finish before time is called, you may check your work on this test only.

Do not turn to any other test in this book.

Answer Key

1. C
2. E
3. B
4. A
5. D
6. D
7. E
8. A
9. C
10. B
11. B
12. A
13. E
14. C
15. D
16. E
17. B
18. E
19. B
20. D
21. E
22. C
23. E
24. B
25. C
26. A
27. E
28. A
29. B
30. C
31. E
32. B
33. C
34. E
35. C
36. B
37. D
38. D
39. D
40. B
41. E
42. D
43. B
44. D
45. A
46. C
47. B
48. D
49. C
50. A
51. C
52. A
53. B
54. E
55. C
56. C
57. E
58. C
59. A
60. D
61. B
62. C
63. B
64. C
65. A
66. C
67. E
68. D
69. E
70. B
71. A
72. E
73. C
74. B
75. B

Answers and Explanations

Part A

1. **(C)** This formula can be rewritten as

$$\frac{1}{2}mv^2 - \frac{1}{2}mv_o^2,$$

which is the change in kinetic energy. The net work done on an object is equal to its change in kinetic energy.

2. **(E)** This formula can be used to determine the acceleration of an object in circular motion.

3. **(B)** This formula describes the potential energy of charge q_2 in the presence of charge q_1 a distance r away.

4. **(A)** This rearrangement of the ideal gas law solves for pressure (P), which is equal to the product of the number of moles (n), the constant of proportionality (R), and temperature (T) divided by volume (V).

5. **(D)** This formula can be rewritten as either

$$\frac{\Delta p}{t},$$

where p represents momentum and t represents time, or

$$\frac{m}{t}(at) = ma,$$

where m represents mass, a represents acceleration, and t represents time. According to Newton's second law of motion, both of these are equal to force.

6. **(D)** The gravitational force between the masses is determined by

$$F = G\frac{m_1m_2}{r^2}.$$

The force is directly proportional to mass, so if one mass is doubled, the force is also doubled.

7. **(E)** The gravitational force between the masses is determined by

$$F = G\frac{m_1m_2}{r^2}.$$

The force is directly proportional to mass, so if both masses are doubled, the force is multiplied by 4.

8. **(A)** The gravitational force between the masses is determined by

$$F = G\frac{m_1m_2}{r^2}.$$

The force is inversely proportional to the square of the distance between the mass, so if the distance is doubled, the force is divided by 1/4.

9. **(C)** The gravitational force between the masses is determined by

$$F = G\frac{m_1m_2}{r^2}.$$

The force is directly proportional to mass, so if one mass is doubled but the other is halved, there is no overall change in the force.

10. **(B)** The gravitational force between the masses is determined by

$$F = G\frac{m_1m_2}{r^2}.$$

The force is directly proportional to mass, so if one mass is divided by two, the force is also divided by two.

11. **(B)** The direction of the force is perpendicular to the magnetic field and to the velocity of the particle. Using the right-hand rule, orient your hand so that your outstretched fingers point in the direction of motion of the particle. When you bend your fingers, they should point downward in the direction of the magnetic field. Your thumb will point to the right, in the direction of the force.

12. **(A)** The direction of the force is perpendicular to the magnetic field and to the velocity of the particle. Using the right-hand rule, orient your hand so that your palm is flat with your fingers pointing toward the right. Your thumb will then point out of the page, in the direction of the force.

13. **(E)** The force is zero if the particle moves parallel to the field lines.

14. **(C)** The direction of the force is perpendicular to the magnetic field and to the velocity of the particle. Using the right-hand rule, orient your hand so that your palm faces you and your fingers point toward the left. Your thumb will point upward, in the direction of the force.

Part B

15. **(D)** Consider X as $^{a}_{b}X$. *Particle X* must balance the equation. Solve $27 + 4 = 30 + a$, so $a = 1$. Solve $13 + 2 = 15 + b$, so $b = 0$. Thus, X must be $^{1}_{0}n$.

16. **(E)** Two pulses traveling in opposite directions on the same string pass each other without being changed.

17. **(B)** The heat loss to cool the steam from 120°C to liquid at 110°C is found by Q = (10 kg) (10°C) × (2.01 × 10^3 J/kg°C) = 2.01 × 10^5 J.

18. **(E)** Photoelectricity is produced when electrons are emitted from a surface as a result of photons of incident light. According to the Compton effect, scattered light has a slightly shorter wavelength than incident light because of a loss of energy that occurs as photons collide with electrons in the material. Polarization occurs when light waves in specific directions are blocked from passing through a material.

19. **(B)** Both vectors have the same magnitude. The westward vector combines with the southward vector to form a vector in the southwest direction.

20. **(D)** As the light exits the prism, it is bent at the same angle at which it entered the prism.

21. **(E)** The voltage is equal to the current measured at the ammeter multiplied by the resistance of R_3. $V = (5)(10) = 50$ V.

22. **(C)** The 5.0 A enters the junction leading to R_1 and R_2. Because $R_1 = R_2$, the current divides evenly.

23. **(E)** The distance $d = 1/2\ at^2$ where a is the acceleration and t is the time. For this object, $d = 1/2(6.0 \text{ m/s}^2)(4.0 \text{ s})^2 = 48$ m.

24. **(B)** The efficiency is found by $1 - \dfrac{30 \text{ J}}{40 \text{ J}} \times 100\% = 25\%$.

25. **(C)** Electrons are negatively charged subatomic particles. Protons are positively charged subatomic particles. Neutrons are neutral subatomic particles. In an atom, the electrons and protons are balanced. An ion is an atom that has gained or lost electrons. The difference between the number of electrons and protons determines the charge of the ion.

26. **(A)** According the Newton's second law of motion, the net force acting on an object is proportional to the product of mass and acceleration. If the mass is doubled, the force must, therefore, be halved.

27. **(E)** The wavelength is the distance between two consecutive points on the wave. For example, the first cycle of the wave begins at 0 meter and is complete at 3.0 meters.

28. **(A)** The amplitude is the maximum disturbance from the resting position. It is the distance from rest to crest or to trough, which is 0.5 meter.

29. **(B)** The energy provided is found by $E = Pt$. In this case, $E = (1.2 \text{ kw})(2 \text{ hr}) = 2.4$ kwhr. The cost is found by multiplying the number of kilowatt-hours by the price per kilowatt-hour. The total cost, $C = (2.4 \text{ kwhr})(\$0.10/\text{kwhr}) = \$0.24$.

30. **(C)** Using Newton's second law, $\Sigma F = ma$, $(M + m)g - Mg = (M + m)a$. Therefore, $a = mg/(M + m)$.

31. **(E)** On average, the molecules of a gas are far apart from one another. Their average separation is much greater than the diameter of each molecule.

32. **(B)** The magnetic field lines form circles around the current-carrying wire.

33. **(C)** As the wave goes through the slit, it spreads out behind the obstacle. The bending of a wave around an obstacle is diffraction.

34. **(E)** Resistance generally increases with temperature and length. It decreases as the cross-sectional area increases.

35. **(C)** First find the change in the volume of the cylinder: $\Delta V = (0.010 \text{ m}^2)(0.020 \text{ m}) = 2 \times 10^{-4} \text{ m}^3$. Then use the change in volume to calculate the work done: $W = P\Delta V = (5.0 \times 10^5 \text{ N/m}^2)(2 \times 10^{-4} \text{ m}^3) = 100 \text{ J}$.

36. **(B)** Entropy is a measure of a system's disorder. As the ice cube melts, the entropy of the molecules increases. According to the second law of thermodynamics, the entropy of the universe increases through all natural processes.

37. **(D)** Approximately four half-lives occur in 32 days.

$$\left(\frac{1}{2}\right)^4 = \frac{1}{16}$$

38. **(D)** The same current passes through each resistor in a series circuit. If this were not true, charge would have to accumulate somewhere in the circuit.

39. **(D)** The acceleration due to gravity is independent of mass and therefore depends on the speed and height of the airplane only.

40. **(B)** The diffraction pattern would be a series of rectangular, parallel bands of light. The central band would be brightest, and the bands to the side decrease in brightness in either direction.

41. **(E)** The horizontal portions of the diagram indicate that heat is being added to or removed from the sample without resulting in a temperature change. These changes in heat result in phase changes. The length of the line is related to the amount of heat loss required for the change. The length of the steam to liquid water line is considerably longer than the length of the liquid water to ice line.

42. **(D)** The string naturally vibrates at certain frequencies. The lowest of these natural frequencies is the fundamental frequency. At this frequency, the wavelength in the string is twice the length of the string.

43. **(B)** Reading the graph shows that the vehicle accelerated at a constant rate from 5 m/s to 35 m/s in 6 seconds. Acceleration is equal to the change in velocity divided by the elapsed time,

$$a = \frac{30 \text{ m/s}}{6 \text{ s}} = 5 \text{ m/s}^2.$$

44. **(D)** I and II. According to Lenz's law, a current is produced by a changing magnetic flux. Magnetic flux is the dot product of magnetic field and area. An increasing magnetic field produces a current because the magnetic flux is increasing. A rotating loop has a changing magnetic flux because the area vector is changing with respect to the magnetic field. A loop moving parallel to a magnetic field does not have a changing magnetic flux.

45. **(A)** A concave lens always forms a virtual image that is smaller than the object and is oriented in the same direction as the object.

46. **(C)** The voltage of 60 V is applied across both branches of the parallel circuit. The total resistance across the branch with the 30-ohm resistor is 40 ohms. For that branch, $I = V/R =$ $60\ V/40\ \Omega = 1.5$ A.

47. **(B)** The straight line with a positive slope indicates that the velocity is increasing at a constant rate. This means that the object is moving with constant acceleration as indicated by a horizontal line on the acceleration graph.

48. **(D)** The release of a beta particle occurs when a neutron breaks into an electron and a proton. The electron is emitted, and the proton remains in the nucleus. The atomic number depends on the number of protons in the nucleus, so the atomic number increases by 1.

49. **(C)** If the meter stick is along the x-axis, and the motion is in the y-direction, there will not be any observed change in length.

50. **(A)** As the truck moves forward, the sound waves bunch together. An observer hears a sound with a higher frequency than the sound produced by the siren.

51. **(C)** The acceleration of the ball is determined by dividing the square of the velocity by the radius of the circle. It is independent of the mass of the ball.

52. **(A)** Frequency is the number of vibrations per second. Therefore, $\frac{21}{30} = 0.7$ Hz.

53. **(B)** According to Kepler's second law, planets sweep out equal areas in equal periods of time. Therefore, planets must move faster when they are closer to the sun than when they are farther away.

54. **(E)** The force between two point charges is inversely proportional to the square of the distance between them. If the distance is doubled, the force is divided by $\frac{1}{4}$.

55. **(C)** The number of turns is related to the voltage by

$$N_p = N_s\left(\frac{V_p}{V_s}\right) = 10{,}000\left(\frac{12\ V}{24{,}000}\right) = 5.$$

56. **(C)** Speed is increasing when the graph has a positive slope.

57. **(E)** The acceleration is constant but not zero when the graph shows a straight, but not horizontal, line.

58. **(C)** Wavelength is equal to speed divided by frequency. For this wave, wavelength equals 3.9×10^3 meters per second divided by 2.6×10^4 hertz.

59. **(A)** The negative charge would move in a clockwise circle. A positive charge would move in a counterclockwise direction.

60. **(D)** Once the CD is in motion, its speed does not change. Because it is in circular motion, the direction is constantly changing. Therefore, a point on the edge of the disc accelerates at a constant speed, but changing direction.

61. **(B)** A generator is the opposite of a motor in that it uses the mechanical energy of a moving armature to produce an electric current.

62. **(C)** Calculate the net electric force on the particle due to the other two charges.

$F_{31} = \frac{k(-2Q)(-Q)}{(3d)^2} = \frac{2kQ^2}{9d^2}$ to the right because like charges repel.

$F_{32} = \frac{k(Q)(-Q)}{(d)^2} = \frac{-kQ^2}{d^2}$ to the left because opposite charges attract.

$F = F_{31} + F_{32} = \frac{2kQ^2}{9d^2} - \frac{-kQ^2}{d^2} = -\frac{7kQ^2}{9d^2}$ or $\frac{7}{9}\frac{kQ^2}{d^2}$ to the left.

63. **(B)** The force and mass are constant during this interval. Therefore, the acceleration is also constant. The block continues to speed up, but at a constant rate.

64. **(C)** At $t = 3$ seconds, the force is 6 N. According to Newton's second law of motion, $F = ma$. 6 N = 2 kg (a), which means that a = 3 m/s^2.

65. **(A)** The magnification equals the distance from the image to the lens, 15 centimeters, divided by the distance from the object to the lens, 30 centimeters.

66. **(C)** The expansion ΔL can be found by $\Delta L = \alpha L_o \Delta T$, where α is the coefficient of linear expansion, L_o is the original length, and ΔT is the change in temperature. In this situation, $\Delta L = (12 \times 10^{-6}/°C)(10\ m)(30°C) = 3.6 \times 10^{-3}$ m.

67. **(E)** In addition to his many other contributions to science, de Broglie is credited with recognizing that matter can be described by wave properties just as light can be described by properties of matter.

68. **(D)** The maximum range occurs at an angle of 45° for a given velocity. Therefore, the ball kicked at this angle will travel farther than the ball kicked at a smaller angle.

69. **(E)** All nonzero digits are significant. Zeros appearing between nonzero digits are significant. Zeros at the end of a number with no decimal point written are not significant.

70. **(B)** The frequency is proportional to its energy, $E = hf$, or $f = E/h$. The energy is determined by subtracting the energy of the first state from the energy of the second state: $E_2 - E_1 = -21$ eV – (–13 eV) = –7 eV. The energy is negative because it is emitted. Therefore, frequency = 7 eV/h.

71. **(A)** Compare the energy differences for each transition by subtracting the two values.

For $\Delta E_{4\to3}$ = –3 eV – (–7 eV) = 4 eV
For $\Delta E_{3\to2}$ = –7 eV – (–13 eV) = 6 eV
For $\Delta E_{5\to3}$ = –1 eV – (–7 eV) = 6 eV
For $\Delta E_{4\to2}$ = –3 eV – (–13 eV) = 10 eV
For $\Delta E_{3\to1}$ = –7 eV – (–21 eV) = 13 eV

72. **(E)** Refraction is the bending of light when it moves from one medium to another at an angle. This bending will cause the stem to appear bent.

73. **(C)** If these forces act in opposite directions, the net force is the difference between them (31 N – 13 N = 18 N).

74. **(B)** For small-angle displacements, the period of a simple pendulum depends only on the length of the pendulum.

75. **(B)** Rutherford expected the particles to go directly through the foil. However, about 1 in every 8,000 alpha particles was reflected back to the source. This observation led Rutherford to conclude that an atom contains a small, central core with a positive charge.

Score Sheet

Number of questions correct: ____________________

Less: 0.25 × number of questions wrong: ____________________

(Remember that omitted questions are not counted as wrong.)

Raw score: ____________________

RAW SCORE	SCALED SCORE	RAW SCORE	SCALED SCORE	RAW SCORE	SCALED SCORE	RAW SCORE	SCALED SCORE	RAW SCORE	SCALED SCORE
75	800	52	740	29	600	6	470	–17	300
74	800	51	730	28	590	5	460	–18	290
73	800	50	730	27	590	4	460	–19	290
72	800	49	730	26	580	3	460		
71	800	48	720	25	580	2	450		
70	800	47	720	24	570	1	450		
69	800	46	710	23	560	0	440		
68	800	45	710	22	560	–1	400		
67	800	44	700	21	550	–2	390		
66	800	43	690	20	540	–3	390		
65	800	42	680	19	540	–4	380		
64	800	41	680	18	530	–5	380		
63	790	40	670	17	520	–6	370		
62	790	39	670	16	510	–7	370		
61	780	38	660	15	510	–8	360		
60	780	37	650	14	500	–9	360		
59	770	36	650	13	490	–10	350		

PART III

Physics Topic Review

CHAPTER 1

Measurements and Data Displays

It is often said that mathematics is the language of physics. The reason for this conclusion is that so much of physics deals with measurements and calculations. It is therefore useful to review some basic information involving measurements and data.

Measurements

On the SAT Physics exam, you will be required to interpret and use measurements provided in a variety of formats. Perhaps one of the most important aspects of measurements for you to know to be able to deal with measurements correctly is to be familiar with units of measurements. The units used by physicists and in questions in SAT Physics are those of the International System of Units (SI). The SI has seven base quantities that are assumed to be independent. The base quantities, along with their names and symbols, are shown in Table 1.

TABLE 1. SI Base Units

QUANTITY	UNIT NAME	SYMBOL
Length	meter	m
Mass	kilogram	kg
Time	second	s
Electric current	ampere	A
Thermodynamic temperature	kelvin	K
Amount of substance	mole	mol
Luminous intensity	candela	cd

Derived quantities are those defined in terms of the seven base quantities. Common derived quantities are shown in Table 2. Notice how the unit for each derived quantity consists of a combination of base units.

TABLE 2. Derived Quantities

QUANTITY	UNIT NAME	SYMBOL
Area	square meter	m^2
Volume	cubic meter	m^3
Velocity	meter per second	m/s
Acceleration	meter per second squared	m/s^2
Mass density	kilogram per cubic meter	kg/m^3
Current density	ampere per square meter	A/m^2
Luminance	candela per square meter	cd/m^2

As you complete your review, you will encounter many other SI units. Note that most have more than one expression in terms of SI units. Several examples are listed in Table 3.

TABLE 3. Equivalent Units

QUANTITY	UNIT NAME	SYMBOL	EQUIVALENT
Force	Newton	N	$kg \cdot m/s^2$
Pressure	Pascal	Pa	N/m^2
Energy	Joule	J	$N \cdot m = kg \cdot m^2/s^2$
Cyclic frequency	Hertz	Hz	s^{-1}

Significant Figures

No measuring device can give perfectly accurate measurements without some degree of uncertainty. The rightmost digit is considered to be uncertain. For example, a measured mass of 84.6 g is considered to have an uncertainty of plus or minus 0.1 g. A measured mass of 84.60 has an uncertainty of plus or minus 0.01 g. Digits in a number that are known with some degree of certainty are said to be **significant figures**. There are several basic rules for deciding the number of significant figures:

- All nonzero digits are significant.
  5.648 cm has 4 significant figures.
  3.1 cm has 2 significant figures.
- Zeros within a number are significant.
  2509 m has 4 significant figures.
  32.06 has 4 significant figures.

- Zeros to the left of the first nonzero digit are not significant.
  0.018 g has 2 significant figures.
  0.009 g has 1 significant figure.
- Trailing zeros that are to the right of the decimal point are significant.
  0.50 has 2 significant figures.
  0.0310 has 3 significant figures.
- Trailing zeros are not significant unless the decimal point is indicated.
  540 has 2 significant figures.
  540. has 3 significant figures.

Adding and Subtracting

When you combine measurements, the sum can be no more accurate than the least accurate measurement. Therefore, the sum or difference of measurements can contain no more decimal places than the least accurate measurement. Notice in the following example that the sum or difference is rounded to the correct number of decimal places.

$$\begin{array}{r} 186.0 \\ \underline{+0.423} \\ 186.423 \rightarrow 186.4 \end{array} \qquad \begin{array}{r} 586 \\ \underline{+32.924} \\ 618.924 \rightarrow 619 \end{array} \qquad \begin{array}{r} 254.450 \\ \underline{-192.1} \\ 62.35 \rightarrow 62.4 \end{array}$$

Multiplying and Dividing

As with adding and subtracting, the result can be no more accurate than the least accurate measurement. When multiplying and dividing, it is the number of significant figures you count rather than the decimal places. The product or quotient of measurements can have no more significant figures than the least accurate measurement. Again notice that the product or quotient is rounded to the correct number of significant figures.

$$\begin{array}{r} 4.0 \\ \underline{\times 18.40} \\ 73.60 \rightarrow 74 \end{array} \qquad 12.2\overline{)228.1156} = 18.698 \rightarrow 18.7$$

Scientific Notation

Any ambiguity presented by zeros in measurements can be avoided by using **scientific notation**, which is also known as standard notation or exponential notation. In this form, the number of significant figures is clearly indicated. Scientific notation is generally a method of writing very large or very small numbers.

Numbers written in scientific notation have three parts: the base, the coefficient, and the exponent.

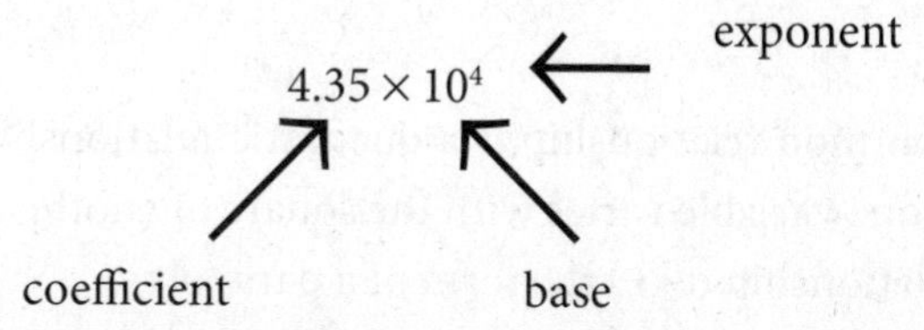

To write a number in scientific notation, write the significant digits as a number between 1 and 10. If the original number was greater than 1, write the number of spaces the decimal point moved as a positive exponent of 10. If the original number was less than 1, write the number of spaces the decimal point moved as a negative exponent of 10.

Example:

Write 12,250,000 in scientific notation.

Move the decimal 7 places to the left. 1.225×10^7

Write 0.000056

Move the decimal 5 places to the right 5.6×10^{-5}

Graphing

Scientific data is often presented in a graph. Although SAT Physics will not require you to draw a graph, you will need to be able to recognize and interpret the information in different kinds of graphs. Most data will be plotted using Cartesian coordinates where the independent variable is plotted along the x-axis and the dependent variable is plotted along the y-axis.

While specific data will vary, there are some general trends that you should be able to recognize easily. A **direct relationship** is one in which two variables increase or decrease simultaneously. In other words, if the independent variable increases or decreases, so does the dependent variable. An **indirect relationship** is one in which the dependent variable changes opposite to the independent variable. For example, if the independent variable increases, the dependent variable decreases.

One type of relationship is a linear relationship. This type of relationship can be described by an equation in the form $y = mx + b$ where x is the independent variable, y is the dependent variable, m is the slope, and b is the y-intercept.

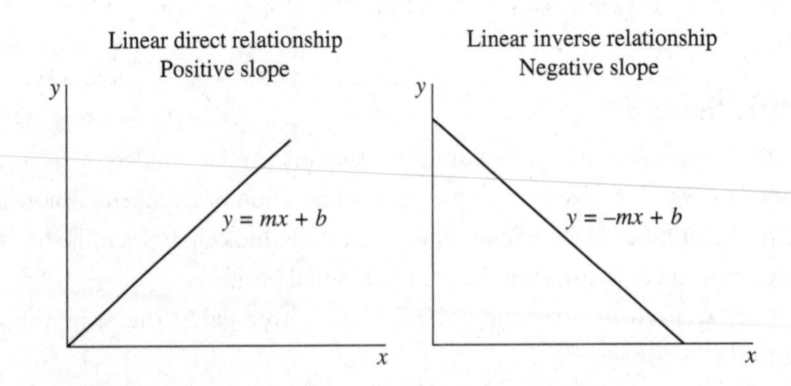

Another common relationship is a quadratic relationship, which is a relationship in which one variable varies with the square of another variable. The graph of a quadratic relationship takes the form of a parabola.

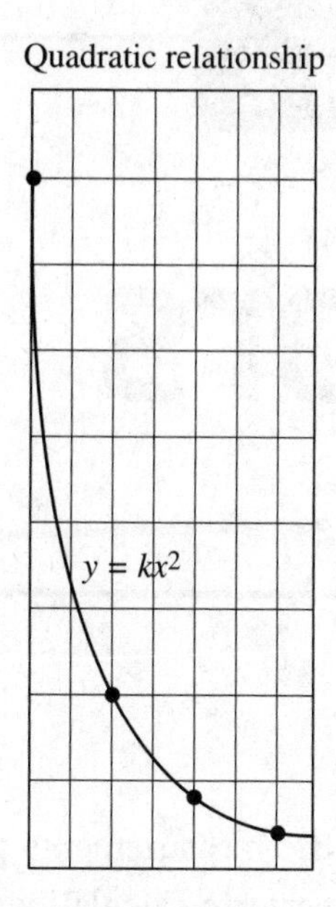

A third relationship you should recognize is the graph of a square root relationship. It takes the shape of the graph shown.

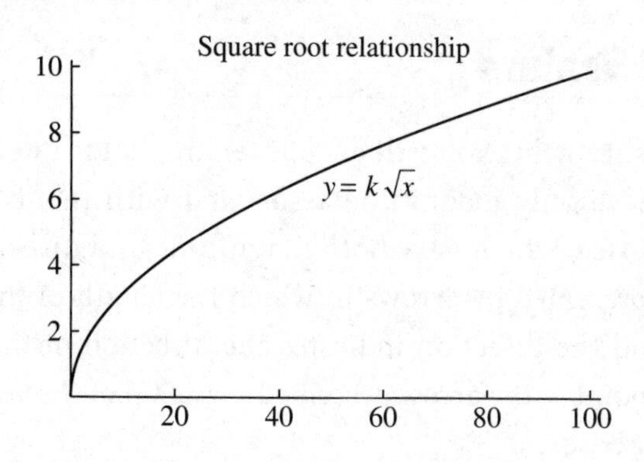

TEST-TAKING HINT

When performing calculations, keep all of the digits until the last step. Then round to the correct number of significant figures.

Finally, there is one graph that may be presented that takes the form of a horizontal line. This type of graph represents a relationship in which the dependent variable remains the same, despite changes in the independent variable. For these types of graphs, pay attention to the variables to be plotted. For example, if the dependent variable is distance, the graph shows that distance (position) stays the same. If the dependent variable is speed, the graph shows that speed stays the same.

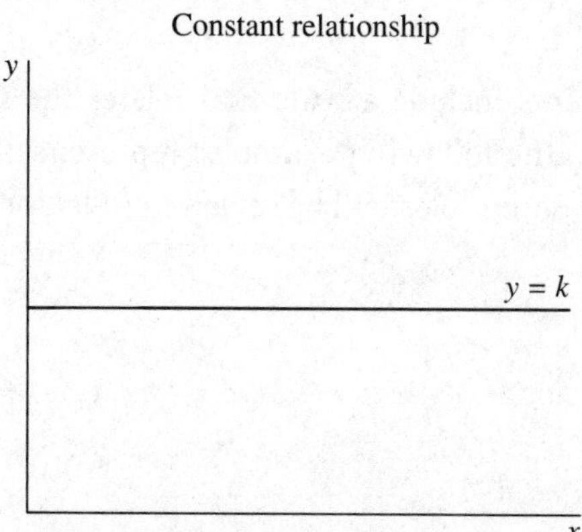

CHAPTER 2

Vectors

Understanding the nature of the measurements presented on SAT Physics will help you relate quantities and perform calculations correctly. In this chapter, you will review specific types of measurements known as vectors.

Vectors and Scalars

Some of the measurements you will encounter are **scalar quantities**, which are quantities that have only magnitude associated with them. Other quantities are **vector quantities**, which have both magnitude and direction.

Vectors are represented by arrows in which the length of the arrow relates to the magnitude and the direction indicates the direction of the vector. You will generally see the point of the arrow described as the tip of the vector and the other ends as the tail, or base.

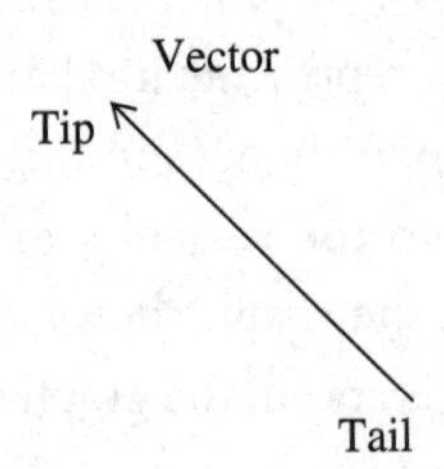

Some vector diagrams include a scale that relates the size of the drawing to actual measurements. The following examples represent displacement in meters and force in newtons. Both topics will be reviewed in later chapters.

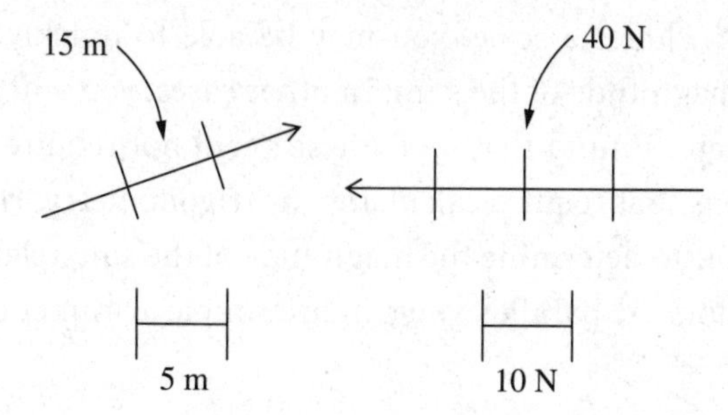

Vectors can be represented in several different ways. One common method is as a letter with an arrow above it, such as $\vec{F}$. SAT Physics uses a boldface letter written in italic print, such as ***F***. The magnitude of a vector is also written in italic print, but not in boldface.

Vector Addition

Two or more vectors can be combined to find the resultant ***R*** through addition. There are two basic methods of vector addition that each yield the same result: the parallelogram method and the tip-to-tail method.

Parallelogram Method

To use this method, draw the vectors so that they extend from the same point and use the same scale. Use the vectors as two sides of a parallelogram and complete the figure by adding the two missing sides. The diagonal of the parallelogram is the sum of the vectors as shown in the diagram.

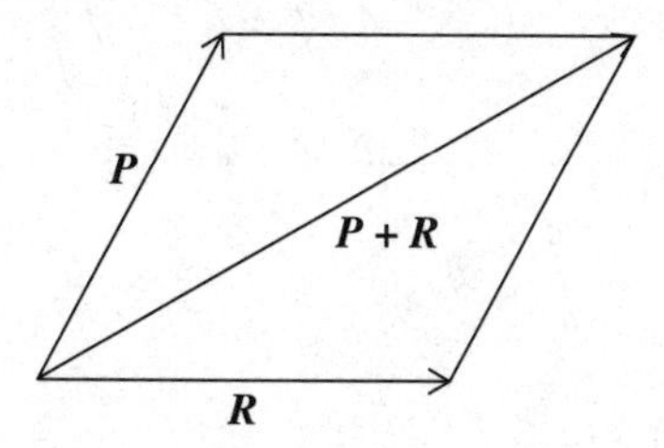

Tip-to-Tail Method

To use this method, place the tail of one vector at the tip of the other. Then draw a vector from the tail of the first vector to the tip of the other. The vector you draw is the sum.

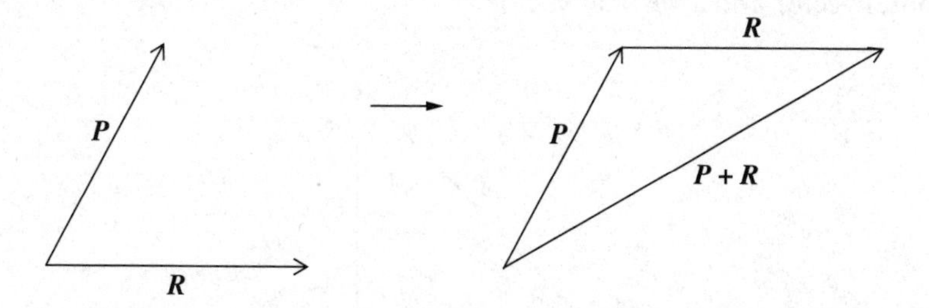

If the diagram includes a scale, you may be able to quickly determine or at least estimate the magnitude of the sum. In other cases, you will need to calculate the magnitude. Keep in mind that SAT Physics will not require you to complete complex calculations that require calculators or trigonometry. However, in many cases you will be able to determine the magnitude of the sum relatively simply. For example, if the vectors are parallel, you can use simple arithmetic.

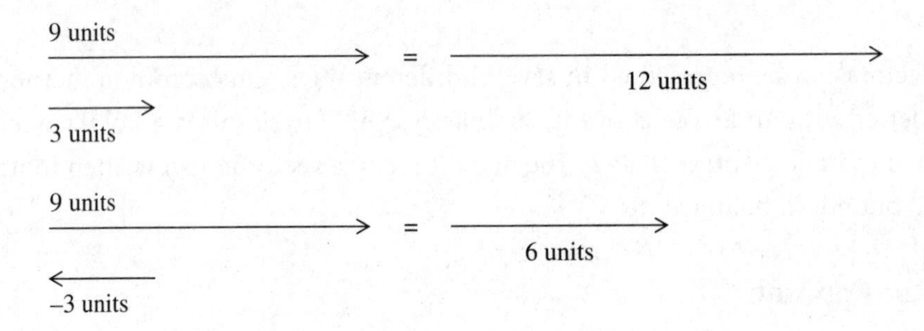

If vectors are perpendicular, you can use the Pythagorean Theorem to find the magnitude of the resultant.

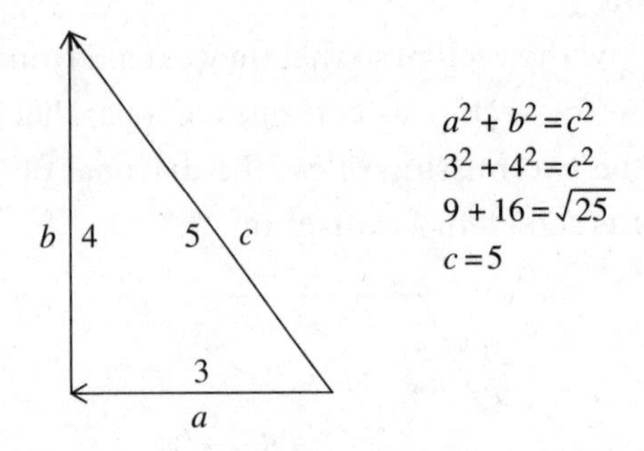

Resolution of Vectors

A single vector may be resolved, or broken down, into its components. Vectors are commonly resolved into horizontal and vertical components so that trigonometric relationships can be used. The sum of the components is the original vector.

Consider a vector $\boldsymbol{F}$ at an angle to the horizontal. To find the components, draw a horizontal vector and a vertical vector.

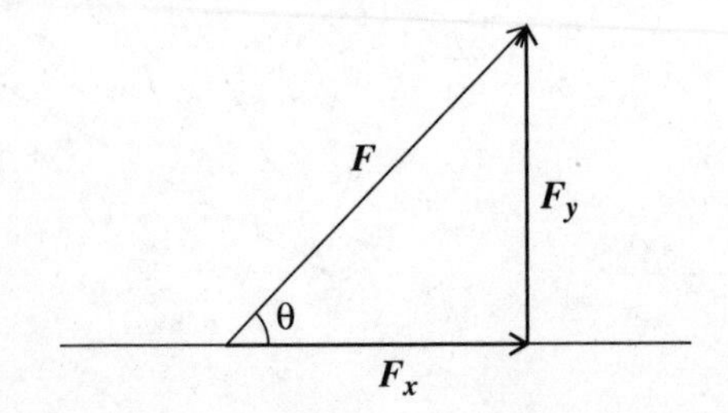

The component vectors form a right triangle. Therefore, you can solve for the magnitude of each component.

$$\sin\theta = \frac{Fy}{F} \qquad \text{therefore } \boldsymbol{F_y} = \boldsymbol{F}\sin\theta$$

$$\cos\theta = \frac{Fx}{F} \qquad \text{therefore } \boldsymbol{F_x} = \boldsymbol{F}\cos\theta$$

Example:

A force is exerted at an angle of 30° with the horizontal. If the magnitude of the force vector is 20 newtons, what are the vertical and horizontal components?

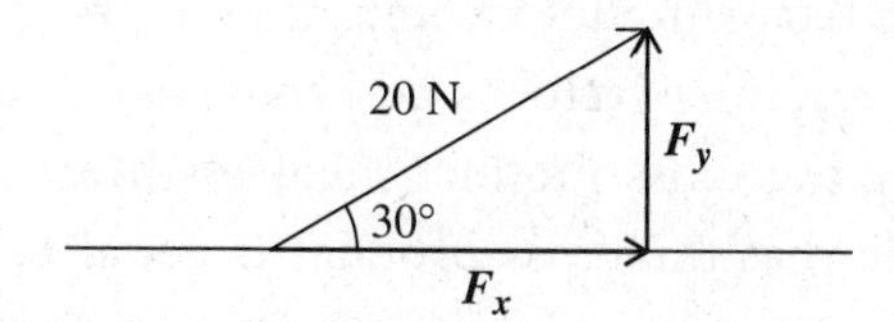

$$\boldsymbol{F_y} = \boldsymbol{F}\sin\theta = (20\text{ N})(0.5) = 10\text{ N}$$

$$\boldsymbol{F_x} = \boldsymbol{F}\cos\theta = (20\text{ N})(0.866) = 17\text{ N}$$

Vector Multiplication

You may encounter problems that involve vector multiplication. There are two basic forms of vector multiplication: dot product and cross product.

Dot Product

The **dot product,** or scalar product, combines vectors in such a way that the result is a scalar. Essentially, the dot product involves multiplying the component of one vector in the direction of the other by the magnitude of the other vector. Notice in the diagram that the dot product of the vectors equals the product of the magnitude of the vectors multiplied by the cosine of the angle between them.

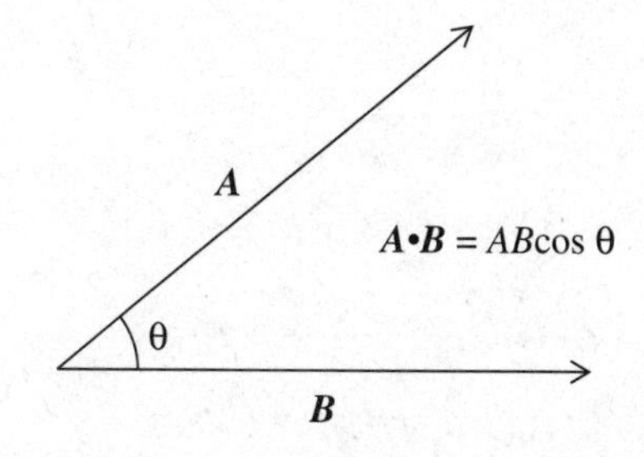

Example:

Find the dot product of the vectors F_1 and F_2.

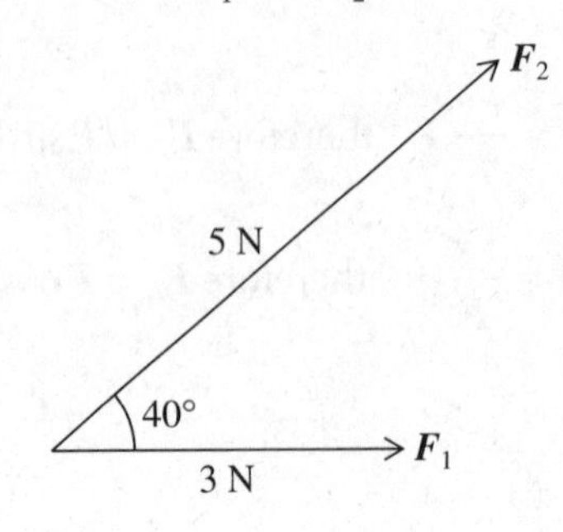

$$F_1 \cdot F_2 = (3)(5)\cos 40° = 11.49$$

Cross Product

The **cross product,** or vector product, combines vectors in such a way that the result is a vector. The resulting vector is perpendicular to both of the original vectors. Thus, finding the cross product involves three dimensions. Note in the following example that the cross product is equal to the product of the magnitudes of the vectors times the sin of the angle between them and $\hat{n}$, which is a unit vector perpendicular to both vectors.

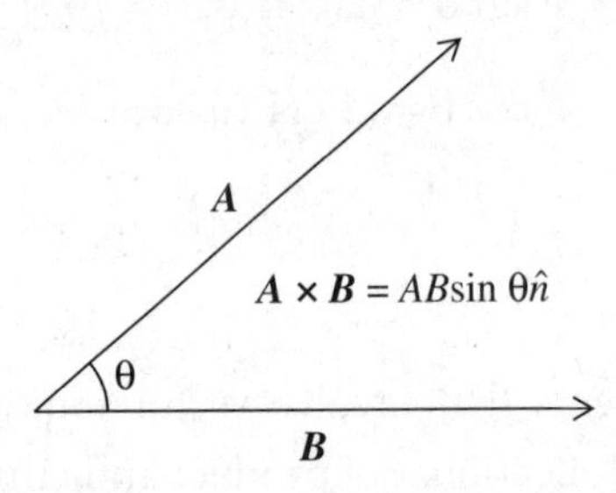

TEST-TAKING HINT

If you are right-handed, you may be tempted to use your left hand as you are writing on the SAT Physics exam. However, when directions are chosen according to the right-hand rule, you must use your right hand to determine relationships. Using your left hand will result in incorrect answers.

To determine the direction of $\hat{n}$, you need to use the **right-hand rule**. According to this rule, point your forefinger in the direction of **A** and your middle finger in the direction of **B**. Your thumb will point in the direction of vector $\hat{n}$.

Example:

Find the cross product of the vectors F_1 and F_2.

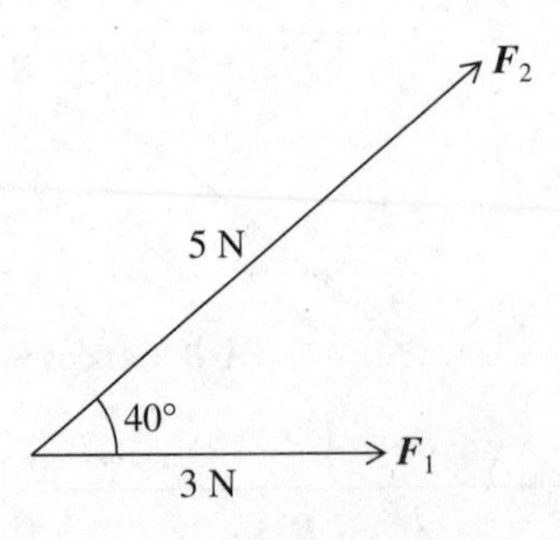

$$F_1 \times F_2 = (3)(5)\sin 40° \hat{n} = 9.64\,\hat{n}$$

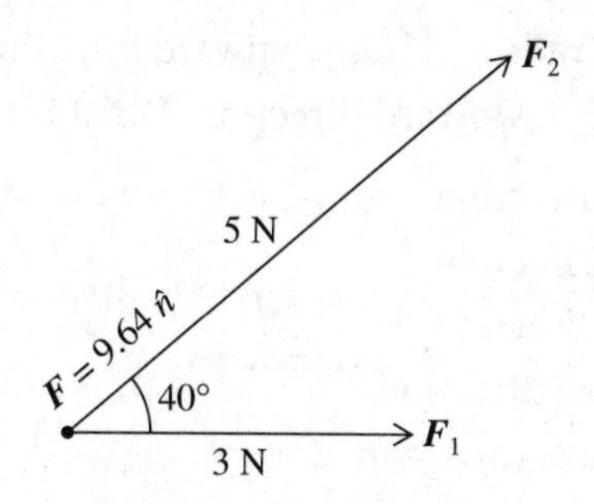

REVIEW QUESTIONS

Questions 1 and 2 refer to the following force values.

(A) 50.0 N
(B) 32.5 N
(C) 43.0 N
(D) 25.0 N
(E) 50.6 N

The diagram below shows a person exerting a force of 50.0 N on a sled at an angle of 30.0°.

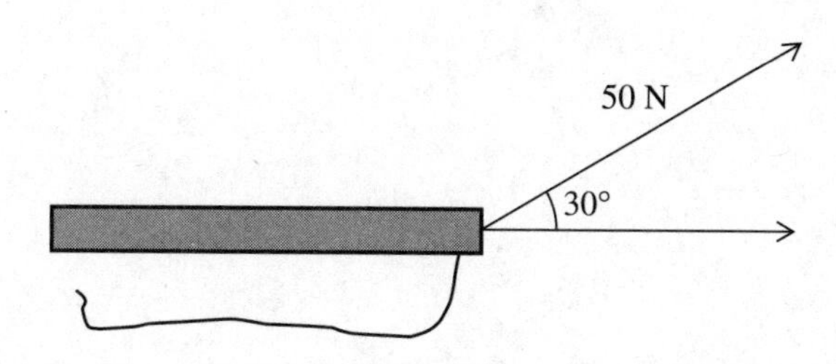

1. What is the horizontal component of the force?

2. What is the vertical component of the force?

3. Which of the following examples describes a vector quantity?

(A) It took 9.2 minutes for the water to boil.
(B) She exerted a force of 5 newtons downward on the board.
(C) The object has a mass of 6.5 kilograms.
(D) An acorn moved about 10 meters.
(E) The temperature rose 23°C.

4. Vector $\boldsymbol{A}$ has a magnitude of 4 in the upward direction and $\boldsymbol{B}$ has a magnitude of 3 in the downward direction. What is the value of $2\boldsymbol{A} + \boldsymbol{B}$?

(A) 2 in the upward direction
(B) 6 in the upward direction
(C) 5 in the upward direction
(D) 5 in the downward direction
(E) 11 in the downward direction

5. Which vector represents $\boldsymbol{A} + \boldsymbol{B}$?

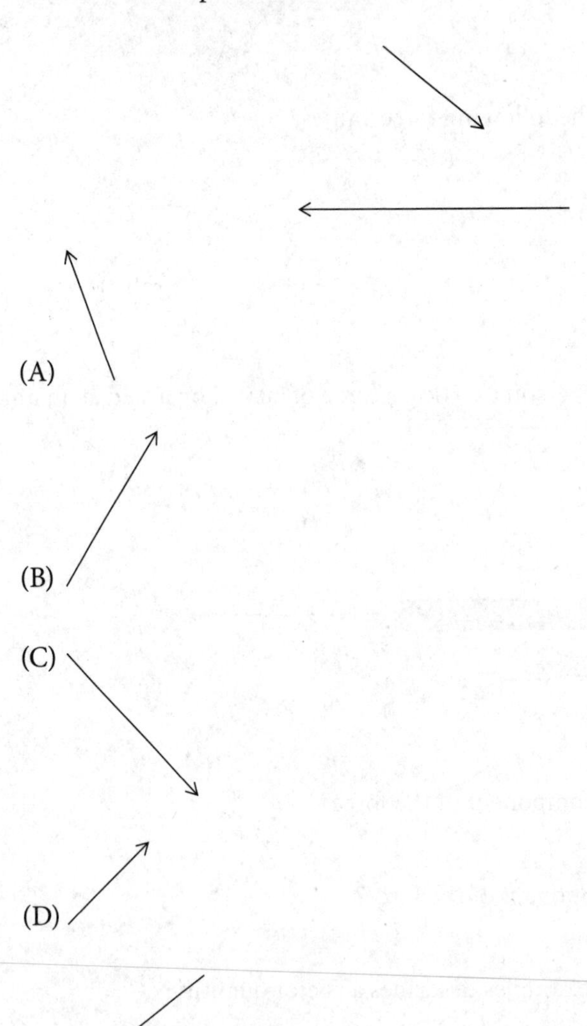

6. Vector $\boldsymbol{A}$ has a magnitude of 12.0 and Vector $\boldsymbol{B}$ has a magnitude of 3.0. If the vectors are at an angle of 60.0°, what is the dot product $\boldsymbol{A} \bullet \boldsymbol{B}$?

(A) 2
(B) 4
(C) 15
(D) 18
(E) 36

7. Vector $\boldsymbol{A}$ has a magnitude of 9.0 and Vector $\boldsymbol{B}$ has a magnitude of 3.0. If the vectors are at an angle of 30.0°, what is the magnitude of the cross product $\boldsymbol{A} \times \boldsymbol{B}$?

(A) 13.5
(B) 16.2
(C) 23.4
(D) 27.0
(E) 32.0

8. Vector $\boldsymbol{A}$ points into the page. Vector $\boldsymbol{B}$ points downward, parallel to the plane of the page. Which direction describes the cross product $\boldsymbol{A} \times \boldsymbol{B}$?

(A) out of the page
(B) to the left, parallel to the plane of the page
(C) to the right, parallel to the plane of the page
(D) upward, parallel to the plane of the page
(E) into the page

9. Which statement about $\boldsymbol{A} \times \boldsymbol{B}$ is true?

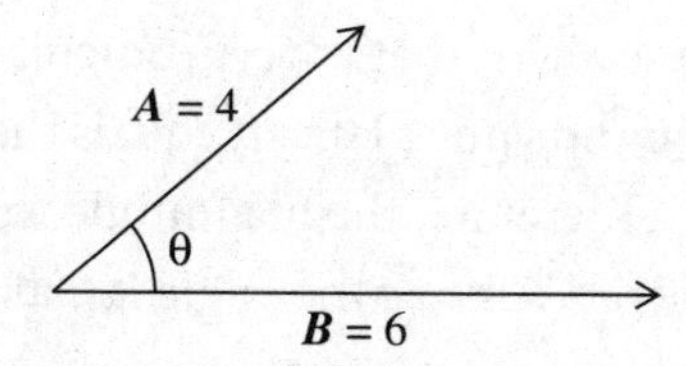

(A) It is in the plane of the page.
(B) Its magnitude must be greater than 24.
(C) It makes a greater angle with $\boldsymbol{B}$ than with $\boldsymbol{A}$.
(D) It is a vector that points into the page.
(E) It is equal in magnitude to the sum of the magnitudes of $\boldsymbol{A}$ and $\boldsymbol{B}$.

10. An airplane flies with a velocity of 100 km/h directly east. A wind blows on the airplane with a velocity of 30 km/h directly north. What is the magnitude of the resultant velocity of the airplane?

(A) 0 km/h
(B) 30 km/h
(C) 70 km/h
(D) 90 km/h
(E) 100 km/h

ANSWERS AND EXPLANATIONS

1. **(C)** The horizontal component is found by $50 \cos 30 = 50 \times \frac{\sqrt{3}}{2} = 25\sqrt{3} = 43$ N.
2. **(D)** The vertical component is found by $50 \sin 30 = 50 \times \frac{1}{2} = 25$ N.
3. **(B)** A vector quantity has both magnitude and direction. All of the answer choices except for B include magnitude only. The force is described by its magnitude, or 5 N, and direction, or downward.
4. **(C)** The magnitude of the vector in the upward direction is doubled, 8. Adding the magnitude of the vector in the downward direction is the same as subtraction. $8 - 3 = 5$.
5. **(E)** Placing the vectors tip-to-tail shows the resulting vector.

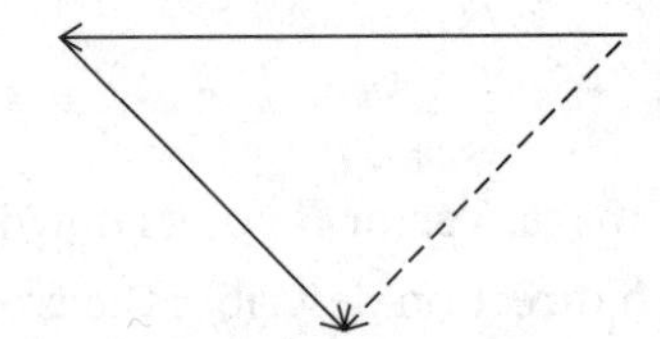

6. **(D)** The dot product $\boldsymbol{A} \bullet \boldsymbol{B} = (3.0)(12.0) \cos 60.0° = 18.0$.
7. **(A)** The cross product $\boldsymbol{A} \times \boldsymbol{B} = (9.0)(3.0) \sin 30.0° = 13.5$.
8. **(B)** According to the right-hand rule, if the forefinger points in the direction of $\boldsymbol{A}$ and the middle finger points in the direction of $\boldsymbol{B}$, the thumb points in the direction of the cross product.
9. **(D)** The cross product is a vector that is perpendicular to both of the original vectors so it cannot be in the same plane. It equals the product of 24 and sin θ, where θ is less than 90°. Therefore, the magnitude would be less than 24. The angle it makes with $\boldsymbol{A}$ is the same as the angle it makes with $\boldsymbol{B}$. It is not found by adding the magnitudes of the vectors.
10. **(E)** Draw the vectors representing the velocity of the airplane and the wind tip-to-tail. Use the Pythagorean Theorem to find the resultant.

$(100)^2 + (30)^2 = R^2$

$10{,}000 + 900 = R^2$

$100 \approx R$

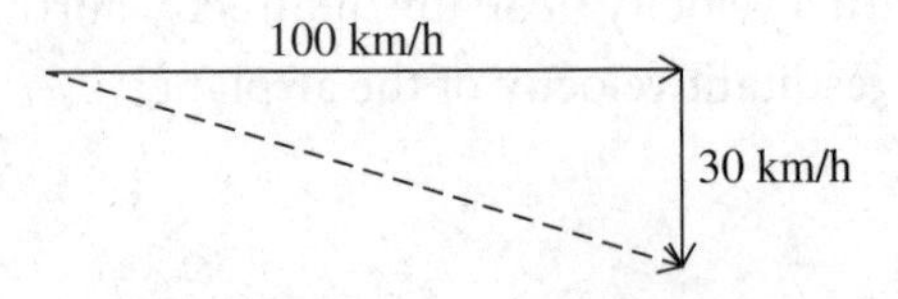

CHAPTER 3

Kinematics

Physical bodies move in many different ways. Kinematics is the study of the motion of objects. Understanding the principles of kinematics will enable you to describe how bodies move so that when you review dynamics in the next chapter, you will be able to explain the causes for motion.

Displacement

A moving object can take many different paths to the same ending point. **Displacement** is the change in position of an object. To evaluate displacement, compare only the starting and ending points. Displacement is a vector quantity because it has both magnitude and direction. As you can see in the diagram, displacement is different from the **distance** traveled. SAT Physics may require you to distinguish between displacement and distance.

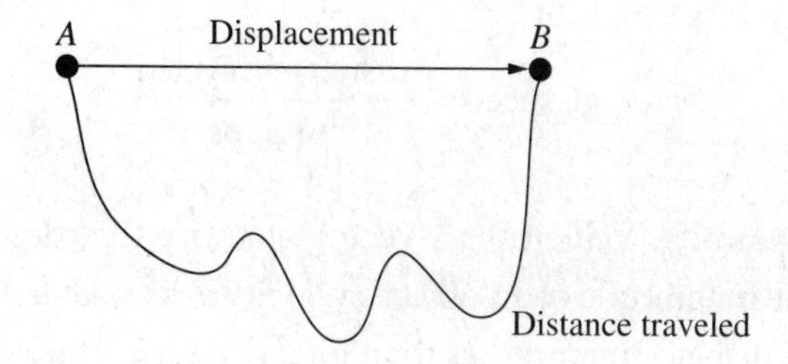

Example:

Tyler and Maria are biologists searching for evidence of sea turtle nests along a shoreline. Tyler walks 30 m directly south along the water. Maria investigates several points up and down the beach. When Tyler and Maria get to the end point, who has walked the farther distance? What are their displacements?

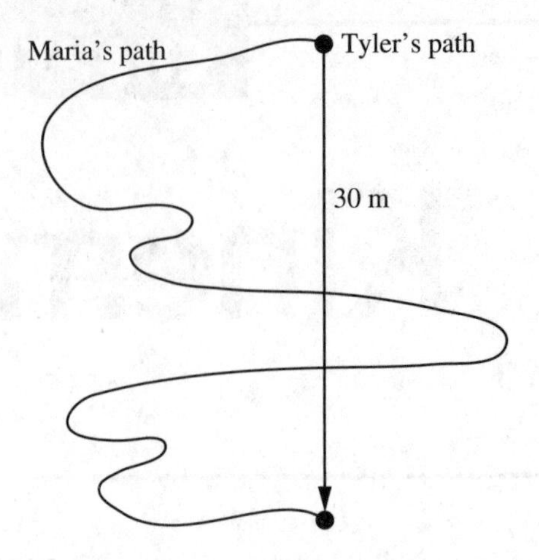

Tyler took a direct path, whereas Maria took an indirect path. As a result, Maria traveled a greater distance. Her path is clearly longer. Both Tyler and Maria started and ended at the same point. Therefore, both biologists have the same displacement of 30 m south. For Tyler only, the distance traveled is equal to the displacement.

Once Tyler and Maria get to the end point, both biologists turn around and walk 30 m directly north to return to their starting point. Now how does the distance Tyler walked compare to his displacement?

Tyler has walked a total distance of 60 m, but his displacement is zero because his ending point is the same as his starting point.

Speed and Velocity

When you observe a moving object, one of the first aspects you may notice is how fast it is moving. The **average speed** of an object is the distance traveled divided by the time it takes to travel this distance.

$$\text{average speed} = \frac{\text{distance traveled}}{\text{time elapsed}}$$

Speed is a scalar quantity. Velocity is a vector quantity that describes speed in a given direction. The magnitude of an object's velocity is its speed. Average velocity is defined in terms of displacement rather than total distance. The **average velocity** (v) of an object is its displacement divided by time.

$$\text{average velocity} = \frac{\text{displacement}}{\text{time elapsed}}$$

You can use the Greek letter delta Δ to represent a change in a quantity. For example, Δx can represent a change in displacement and Δt can represent a change in time. A bar over a letter is commonly used to indicate an average. Thus the average velocity can be written as follows.

$$\bar{v} = \frac{x_2 - x_1}{t_2 - t_1} = \frac{\Delta x}{\Delta t}$$

Example:

A cyclist takes 2.5 h to travel 40.0 km along a straight road toward the east. What is the average velocity of the cyclist?

$$\bar{v} = \frac{\Delta x}{\Delta t} = \frac{40\text{ km}}{2.5\text{ h}} = 16\text{ km/h east}$$

The change in position is 40 km and the change in time is 2.5 h. The result is an average velocity of 16 km/h.

Instantaneous Velocity

The previous discussions involved averages determined over a given time interval. Many problems on SAT Physics involve velocity at a given moment. **Instantaneous velocity** at any moment is the average velocity over an infinitesimally short interval of time. Unless a question specifically mentions average velocity, assume it is looking for instantaneous velocity. Also note that if an object moves at a constant, or uniform, velocity over some time interval, then its instantaneous velocity at any instant is equal to its average velocity.

Accelerated Motion

Motion in which the velocity is constant is considered uniform motion. If the speed, direction, or both speed and direction change, then velocity is not constant. If the velocity changes, the motion is said to be accelerated. **Acceleration** (a) is change in velocity divided by the time it took to make that change. Because acceleration involves velocity, it too is a vector quantity.

$$\text{average acceleration} = \frac{\text{change of velocity}}{\text{time elapsed}}$$

$$\bar{a} = \frac{v_2 - v_1}{t_2 - t_1} = \frac{\Delta v}{\Delta t}$$

You can see from the formula that a unit of velocity is being divided by a unit of time. Therefore, units of acceleration involve a unit of velocity, such as meters per second, and a unit of time, such as seconds. A common unit of acceleration is the meter per second per second (m/s^2).

Example:

A car accelerates from rest to 64 km/h in 4.0 s. What is the magnitude of its average acceleration?

The car starts from rest, so $v_1 = 0$. The final velocity, v_2, is 64 km/h.

$$\bar{a} = \frac{64\text{ km/h}}{4.0\text{ s}} = 16\ \frac{\text{km/h}}{\text{s}}$$

The average acceleration of the car is 16 kilometers per hour per second.

Uniformly Accelerated Motion

When acceleration is constant and in one dimension, the motion is also known as **uniformly accelerated motion**. In other words, the object is moving in a straight line and acceleration stays the same. For this type of motion, you should know the following kinematic equations that relate position x, initial velocity v_o, final velocity v_f, acceleration a, and elapsed time t. SAT Physics questions will often give you three of the five variables and ask you to solve for the fourth. The fifth will not be involved in the question. Knowing this, you'll need to choose the equation that omits the variable that is not involved in the question.

$$v = v_o + at$$

$$x = x_o + v_o t + \frac{1}{2}at^2$$

$$v^2 = v_o^2 + 2a(x - x_o)$$

$$\bar{v} = \frac{v + v_o}{2}$$

Remember that these equations are valid only when acceleration is constant. In many questions, x_o, which is the object's position at $t = 0$, is zero, so the equations are simplified even further. Also make sure you recognize that x represents position and not distance. Therefore, $x - x_o$ is displacement.

Example:

A driver steps on the brakes of a car, causing the car to slow to a stop at a constant acceleration of −6.0 m/s². If the car is moving at an initial velocity of 32 m/s, estimate the displacement of the car as it comes to a stop.

List the known information and the unknown information.

KNOWN	UNKNOWN
$v_i = 32$ m/s	$x - x_o$
$v_f = 0$ m/s	
$a = -6.0$ m/s²	

Choose the equation that contains the known and unknown variables.

$$v^2 = v_o^2 + 2a(x - x_o)$$

Substitute the values into the equation.

$$0 = (32 \text{ m/s})^2 + 2(-6.0 \text{ m/s}^2)(x - x_o)$$

$$0 = 1024 \text{ m}^2/\text{s}^2 + (-12.0 \text{ m/s}^2)(x - x_o)$$

$$(12.0 \text{ m/s}^2)(x - x_o) = 1024 \text{ m}^2/\text{s}^2$$

$$(x - x_o) = 1024 \text{ m}^2/\text{s}^2/(12.0 \text{ m/s}^2)$$

$$(x - x_o) = 85.3 \text{ m} \approx 85 \text{ m}$$

The car travels about 85 m as it comes to a stop.

Graphing Motion

Many of the SAT Physics questions involving kinematics will provide graphs plotting position velocity, or acceleration against time. Being prepared to read these kinds of graphs will help you to solve problems quickly and efficiently.

Position versus Time Graph

First consider a position versus time graph. Examine the following graph, which represents the motion of a car moving at a constant velocity of 12 m/s. You can see that x increases by 12 m each second. Because the position increases linearly with time, the graph of x versus t is a straight line. Each point on the graph represents the position of the car at a particular time. For example, at 3.0 s, the car is at 36 m. At 4.0 s, the car is at 48 m.

The shaded triangle indicates the slope of the line, which is the change in the dependent variable (Δx) divided by the corresponding change in the independent variable (Δt).

$$\text{slope} = \frac{\Delta x}{\Delta t}$$

If you look at the equation for slope, you will notice that it is the same as the equation for velocity. For this graph, $\Delta x/\Delta t$ = (12 m)/(1.0 s) = 12 m/s, which was the velocity given at the beginning of the example.

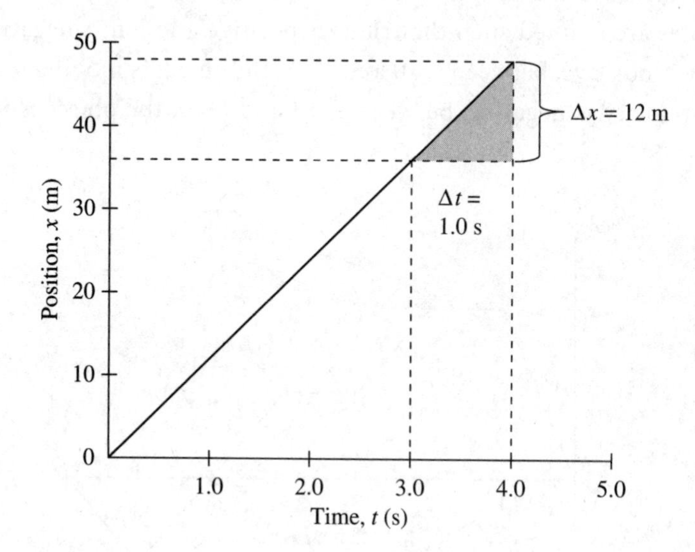

Calculating the velocity of a straight-line graph is relatively straightforward. However, if a graph is curved because velocity changes, calculating velocity requires calculus. Fortunately, SAT Physics does not require you to use calculus. If you encounter a curved position versus time graph, you will need to consider qualitative questions about the motion. Most likely, you will need to identify the point with the greatest or least velocity. Keep in mind that points at the top of peaks or the bottom of valleys have a slope of zero and therefore have zero velocity.

In the following graph, the velocity is zero at points A, B, and D. The velocity is greatest at point C and least at point E. Note also that the velocity at point E is negative, whereas the velocity at point C is positive.

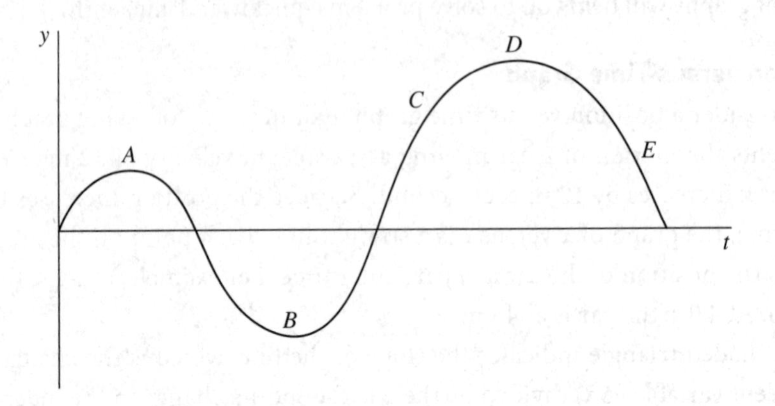

Velocity versus Time Graph

Now consider a velocity versus time graph. This type of graph indicates the velocity of an object at a given time. By looking at the following graph, you can see the velocity at any given time. For example, at 1 second after it started moving, the object is moving at 2 m/s. At 2 seconds the object is still moving at 2 m/s, but at 3 seconds, the object is moving at 1 m/s.

You can also use the graph to determine the direction of the object. The y-coordinates are defined such that right is positive and left is negative. When the velocity is positive, between $t = 0$ and $t = 5$, the object is moving to the right. When the velocity is negative, between $t = 5$ and $t = 7$, the object is moving to the left.

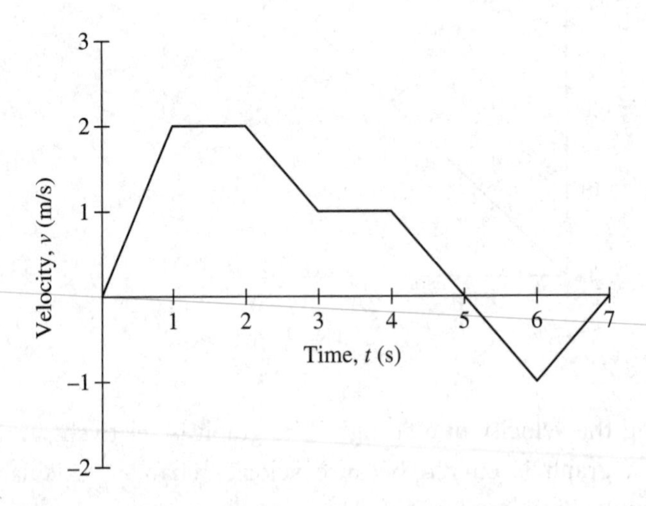

Just as you found velocity as the slope of the positive versus time graph, you can calculate acceleration from a velocity versus time graph. Where the graph is horizontal, the acceleration is zero. Therefore, acceleration is zero between $t = 1$

and $t = 2$ and between $t = 3$ and $t = 4$. The acceleration between $t = 4$ and $t = 6$ is constant, so you can find the acceleration at $t = 5$ by calculating the average acceleration between $t = 4$ and $t = 6$.

$$\text{average acceleration} = \frac{v_{final} - v_{initial}}{t_{final} - t_{initial}} = \frac{-1-(1)\text{ m/s}}{6-4\text{ s}} = \frac{-2\text{ m/s}}{2\text{ s}} = -1\text{ m/s}^2$$

You can also use a velocity versus time graph to calculate displacement. To do so, divide the time axis into subintervals. The displacement in a given time interval equals the area under the graph during that interval. For example, in the following graph, the displacement between $t = 1$ and $t = 2$ is the area of the rectangle shown ($\Delta t \times v$), or 1 s × 2 m/s = 2 m to the right.

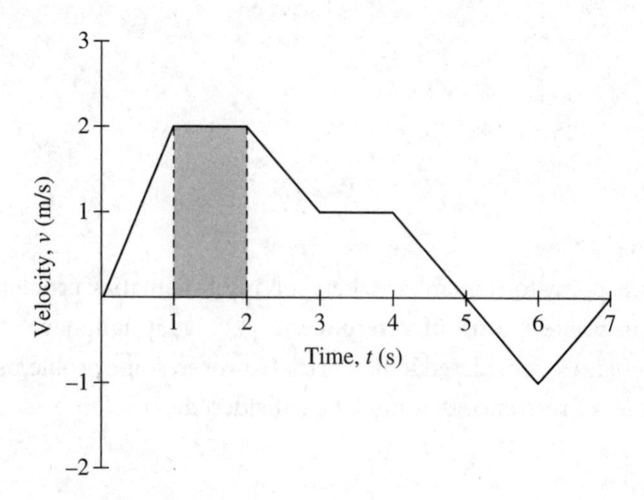

Acceleration versus Time Graph

Examine acceleration versus time graphs in much the same way as the previous graphs. You can see that the object represented by the following graph has positive acceleration between $t = 0$ and $t = 1$ and negative acceleration between $t = 4$ and $t = 5$. It accelerates at a constant rate of 1 m/s² between $t = 1$ and $t = 4$. The object is not accelerating between $t = 5$ and $t = 7$, which means that its velocity is constant.

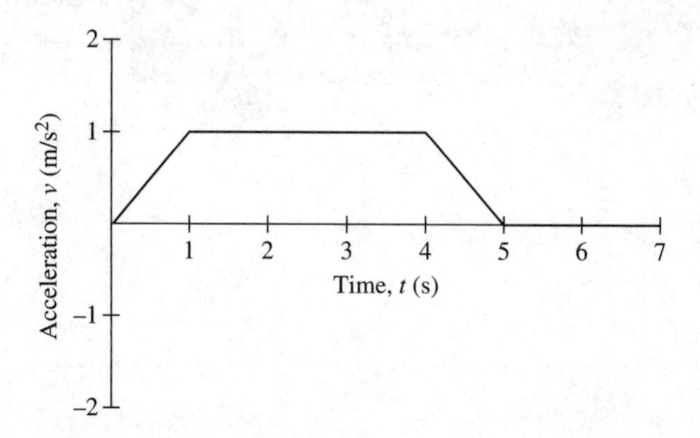

Just as you used a velocity versus time graph to calculate displacement, you can use an acceleration versus time graph to calculate the change in velocity. The change in velocity in a given time interval equals the area under the graph during that interval. The acceleration is constant between $t = 1$ and $t = 4$, so the change in velocity during that interval is equal to the area of the shaded rectangle: $3\ \text{s} \times 1\ \text{m/s}^2 = 3\ \text{m/s}$. The velocity at $t = 4$ is 3 m/s greater than it was at $t = 1$.

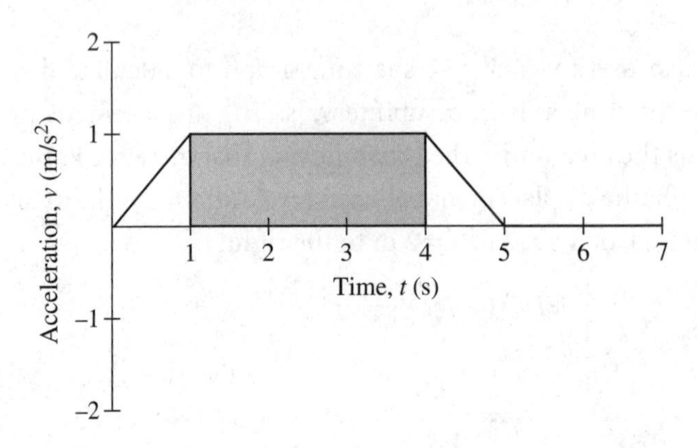

TEST-TAKING HINT

You might think that some questions are missing information. However, there may be clues that will help you translate information in questions into values for kinematic equations. For example, the phrase *starts from rest* tells you that $v_o = 0$. The phrase *comes to rest* tells you that $v_f = 0$, and the phrase *moves at constant velocity* tells you that $a = 0$.

Frame of Reference

One important point to remember when solving kinematics problems is to consider the appropriate **frame of reference**. In everyday language, the frame of reference is generally considered to be Earth. However, some problems will include alternate frames of reference that must be considered.

Example:

Imagine a passenger on a train traveling at 65 km/h with respect to the ground. Another passenger walks to the front of the train at 4 km/h with respect to the train. The walking passenger is actually moving at 69 km/h (65 km/h + 4 km/h) with respect to the ground. As you can see, it is always important to look for clues that describe frame of reference in both problems and solutions.

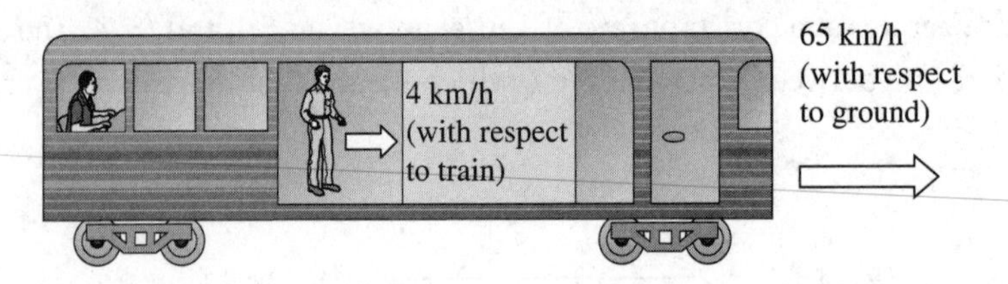

REVIEW QUESTIONS

Questions 1 and 2 refer to the following speeds.

(A) 5.2 m/s
(B) 17.0 m/s
(C) 12.0 m/s
(D) 10.4 m/s
(E) 45.2 m/s

1. Usain Bolt ran straight from the start to the finish line of the 100-meter dash in 9.58 s. What was his average speed?

2. The graph below shows the acceleration of a particle. At $t = 0$, the speed of the particle is 1 m/s. What is the speed of the particle at $t = 5$?

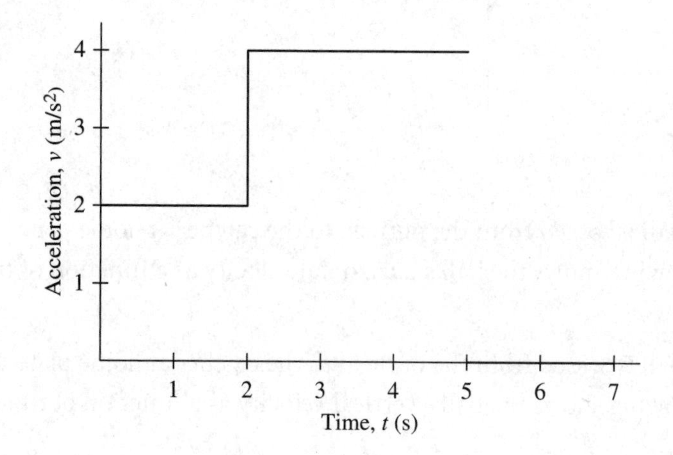

3. A shopper in a hardware store pushes a cart 25 m north up one aisle. She then moves 12 m west before turning south and moving 30 m. She moves 12 m east in order to pay for her purchase. What is her displacement?

(A) 15 m north
(B) 5 m south
(C) 5 m west
(D) 55 m south
(E) 79 m east

Questions 4 and 5

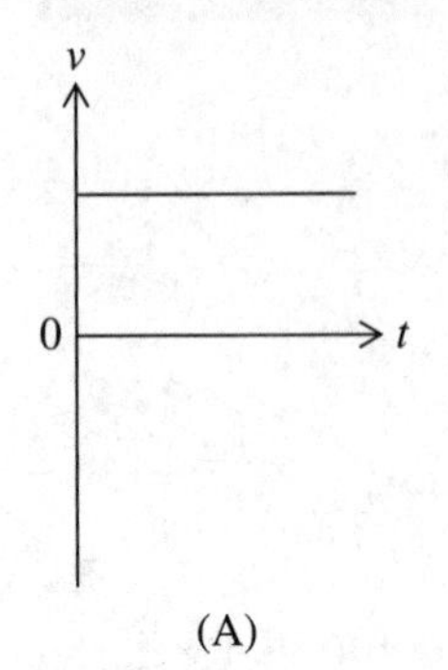

(A)

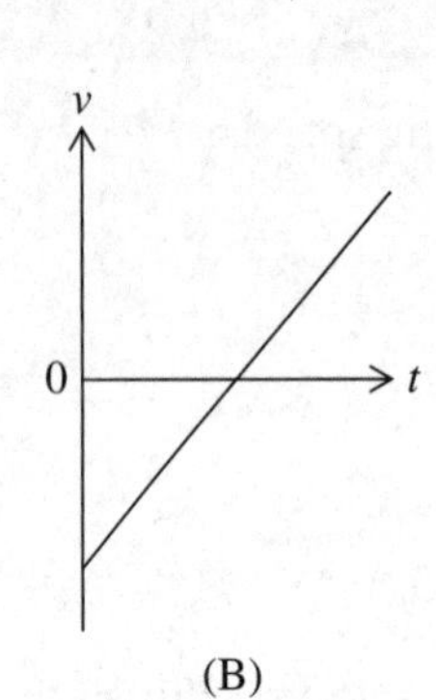

(B)

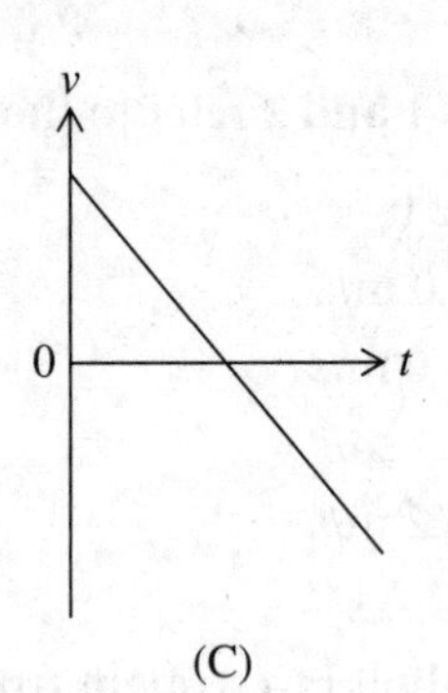

(C)

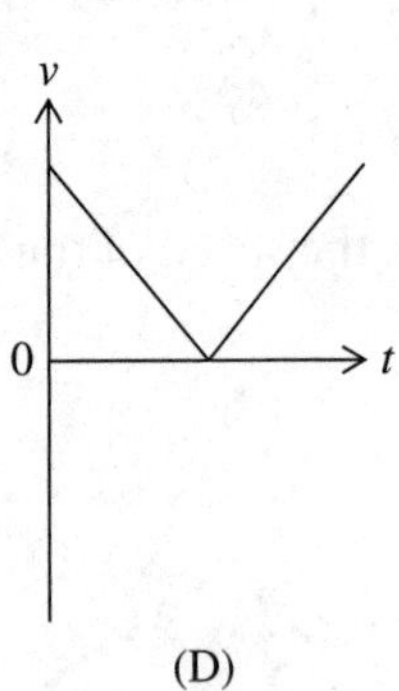

(D)

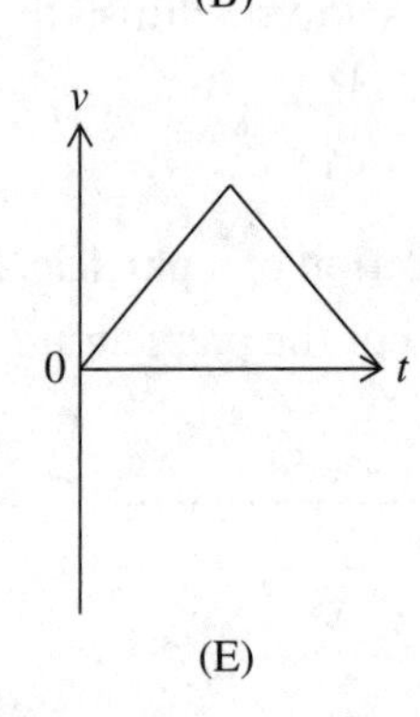

(E)

4. A baseball is tossed from the outfield to the catcher at home plate. Which of the following shows the ball's horizontal velocity as a function of time?

5. A baseball is tossed from the outfield to the catcher at home plate. Which of the following shows the ball's vertical velocity as a function of time?

6. How long does it take a car to travel 50.0 m if it accelerates from rest at a constant rate of 3.00 m/s^2?

(A) 3.05 s
(B) 4.08 s
(C) 5.77 s
(D) 12.2 s
(E) 16.7 s

Questions 7–10 relate to the graph below, which represents the motion of Cars A and B on straight parallel paths. Car B passes Car A at the same instant that Car A starts from rest at $t = 0$ s.

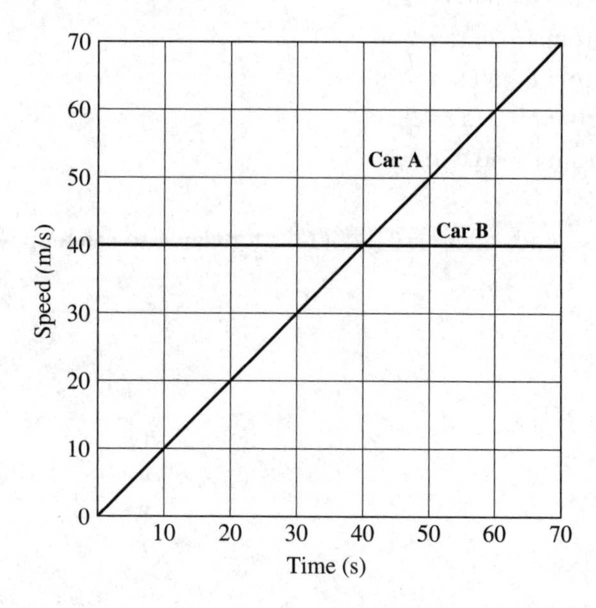

7. What is the acceleration of Car A during the interval between $t = 0$ and $t =$ 70.0 seconds?

(A) 0 m/s^2
(B) 1 m/s^2
(C) 10 m/s^2
(D) 40 m/s^2
(E) 70 m/s^2

8. How far did Car A travel between $t = 0$ and $t = 70.0$ seconds?

(A) 70 m
(B) 110 m
(C) 280 m
(D) 2450 m
(E) 4900 m

9. Which choice describes the car that traveled the greatest distance during the time intervals listed?

(A) Car A from $t = 0$ to $t = 20$
(B) Car A from $t = 20$ to $t = 40$
(C) Car A from $t = 50$ to $t = 70$
(D) Car B from $t = 0$ to $t = 20$
(E) Car B from $t = 30$ to $t = 50$

10. How many seconds after $t = 0$ did it take for Car A to catch up with Car B?

(A) 0
(B) 20
(C) 40
(D) 60
(E) 80

ANSWERS AND EXPLANATIONS

1. **(D)** Because he ran straight, his displacement is the same as his distance.

$$\bar{v} = \frac{\Delta x}{\Delta t} = \frac{100.0 \text{ m}}{9.58 \text{ s}} = 10.4 \text{ m/s}$$

2. **(B)** The acceleration is constant from $t = 0$ to $t = 2$ and again from $t = 2$ to $t = 5$. The constant acceleration means that you can use of your kinematic equations: $v = v_o + at$. At $t = 2$ s, the speed of the particle is $v_{2s} = v_o + 2 \text{ m/s}^2(2 \text{ s}) = 1 \text{ m/s} + 4 \text{ m/s} = 5 \text{ m/s}$. At $t = 4$ s, the speed of the particle is $v_{4s} = v_{2s} + 4 \text{ m/s}^2(3 \text{ s}) = 5 \text{ m/s} + 12 \text{ m/s} = 17 \text{ m/s}$.

3. **(B)** This is a problem that will be easier to solve by drawing a diagram. In doing so, you will see that the distance moved west and east cancel out. The southward movement is 5 m longer than the northward movement, so the total displacement is 5 m south. Be sure to differentiate displacement from distance, because the shopper moved a total distance of 79 m.

4. **(A)** No forces act horizontally on the baseball, so according to Newton's first law, the baseball will travel at a constant horizontal velocity.

5. **(C)** The vertical force acting on the baseball is gravity, so the vertical force will have a negative slope. The line crosses the *x*-axis because at the top of the path, the baseball has zero velocity.

6. **(C)** Solving this problem comes down to selecting the correct kinematic equation. The known values are $x_o = 0$, $x = 50.0$ m, $a = 3.00 \text{ m/s}^2$, and $v_o = 0$. The unknown is the time *t*. The equation you need is $x = x_o + v_o t + \frac{1}{2}at^2$. Once you substitute x_o and v_o into the equation, it simplifies to $x = \frac{1}{2}at^2$. By rearranging to solve for *t*, the equation becomes $t = \sqrt{\frac{2x}{a}}$.

$$t = \sqrt{\frac{2(50.0 \text{ m})}{3.00 \text{ m/s}^2}} = \sqrt{33.3} \approx 6 \text{ s}$$

7. **(B)** Use the change in velocity and the elapsed time to calculate the acceleration. Car A starts from rest so its initial velocity is zero. At the end of 70 s, its velocity is 70 m/s according to the graph.

$$a = \frac{70 \text{ m/s}}{70 \text{ s}} = 1 \text{ m/s}^2$$

8. **(D)** Car A starts at rest and moves with constant acceleration. The distance traveled is therefore one-half the product of its acceleration and the square of the time traveled.

$$d = 1/2at^2$$
$$= 1/2(1\ \text{m/s}^2)(70\ \text{s})^2 = 2{,}450\ \text{m}$$

9. **(C)** For a graph of speed versus time, you can use the area under the graph to find the distance traveled during that interval. Car B travels at a constant speed so the area under the graph for each 20-second interval is 8 squares. For Car A, you must examine each individual interval. Between $t = 0$ to $t = 20$, there are 2 squares under the graph. Between $t = 20$ to $t = 40$, there are 6 squares under the graph. Between $t = 50$ to $t = 70$, there are 12 squares under the graph.

10. **(E)** Car B travels at a constant speed of 40 m/s. The distance traveled is the product of the speed and the elapsed time, $40t$. The cars meet when they travel the same distance. The distance Car A travels is $1/2at^2$. Set the two expressions equal to each other and solve for t.

$$1/2at^2 = 40t$$
$$\tfrac{1}{2}(1\ \text{m/s}^2)t^2 = 40t$$
$$t = 80\ \text{s}$$

CHAPTER 4

Dynamics

Describing the motion of an object, or *kinematics*, is closely related to understanding why something moves the way it does, or *dynamics*. The key to dynamics is a thorough understanding of forces, and, in particular, Newton's laws of motion.

Introduction to Forces

You exert a force whenever you lift a book, open a door, or throw a ball. A **force** is a push or pull. Forces are vector quantities with both magnitude and direction. The SI unit of force is the newton. Therefore, a force might be 10 newtons acting to the north or 2 newtons acting to the east.

As vectors, forces can be added and combined using the rules of vectors presented in Chapter 2. Recall that vectors could cancel out. If the forces acting on an object cancel out, the forces are said to be **balanced forces**. The overall, or net force, is zero. **Unbalanced forces** exist when the forces on an object do not cancel out. In this case, the net force is not zero.

It is helpful to be familiar with several common forces. Even if you are not asked to describe them explicitly, an understanding of their nature will be expected within several questions of SAT Physics. Other forces will be discussed as they are reviewed in later chapters.

Gravitational Force

As you read these words, you are being pulled by the **gravitational force**, which is the force of attraction between every pair of masses in the universe. The measure of the gravitational force exerted on an object is **weight**. Your weight, for example, is a measure of the gravitational force Earth exerts on you. Make sure you do not confuse weight with mass, as often occurs. Mass is the amount of matter in an object. It is a scalar quantity that is independent of location. Weight, however, is a vector quantity that varies with the gravitational force acting on an object. The mass of a ball on Earth is the same as its mass on the moon because the amount of matter in it does not change. The weight of the ball on the moon, however, is much less than its weight on Earth because the gravitational force exerted by the moon is about one-sixth the gravitational force exerted by Earth.

The weight of an object is the product of the mass of the object and the acceleration due to gravity, *g*. For example, the following equation shows the calculation for the weight of a 50-kg mass on Earth. Note that weight can be represented by $\boldsymbol{F}_{\textbf{weight}}$ or *W*.

$$\begin{aligned} F_{\text{weight}} &= mg_{\text{Earth}} \\ &= (50\text{ kg})(9.8\text{ m/s}^2) \\ &= 490\text{ N (toward the center of Earth)} \end{aligned}$$

Normal Force

In everyday language, the word *normal* might mean ordinary or usual. In physics, however, the word has a very specific meaning. The normal is a line perpendicular to a surface. The **normal force**, therefore, is a force that acts perpendicular to a surface of contact between two objects. The normal force is equal in magnitude, but opposite in direction, to the weight of an object. The normal force can be represented as $\boldsymbol{F}_{\textbf{normal}}$ or ***N***.

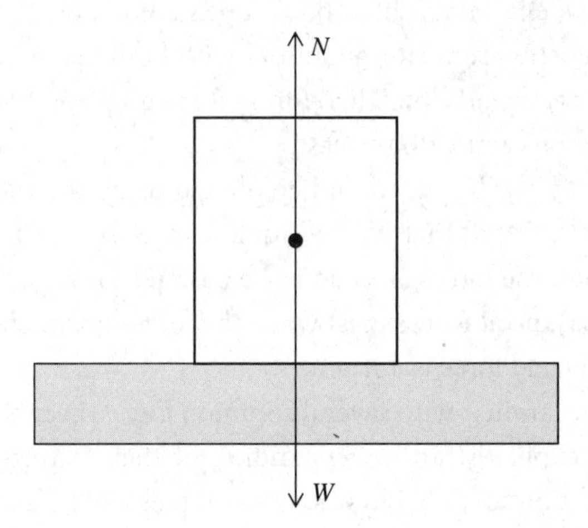

Friction

When you rub sandpaper across a piece of wood, friction is created. **Friction** is a force between materials that are in contact with one another. The force of friction acts parallel to the plane of contact between the two materials and acts in the direction opposite to the direction of the object being moved. The surfaces do not have to be as rough as sandpaper for friction to be created. Even surfaces that seem smooth to the unaided eye are rough at the microscopic level and establish friction.

There are two basic types of friction: static friction and kinetic friction. **Static friction** is the force that arises between two surfaces that are not moving relative to one another. For example, suppose you push a dresser that is at rest on a floor. Static friction resists your force and holds the dresser in place. In the following diagram, you can see that the force of your push, F_{push}, is equal and opposite to the force of static friction, F_{static}.

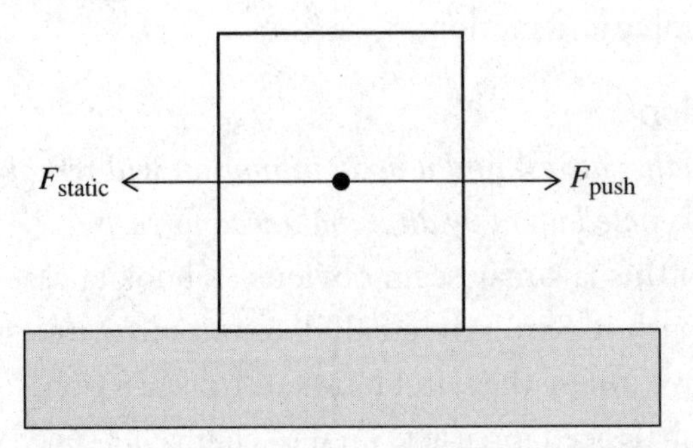

Static friction occurs when the net force on the object is zero, and therefore, the object is at rest. Once the force of the push overcomes the force of static friction, the net force is no longer zero and the object is set in motion. **Kinetic friction** is the force that arises between two surfaces that are moving relative to one another.

Once static friction is overcome and the object begins to move, the relationship between the force of kinetic friction, $F_{kinetic}$, and the force of the push, F_{push}, determines the specifics of the motion of the object. If F_{push} is equal in magnitude to $F_{kinetic}$, the net force again becomes zero and the object moves at constant velocity. If F_{push} is greater in magnitude than $F_{kinetic}$, as in the following diagram, there is a net force in the direction of motion so the object accelerates. If F_{push} is lower in magnitude than $F_{kinetic}$, there is a net force, but it is opposite to the direction of motion. Therefore, the object will slow down and come to a stop.

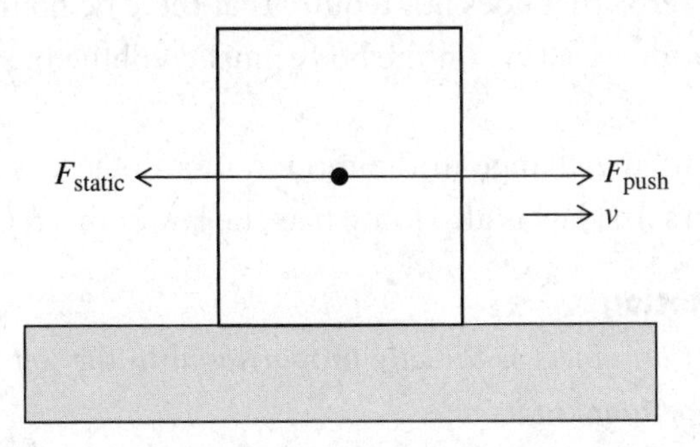

The magnitude of the friction force depends on the nature of the surfaces in contact and the normal force. The ratio of the force of friction between two bodies is known as the **coefficient of friction** and is represented by the Greek letter μ. The coefficients are constants that vary with the material and have both static (μ_s) and kinetic components (μ_k). The value of μ always falls between 0 and 1, with lower values indicating smoother surfaces. In addition, the coefficient of kinetic friction will always be less than the coefficient of static friction for a given surface. Questions on SAT Physics may provide you with the coefficient of friction and ask you to calculate the resulting friction force.

$$F_{static} = \mu_s N$$
$$F_{kinetic} = \mu_k N$$

Newton's Laws of Motion

In 1686, Isaac Newton presented three laws of motion that summarize the relationship between force and motion.

First Law of Motion

A body at rest remains at rest and a body in motion will remain in motion at constant velocity unless acted upon by an unbalanced force.

The first part of this law may seem obvious. A book at rest on a desk does not move unless you push it. Similarly, a ball does not move unless you throw it, and a leaf does not move unless the wind blows it. In other words, an object does not move unless an unbalanced force acts on it to change its speed or direction.

The second part of this law requires a little more consideration. It suggests that once an object is moving, it will continue to move in the same way unless an unbalanced force acts on it. Yet you know that a soccer ball kicked across a field and a toy car pushed across a floor eventually come to a stop. Now you know that if an object in motion comes to a stop, an unbalanced force must be acting on it. In the case of the soccer ball and the toy car, that force is friction. If friction were not present, objects would continue to travel with constant motion forever.

Another way to state Newton's first law of motion is that the velocity of the object does not change if the net force acting on the object is zero. Remember that velocity involves both speed and direction, so when a body is at rest or moving with constant speed in a straight line, the resultant of all the forces acting on the body must be zero. That does not require that there be no forces acting on the body. Instead, any forces acting on the body must combine in such a way that the net force is zero.

The resistance to any change in the motion of an object is known as **inertia**. Therefore, Newton's first law is also known as the law of inertia.

Second Law of Motion

The acceleration of an object is directly proportional to the net force exerted on it and inversely proportional to its mass.

What this law means is that if the net force acting on an object is not zero, the object will be accelerated. The direction of acceleration will be the same as the direction of the force. The magnitude of the acceleration will depend on the magnitude of the force and the mass of the object.

You may know from experience that you have to use more force to accelerate a shopping cart by the same amount when it is filled than when it is empty. That's because the mass of the cart increases as it is filled, so a greater force is required to produce the same amount of acceleration. If you increase the force you exert without changing the mass, the acceleration increases.

Newton's second law can be represented by the following equation.

$$a = \frac{F}{m}$$

where F is force, m is mass, and a is acceleration. In the SI system, force is measured in newtons, mass in kilograms, and acceleration in m/s^2. This equation is often rewritten in a form with which you may be more familiar.

$$F = ma$$

Written this way, you can see that one newton is the force needed to accelerate a 1-kilogram object by 1 meter per second per second.

$$1\text{ N} = 1\text{ kg} \times 1\text{ m/s}^2$$

Example:

A. A car has a mass of 1000 kg. What force is required to accelerate the car 0.05 m/s^2?

$$\begin{aligned} F &= ma \\ &= 1000\text{ kg} \times 0.05\text{ m/s}^2 \\ &= 50\text{ N} \end{aligned}$$

B. What force would be required to accelerate a truck with a mass of 10,000 kg by the same amount?

$$\begin{aligned} F &= ma \\ &= 10{,}000\text{ kg} \times 0.05\text{ m/s}^2 \\ &= 500\text{ N} \end{aligned}$$

A greater force is required to achieve the same acceleration because the mass of the truck is greater than the mass of the car.

C. What force would be required to accelerate the car 0.5 m/s^2?

$$\begin{aligned} F &= ma \\ &= 1000\text{ kg} \times 0.5\text{ m/s}^2 \\ &= 500\text{ N} \end{aligned}$$

A greater force is required to accelerate the same mass, the car, by a greater amount.

Third Law of Motion

For every action, there is an equal and opposite reaction.

This law explains that whenever one object exerts a force on a second object, the second object exerts a force on the first object that is equal in magnitude but opposite in direction. Unlike balanced forces, the two forces do not cancel because they do not act on the same object. As a result, the forces can lead to a change in the motion of the objects. The following diagram shows that a foot exerts a force on a soccer ball (action). The soccer ball exerts an equal and opposite force on the foot (reaction). Be aware that the names of the forces are somewhat arbitrary in that they occur simultaneously rather than sequentially.

There are numerous examples of Newton's third law of motion. For example, consider the rocket in the following diagram. When it burns fuel, gases are expelled out of the bottom. The action force can be considered the force exerted on the gases away from the rocket. The gases, in turn, exert an equal but opposite force on the rocket. This reaction force propels the rocket forward. The force exerted by the rocket F_{rocket} is exactly equal in magnitude and opposite in direction to the force exerted by the gases F_{gases}.

TEST-TAKING HINT

The SAT Physics exam will not ask you to draw diagrams. However, questions might require that you either interpret free-body diagrams or recognize the correct diagram from a group of choices. Make sure you understand all of the information presented in such a diagram so that you do not overlook seemingly minor differences between diagrams.

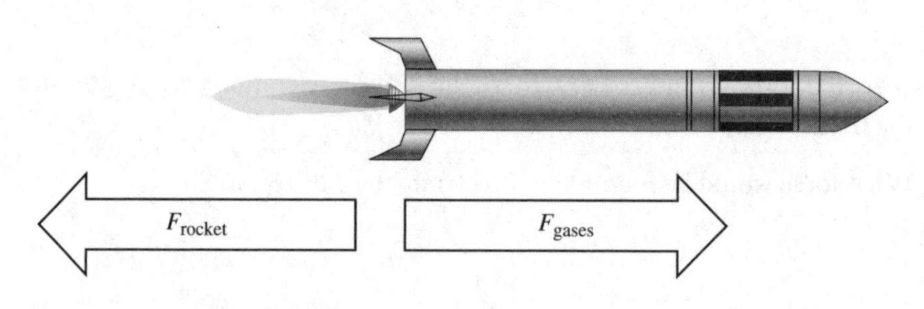

Free-Body Diagrams

Many of the diagrams you have been analyzing are special versions of vector diagrams known as **free-body diagrams**. Diagrams such as these are commonly used to compare the magnitudes and directions of all the forces acting on an object. The length of the arrow represents the magnitude of the force, and the direction of the arrow indicates the direction in which the force is being exerted. Although they can vary somewhat, most free-body diagrams represent the object as a square with arrows extending from its center. The arrows are labeled to indicate the type of force.

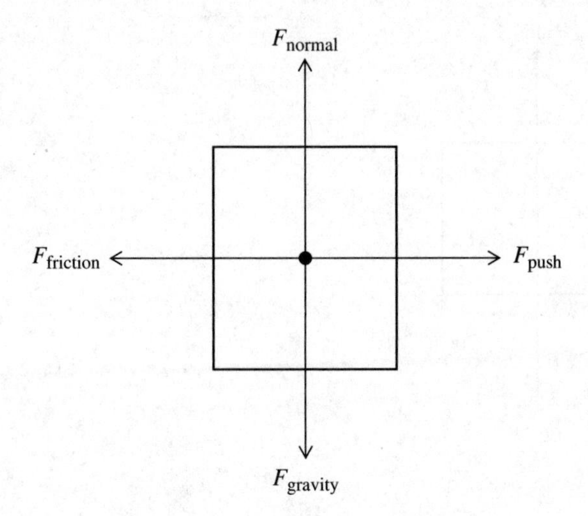

REVIEW QUESTIONS

Questions 1–3 refer to the following net forces.

(A) 3 N northward
(B) 3 N southward
(C) 30 N southward
(D) 30 N northward
(E) Zero

1.

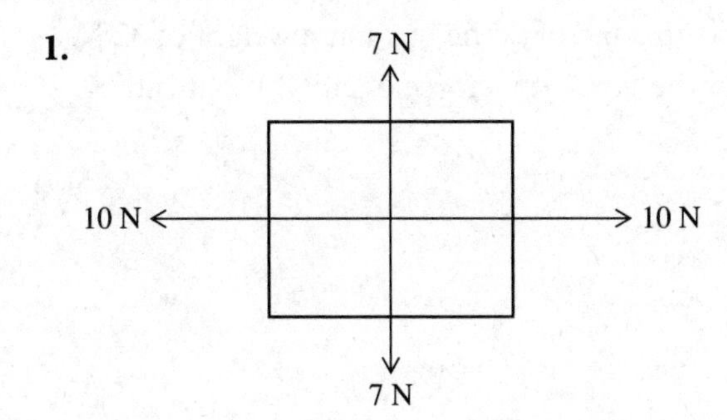

2.

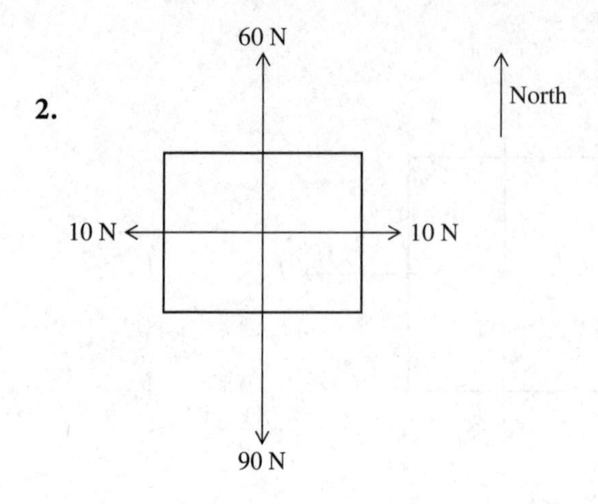

3.

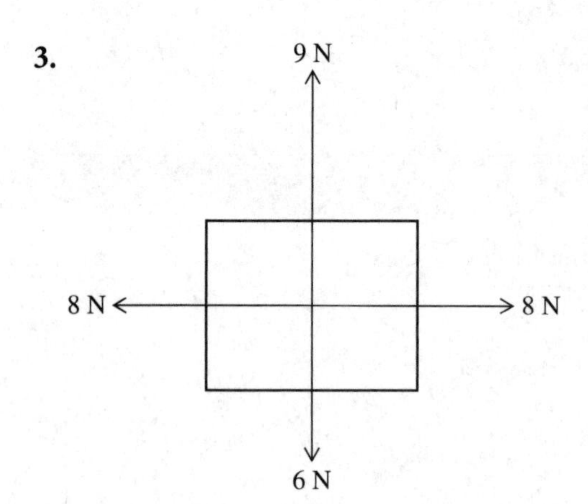

4. A metal box is located on a table in a factory. The box has a weight of 32 N, and a machine pushes down on the box with a force of 20 N. What is the normal force on the box?

(A) 12 N
(B) 20 N
(C) 32 N
(D) 52 N
(E) 70 N

5. The weight of an object on Earth is 147 N. What is its approximate mass?

(A) 5 kg
(B) 10 kg
(C) 15 kg
(D) 20 kg
(E) 25 kg

6.

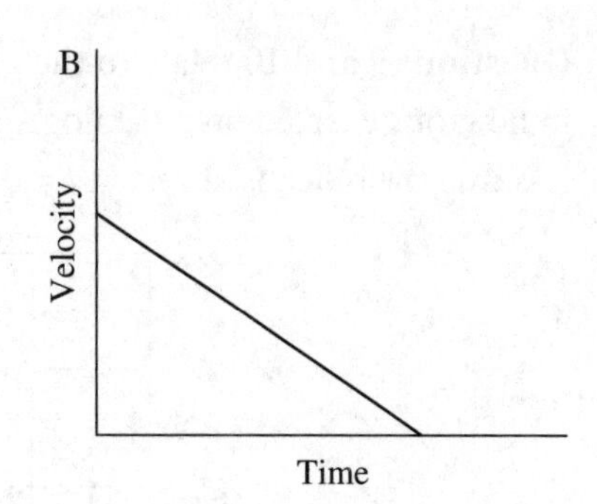

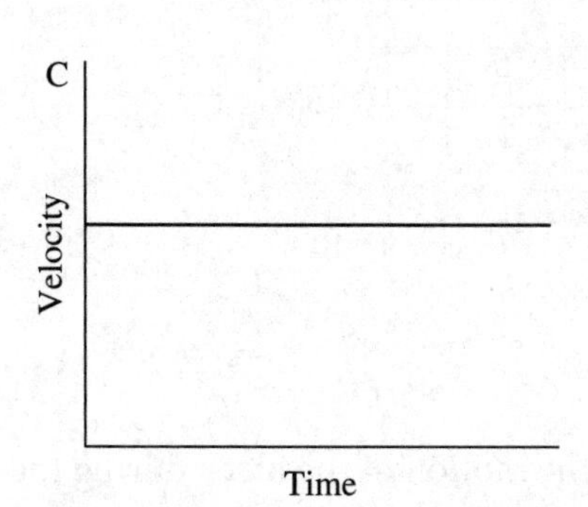

Which of the graphs represent(s) the motion of an object with only balanced forces acting on it?

(A) A only
(B) B only
(C) C only
(D) A and B only
(E) A, B, and C

7. A wagon is accelerating at a rate of 3 m/s^2. If the net force on the wagon is doubled and the mass is tripled, the acceleration becomes

(A) ½ m/s^2
(B) 1 m/s^2
(C) 3/2 m/s^2
(D) 2 m/s^2
(E) 6 m/s^2

8. A student on a skateboard pushes against the ground with her foot. The reaction force is exerted

(A) on the ground by the skateboard
(B) on the air by the skateboard
(C) on the ground by her foot
(D) on the skateboard by her foot
(E) on her foot by the ground

Questions 9 and 10 relate to the graph below, which shows the net force $\boldsymbol{F}$ in newtons exerted on a 6-kilogram block as a function of time in t seconds. Assume the block is at rest at $t = 0$ and that $\boldsymbol{F}$ acts in a fixed direction.

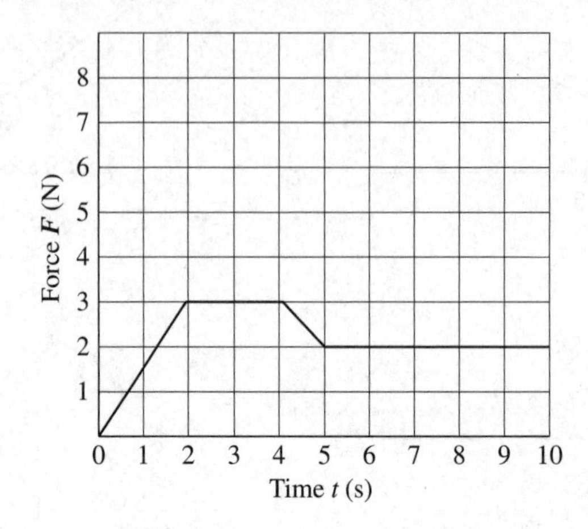

9. Which conclusion can be reached about the motion of the block during the time interval 2 to 4 seconds?

(A) The block is not in motion.
(B) The block is moving with constant acceleration.
(C) The mass of the block is increasing.
(D) The block is moving with constant speed.
(E) The force exerted on the block is decreasing.

10. The acceleration of the block at $t = 6$ sec is

(A) 1/3 m/s^2
(B) 2 m/s^2
(C) 3 m/s^2
(D) 6 m/s^2
(E) 12 m/s^2

ANSWERS AND EXPLANATIONS

1. **(E)** The net force is zero when the forces in all directions cancel out. Because the forces are exerted in the vertical and horizontal directions only, they can be directly added. Remember to assign a negative to one of the forces in each direction. Add the horizontal forces first, and then the vertical forces.

2. **(C)** The horizontal forces are balanced, so they cancel. The vertical forces are unbalanced. The net force is the difference between them, 30 N, in the direction of the greater force, southward.

3. **(A)** The horizontal forces are balanced, so they cancel out. The vertical forces are unbalanced. The net force is the difference between them, 3 N, in the direction of the greater force, northward.

4. **(D)** The normal is equal in magnitude and opposite in direction to the net downward force on the box. The net downward force is the weight of the box, 32 N, plus the additional downward force, 20 N, which is 52 N.

5. **(C)** The weight of an object is the product of its mass and the acceleration due to gravity, which on Earth is approximately 10 m/s^2. Therefore, the mass equals 147 N ÷ 10 m/s^2 = 14.7 kg, which is approximately 15 kg.

6. **(C)** If all the forces acting on an object are balanced, the net force is zero. When the net force is zero, the velocity of an object remains the same. This is true only for the object represented in graph C. The object in graph A is speeding up and the object in graph B is slowing down. Therefore, both objects are accelerating and must have a net force acting on them as the result of unbalanced forces.

7. **(D)** If the net force is doubled, the acceleration is also doubled because these quantities are directly proportional. If this change alone occurred, the new acceleration would become 6 m/s^2. However, the mass is tripled. Because mass and acceleration are inversely related, the acceleration must be divided by this value. The final acceleration becomes 2 m/s^2.

8. **(E)** According to Newton's third law of motion, the action and reaction forces are exerted by objects in contact with one another. Therefore, the foot exerts an action force on the ground and the ground exerts a reaction force back on the foot that is equal in magnitude but opposite in direction.

9. **(B)** The force and mass on the block during this time interval are constant. Therefore, the acceleration is also constant. The block is in motion and its speed is increasing at a constant rate.

10. **(A)** At $t = 6$ sec, the force on the block is 2 newtons, according to Newton's second law of motion, $F = ma$. Therefore, 2 N = 6 kg (a), so $a = 1/3$ m/s^2.

CHAPTER 5

Work, Energy, and Power

The understanding of forces and motion you have developed thus far will enable you to master the concepts of work, energy, and power. Each of the concepts is distinct, yet closely related to the others. Recognizing the meaning of each term in physics, the relationships among them, and how to perform calculations involving them will enable you to solve several different types of questions on SAT Physics.

What Is Work?

A small child lifting a piece of apple to her mouth does more work than a large weightlifter holding a heavy barbell over his head. Why? Unlike the colloquial definitions of work, scientists have a very specific definition of work. In physics, **work** is done only when a force exerted on an object causes that object to move some distance. Even if a force is exerted, work is not done if the force does not cause an object to move.

Calculating Work

Work is equal to the product of the force and the displacement of the object in the direction of the force.

$$W = \text{force} \times \text{displacement}$$

$$= F \times d$$

A common SI unit of work is the newton-meter, which is also known as the **joule** (J). To calculate the work done on or by an object, multiply the force by the displacement the force causes the object to move. If the force is exerted parallel to the direction of movement, multiplication is straightforward.

Example:

A 30-newton force moves a 20-newton object 4 meters. How much work is done by the force?

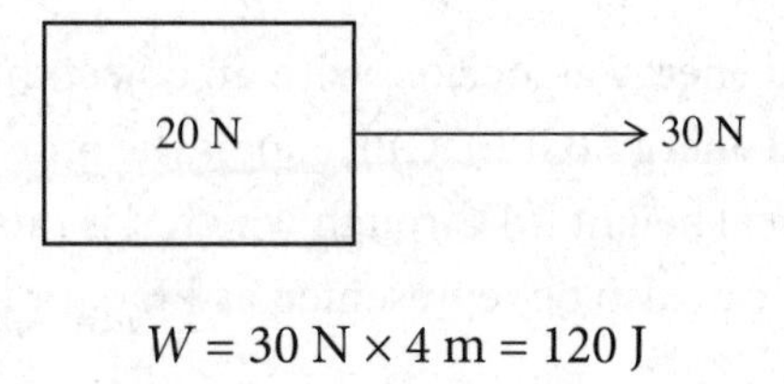

$$W = 30\text{ N} \times 4\text{ m} = 120\text{ J}$$

If the force is exerted at angle to the direction of movement, you must first find the component of the force that is parallel to the movement. In this case,

$$W = F \times d \cos \theta$$

where θ is angle between the force vector and the displacement vector. Actually, you use the same equation to calculate work when the force is parallel to the displacement. In these situations, θ is 0 so the cos 0 is 1. You will not need to perform calculations that require a calculator on SAT Physics, but you may need to use given information to solve a problem.

Example:

A 100-newton force is exerted an angle to move a box a horizontal distance of 8 m as shown.

$\cos 60° = 0.0500$

100 N

20 N

60°

$$W = 100\text{ N} \times 8\text{ m} \times \cos 60° = 40\text{ J}$$

In some situations, the force acts directly opposite to the direction of motion in order to slow it down. The force, therefore, does not cause displacement but instead hinders it. These situations involve negative work because the value of work yields a negative number. The reason is that θ is 180° and cos 180° is –1.

What Is Energy?

The simplest definition of **energy** in physics is the ability to do work. The units of energy, therefore, are the same as the units of work—joules. There are two general types of energy: potential energy and kinetic energy.

Potential Energy

An object has **potential energy** (PE) due to its position or condition. Because of potential energy, an object has the ability (or potential) to do work. When you lift a book from the floor to your desk, for example, you do work against the force of gravity. Once you lift the book, it has a greater ability to do work. This increase in its ability to do work is its potential energy when compared with its original position on the floor.

The type of potential energy associated with an object that is raised is known as **gravitational potential energy** (GPE). GPE equals the product of the weight of the object (w) and the vertical height (h) through which it is raised. [Note that gravitational potential energy can also be represented as PE_{grav} or by U_g.]

$$\text{GPE} = wh = mgh$$

Example:

An object with a mass of 20 kg is raised 5.0 m. What is its gravitational potential energy?

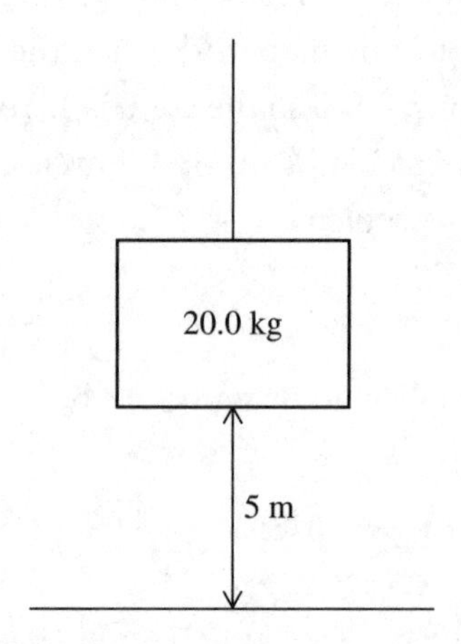

$$\text{GPE} = 20.0\text{ kg} \times 10\text{ m/s}^2 \times 5.0\text{ m} = 980\text{ J}$$

Another form of potential energy commonly tested on SAT Physics is **elastic potential energy**, which is energy stored when an elastic material is stretched or compressed. A compressed spring or a stretched rubber band have elastic potential energy. The amount of potential energy is equal to the work done in compressing or stretching.

Some questions on the SAT Physics test may involve special cases of springs that obey Hooke's Law. According to Hooke's Law, a spring exerts a force on a mass attached to it according to the following equation

$$F_{spring} = kx$$

where k is a constant of proportionality called the spring constant and x is the displacement from equilibrium. The spring constant varies with the spring. A very tight spring, for example, will have a greater spring constant than a looser spring.

You can see from the equation that the further a spring is displaced from equilibrium, the greater the force will be that the spring exerts in the direction of the equilibrium position.

If the potential energy is assigned a value of zero at the equilibrium position, the following equation can be used to relate the elastic potential energy to the spring constant and the displacement.

$$PE_{spring} = \frac{1}{2} kx^2$$

Kinetic Energy

The energy associated with the motion of a moving body is known as **kinetic energy**. A rocket traveling through space, a firefly traveling from one leaf to another, and an atom within a sample of matter all have kinetic energy because they are in motion.

There are different types of kinetic energy, including vibrational (objects moving back and forth), rotational (objects turning on an axis), and translational (objects moving from one place to another). The type of kinetic energy you will encounter most often is translational. This type of kinetic energy is directly proportional to the mass of an object and the square of its velocity according to the following equation.

$$KE = \frac{1}{2} mv^2$$

Example:

A pitcher throws a baseball at 20 m/s. If the baseball has a mass of 0.2 kg, what is the kinetic energy of the ball?

$$\begin{aligned} \text{KE} &= \tfrac{1}{2} \times 0.2 \text{ kg} \times (20 \text{ m/s})^2 \\ &= 40 \text{ J} \end{aligned}$$

You saw that the potential energy of an object was equal to the work done on it. How is kinetic energy related to work? According to the **Work-Energy Theorem**, the net work done on an object is equal to the object's change in kinetic energy.

$$W = \Delta KE$$

Mechanical Energy

Both potential energy and kinetic energy are forms of **mechanical energy**, which is energy acquired by objects upon which work is done. Any object that possesses mechanical energy has the ability to do work. Suppose, for example, you lift a hammer to some height. That hammer has potential energy. If you let the hammer fall, it can exert a force on a nail that causes the nail to move some distance.

In other words, the hammer can do work because of its position. Similarly, suppose a billiard ball moving across the table hits into another billiard ball. The first ball can exert a force on the second ball to cause it to move some distance. It does work because of its motion.

The total mechanical energy (TME) of an object is the sum of its kinetic energy and its potential energy.

$$\text{TME} = \text{PE} + \text{KE}$$

Other Forms of Energy

In addition to mechanical energy, you will encounter other forms of energy throughout SAT Physics. They will be discussed in context; however, this summary gives you an idea of some of the different forms of energy that exist.

FORM	DESCRIPTION
Thermal	Total kinetic energy of the particles in a sample of matter
Electromagnetic	Energy carried as electric and magnetic fields oscillating perpendicular to one another
Nuclear	Energy stored in the center, or nucleus, of an atom
Chemical	Energy stored in the bonds that hold particles of matter together
Electrical	Energy associated with moving charges

Conservation of Energy

A ball thrown into the air has kinetic energy because of its motion. It also has gravitational potential energy because work has been done to change its height. At no time is any of the ball's energy destroyed or new energy created. According to the **law of conservation of energy**, energy can change from one form to another, but it cannot be created or destroyed.

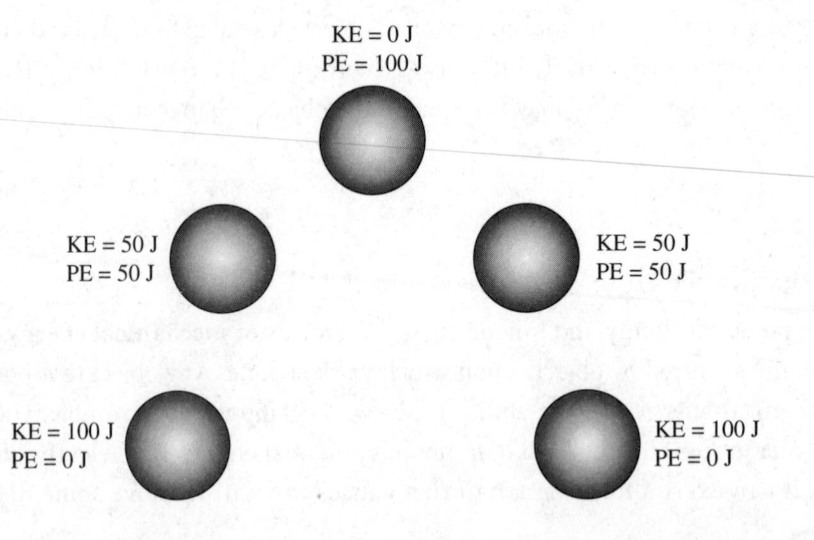

In more advanced topics of physics, it can be shown that mass can be considered a form of energy. Einstein's theory of relativity leads to the famous equation that relates energy to mass and the speed of light c.

$$E = mc^2$$

Power

Another important quantity related to work is **power**, P, which is the rate at which work is done or energy is transformed.

$$P = \frac{\Delta W}{\Delta t} \quad \text{or} \quad P = \frac{\Delta E}{\Delta t}$$

The unit of power is the watt, which is equal to 1 joule per second. If you rewrite the power equation as follows, you can see how power is related to both force and speed. A powerful machine, for example, is both strong and fast.

$$P = \frac{\text{force} \times \text{displacement}}{\text{time}}$$

$$= \text{force} \times \frac{\text{displacement}}{\text{time}}$$

$$= \text{force} \times \text{velocity}$$

Example:

A person with a mass of 60 kg runs up a flight of stairs that has a total height of 5.0 meters. What power is required for the person to run up the stairs in 10 seconds, 20 seconds, 60 seconds? *Use $g = 10\ m/s^2$.*

10 seconds $\quad P = (60\ \text{kg} \times 10\ \text{m/s}^2) \times 5.0\ \text{m}/10\ \text{s} = 300\ \text{J}$

20 seconds $\quad P = (60\ \text{kg} \times 10\ \text{m/s}^2) \times 5.0\ \text{m}/20\ \text{s} = 150\ \text{J}$

60 seconds $\quad P = (60\ \text{kg} \times 10\ \text{m/s}^2) \times 5.0\ \text{m}/60\ \text{s} = 50\ \text{J}$

Notice how the amount of power decreases as the amount of time increases.

TEST-TAKING HINT

While studying for SAT Physics, make a list or prepare individual index cards relating different units of measurement. For example, know that $J = kg \times m^2/s^2 = N \times m$. Recognizing relationships between units will help you check your work and decide if an answer is reasonable.

REVIEW QUESTIONS

Questions 1–3 refer to the following amounts of work.

(A) 15.0 J
(B) 20.4 J
(C) 25.0 J
(D) 49.0 J
(E) 60.0 J

1. A 15-N horizontal force is applied to push a block across a flat, frictionless surface. If the force causes the block to have a displacement of 4.0 m, how much work is done by the force?

2. A ball with a mass of 0.5 kg moves at a velocity of 10.0 m/s. What is its kinetic energy?

3. A cart is pulled at constant speed to the top of an inclined plane. If the mass of the cart is 4.00 kg and the height to which the cart is lifted is 0.52 m, what is the potential energy of the cart at the top of the inclined plane?

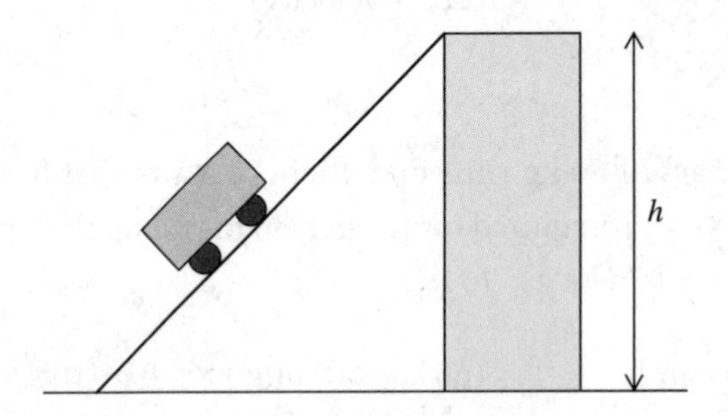

4. The equilibrium position of the spring is located at level I. At which point is the net force acting upward on the mass greatest?

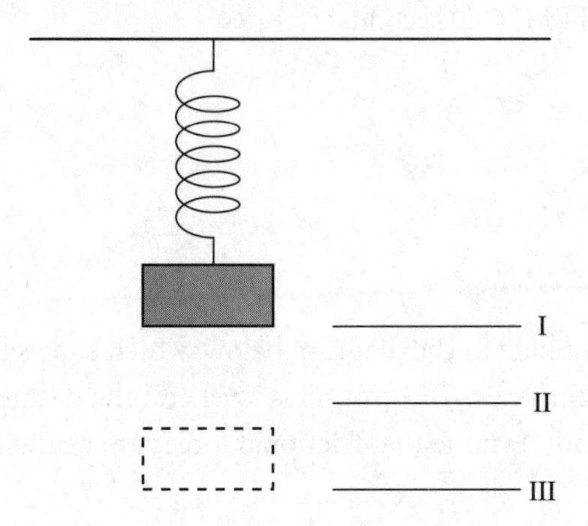

(A) I
(B) Between I and II
(C) II
(D) Between II and III
(E) III

5. An 18-kilogram object has an initial kinetic energy of 36 joules. How far will the object move against a net resisting force of 9 newtons?

(A) 2 m
(B) 4 m
(C) 6 m
(D) 9 m
(E) 30 m

6. If the velocity of an object is doubled, its kinetic energy is

(A) halved
(B) doubled
(C) tripled
(D) quadrupled
(E) unchanged

7. A person with a weight of 340 newtons runs up a flight of stairs to a height of 4.0 meters in 8.0 seconds. The power required for this activity is

(A) 11 W
(B) 43 W
(C) 85 W
(D) 170 W
(E) 680 W

8. A hydraulic ladder lifts a 100.0-newton box to a height of 3.0 meters in 15 seconds. After the box is lifted, how much power is required to hold the box at that height for 60.0 seconds?

(A) 0 W
(B) 5 W
(C) 17 W
(D) 20 W
(E) 33 W

Questions 9 and 10 relate to the diagram below, which shows the path of a heavy ball thrown in the air. The ground is level and the dashed lines are parallel to the ground. Assume that frictional forces are negligible.

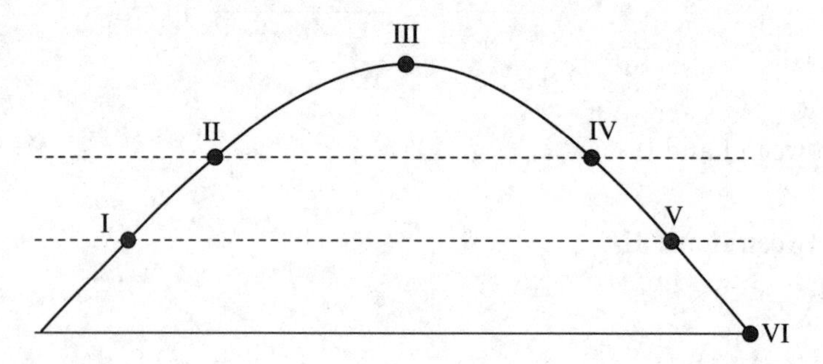

9. The kinetic energy of the ball is least at point

(A) II
(B) III
(C) IV
(D) V
(E) VI

10. At which pair of points is the potential energy of the ball the same?

(A) I and VI
(B) I and II
(C) II and III
(D) II and IV
(E) I and III

ANSWERS AND EXPLANATIONS

1. **(E)** The force is parallel to the displacement, so θ is 0 and cos 0 is 1. Therefore, $W = F \times d \cos \theta = 15 \text{ N} \times 4.0 \text{ m} \times 1 = 60 \text{ J}$.

2. **(C)** Translational kinetic equals one-half the mass of the ball times the square of its velocity. $KE = ½ \times 0.5 \text{ kg} \times (10 \text{ m/s})^2 = 25 \text{ J}$.

3. **(B)** The potential energy at the top of the inclined plane is *mgh*, so PE = approximately $4.0 \text{ kg} \times 10 \text{ m/s}^2 \times 0.52 \text{ m} = 20.8 \text{ J}$, so the closest answer is B.

4. **(E)** The force exerted on the mass is directly proportional to its displacement from equilibrium. Therefore, the force is greatest at the greatest distance from rest.

5. **(B)** The kinetic energy of the object is equal to the work it can do. Therefore, $36 \text{ J} = 9 \text{ J} \times d$, so $d = 4$ m.

6. **(D)** Kinetic energy is directly related to the square of the velocity. If the velocity is doubled, the kinetic energy is then multiplied by a factor of 4 $KE = ½\, m(2v)^2 = 4[½\, mv^2]$.

7. **(D)** The work the person does to climb the stairs is force (weight) times distance. Power is the work done divided by the time during which the work is done. Therefore,

$$P = \frac{340 \text{ N} \times 4 \text{ m}}{8 \text{ s}} = 170 \text{ W}$$

8. **(A)** Power is the rate at which work is done. Work is only done when a force causes an object to move some distance. If the box is being held in one position, work is not done, so no power is involved.

9. **(B)** As the ball rises, the potential energy decreases, and the kinetic energy decreases. At the point where the ball reaches the peak of the curve, it stops moving and changes direction. At this point, kinetic energy is zero.

10. **(D)** The gravitational potential energy of the ball depends on its height. Therefore, the GPE is the same at points I and V, as well as at points II and IV.

CHAPTER 6

Fluid Mechanics

Substances that can flow easily are known as fluids. Gases and liquids are fluids. Unlike solids, fluids do not have fixed shapes. Both liquids and gases adjust to the shape of their container. In addition, a gas expands to fill its container. While liquids have a definite volume, gases do not. You may encounter a few questions on the SAT Physics exam that involve the properties of fluids, so it is best to understand them and recognize changes in them.

Pressure in a Liquid

The particles in a fluid are in constant motion. They collide with each other and the walls of their container. **Pressure** is the force they exert per unit area.

$$P = \frac{F}{A}$$

The pressure of a fluid depends on depth. Pressure increases with depth. Therefore, an object placed in a fluid experiences greater pressure as it is moved deeper within the fluid because the weight of the fluid above it increases. The weight of the fluid is related to the density of the fluid. **Density** is the mass per unit volume.

$$\text{density} = \frac{\text{mass}}{\text{volume}}$$

The denser the fluid above a submerged object is, the more pressure there is exerted on the object. The pressure can therefore be determined by finding the product of the height of the liquid (h), the density (d), and the acceleration due to gravity ($\boldsymbol{g}$). [Note: Density is often represented by the Greek letter rho, ρ, instead of d.]

$$P = hdg$$

If a container of liquid is open, the atmosphere above it exerts additional pressure on the liquid. This added pressure must be included when considering the total pressure of the fluid.

$$P_{\text{total}} = P_{\text{atmosphere}} + P_{\text{fluid}}$$

The SI unit of pressure is the pascal (Pa). This unit is equivalent to 1 N/m^2.

Example:

A biologist is studying a school of fish 11.0 meters below the surface of the ocean. If the density of sea water is 1.00×10^3 kg/m^3 and atmospheric pressure is 1.00×10^5 N/m^2, what is the pressure on the biologist?

$$P_{\text{fluid}} = hdg = (11.0\text{ m})(1.00 \times 10^3\text{ kg/m}^3)(10\text{ m/s}^2) = 1.10 \times 10^5\text{ N/m}^2$$

$$P_{\text{total}} = P_{\text{atmosphere}} + P_{\text{fluid}} = (1.00 \times 10^5\text{ N/m}^2) + (1.10 \times 10^5\text{ N/m}^2)$$

$$= 2.10 \times 10^2\text{ kPa (kiloPascals)}$$

Gas Pressure

Like the particles in a liquid, the particles of a gas are also in constant motion. In fact, they move at greater speeds and travel farther apart. As a result, gases also exert pressure. As with fluids, gas pressure depends on the weight of the gas above a given point. The higher you go in the atmosphere, for example, the less air there is above you. Therefore, atmospheric pressure decreases with altitude. Air is densest near Earth's surface, which explains why air pressure is greatest there.

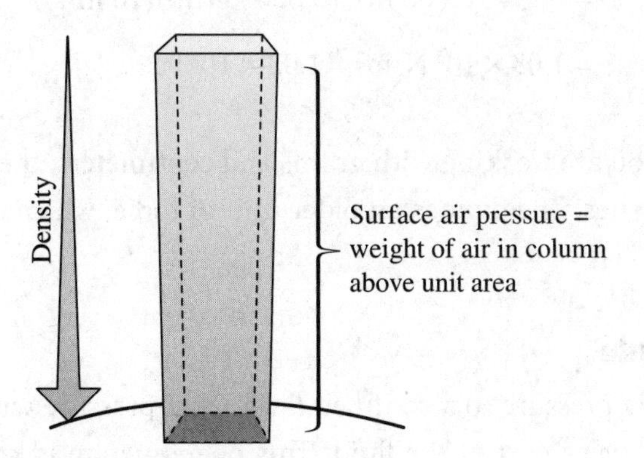

Atmospheric pressure can be measured with a mercury barometer. This device is made by inverting a glass tube from which the air has been removed into a reservoir of mercury. Air will push downward on the mercury in the reservoir, forcing some mercury into the tube. The height of the mercury in the tube will match the pressure of the atmosphere that supports it.

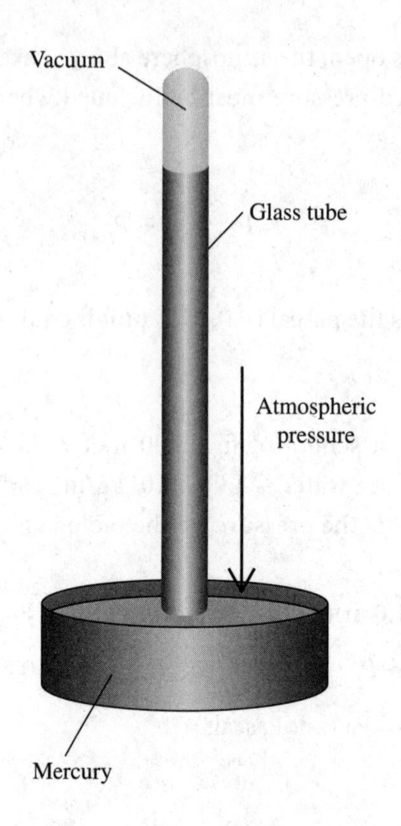

Atmospheric pressure varies with location, altitude, and weather conditions. Standard pressure is the pressure that supports a column of mercury to a height of 760 millimeters. You can use the pressure equation to calculate standard atmospheric pressure. The calculation becomes easier if you convert the height to 0.760 m. The density of mercury is 13.6 g/cm³ or 13,600 kg/m³ in SI units.

$$\mathbf{P} = hdg = (0.760 \text{ m})(13{,}600 \text{ kg/m}^3)(10 \text{ m/s}^2)$$
$$= 1.03 \times 10^5 \text{ N/m}^2 = 1.03 \times 10^5 \text{ Pa}$$

Note that when you are working with grams and centimeters, the resulting unit of power involves dynes. The dyne is an older unit of force, which can be converted as shown.

Pascal's Principle

Suppose you apply pressure to a confined fluid. That pressure will be transmitted undiminished to every part of the fluid. This phenomenon is known as **Pascal's Principle**, which, stated more formally, explains that an increase in pressure to any point of a confined fluid is transmitted equally to every other point in the fluid.

Consider the fluid in the tube shown in the following diagram. If the pressure exerted on the top of the fluid is increased by 3 units, that increase will be observed on all three gauges. No part of the fluid will experience a greater increase than any other part.

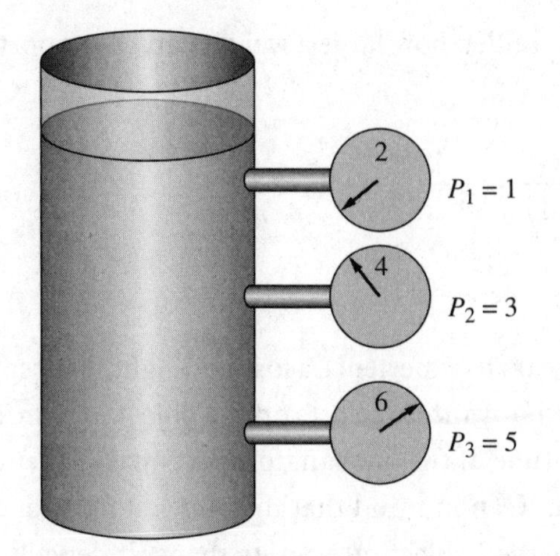

Increase pressure

Gauge 1: $2 + 3 = 5$ units

Gauge 2: $4 + 3 = 7$ units

Gauge 3: $6 + 3 = 9$ units

Pascal's Principle is commonly utilized in hydraulic devices similar to the one shown in the following diagram. A small force F_1 is applied to the piston with the small area A_1. The increase in pressure is transmitted throughout the fluid and therefore acts on the bottom of the larger piston. Because the pressure is exerted over a larger area A_2, the resulting force F_2 is greater than F_1. If the device is designed properly, it can be used to magnify a force in such a way that a heavy load, such as a car, can be lifted.

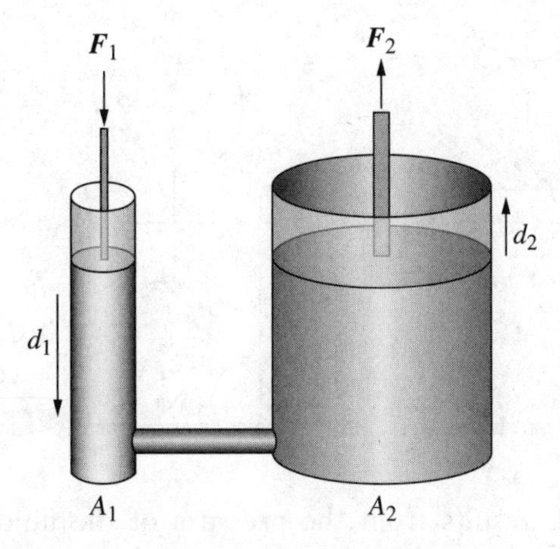

Notice that d_1 is greater than d_2. The reason is that the hydraulic device can be considered to be a machine that does work. The smaller force must be exerted over a longer distance to produce a greater force exerted over a shorter distance. Think of F_1 as the effort and F_2 as the resistance. The mechanical advantage is then the ratio of the resistance to the effort.

$$\text{IMA} = \frac{F_2}{F_1} = \frac{A_2}{A_1}$$

Example:

A hydraulic press has an input cylinder with an area of 4 square centimeters and an output cylinder with an area of 40 square centimeters. If a force of 5 newtons is applied to the piston in the small cylinder, how large a weight can be supported by the piston in the larger cylinder?

$$\frac{F_2}{F_1} = \frac{A_2}{A_1} \rightarrow \frac{F_2}{5\text{ N}} = \frac{40\text{ cm}^2}{4\text{ cm}^2} \rightarrow F_2 = 50\text{ N}$$

Archimedes' Principle

An object submerged in a fluid appears to experience a loss in weight. The reason is that an upward force known as the **buoyant force** acts on the object. According to **Archimedes' Principle**, the magnitude of the buoyant force is equal to the weight of the fluid displaced by the object. Keep in mind that the buoyant force does not depend on the weight of the submerged object or even its shape. It depends only on the weight of the displaced fluid.

According to the following diagram, the object weighs less in the liquid than in air because of the buoyant force. As you can see, the difference in weight equals the weight of the liquid that spills out of the container when the object is submerged.

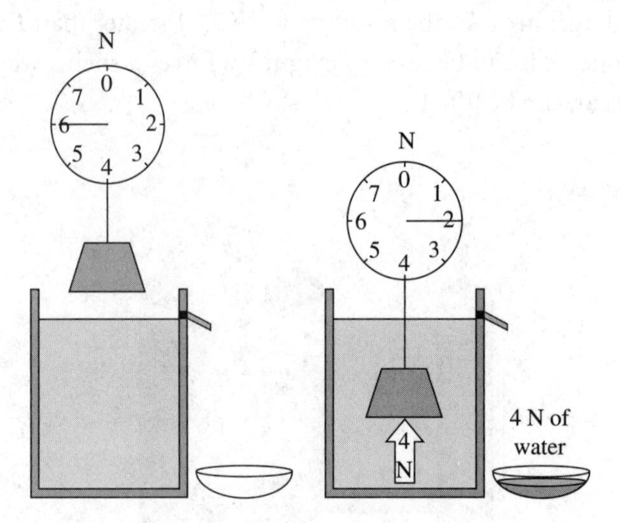

The buoyant force results from the pressure of the fluid. Because pressure increases with depth, the pressure acting on the top of a submerged object is less than the pressure acting on the bottom of the object. Therefore, the force pushing downward on the top of the object is less than the force pushing upward on the bottom of the object. The net force is the buoyant force.

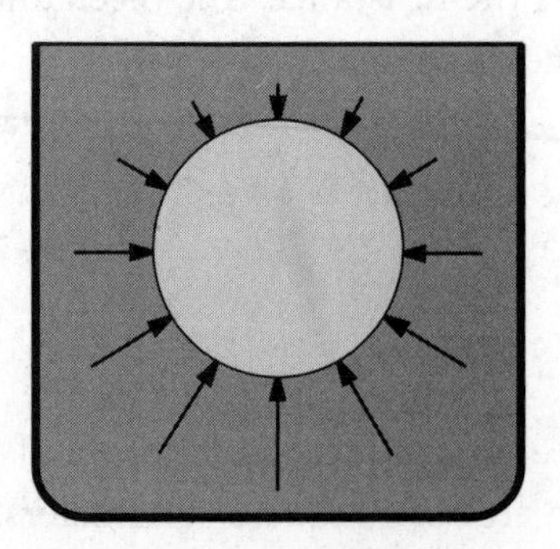

The buoyant force depends on the volume of the object submerged because this is the volume of fluid that will be displaced. Therefore, spheres of different substances that are submerged in water will all displace the same volume of water and therefore experience the same buoyant force. Suppose, for example, spheres of lead, aluminum, and cork are all submerged in water. The spheres have identical volumes. The buoyant force exerted on each sphere will be the same. The difference is that once set free, the lead will quickly sink to the bottom. The aluminum will sink as well, and the cork will rise to the surface. The buoyant force acts opposite to the weight of the sphere. If the weight of the sphere is greater than the buoyant force, as in the cases of lead and aluminum, the sphere will sink. If the weight of the sphere is less than the buoyant force, as in the case of cork, the sphere will float.

Another way to state this relationship is that an object will float in a fluid denser than itself and sink in a fluid that is less dense than itself. The volume of the object is equal to the volume of the displaced fluid. If the object floats, its weight must be less than the weight of the displaced fluid. Weight equals the product of the mass of a sample and the acceleration due to gravity. Because the acceleration due to gravity is the same for the object and the displaced fluid, the mass of the displaced fluid must be greater than the mass of the object. Density is mass divided by volume, so a greater mass divided by the same volume results in a greater density. The fluid must be denser than the object. The same logic explains why an object sinks.

Bernoulli's Principle

An airplane wing experiences lift because the air moving across the top of the wing exerts less pressure than the air moving across the bottom of the wing. The design of a wing is such that air moves faster across the top than across the bottom. **Bernoulli's Principle** states that the pressure of a fluid decreases as the velocity of the fluid increases. In other words, a slow-moving fluid exerts more pressure than a fast-moving fluid.

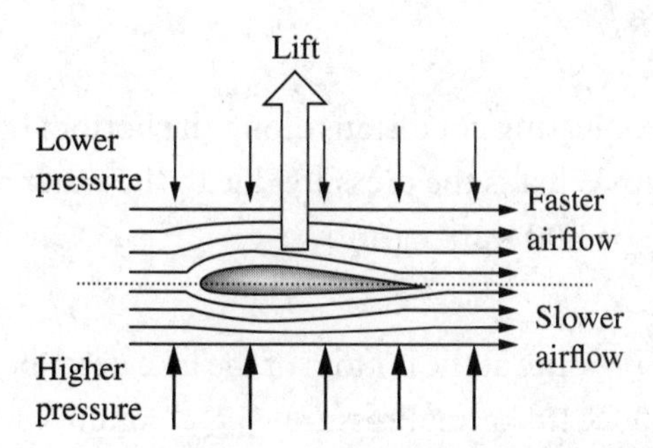

In addition to airplanes, Bernoulli's Principle explains why a shower curtain billows inward. The flow of water causes air around it to move faster than the air outside the shower curtain. The faster-moving air creates a region of lower pressure, allowing the higher pressure outside the shower curtain to push it inward.

TEST-TAKING HINT

Be sure to differentiate between individual quantities and net quantities. You might quickly identify a force on an object, such as the buoyant force on an object submerged in a fluid, but you cannot ignore other forces acting on the same object, such as its downward weight. Only the net force can be used to determine the resulting motion of the object. You can often compare individual quantities by combining arrows used in diagrams.

Bernoulli's Principle also explains how pitchers can cause baseballs to curve. Suppose a ball is thrown from right to left with a clockwise spin. As the ball spins, it drags air with it. As a result, the air flowing over the ball is opposite to the spin, whereas the air flowing under the ball is in the same direction as the spin. Because the air below the ball moves faster than the air above it, the net pressure is downward. Depending on how the pitcher spins the ball, curves in different directions can be achieved.

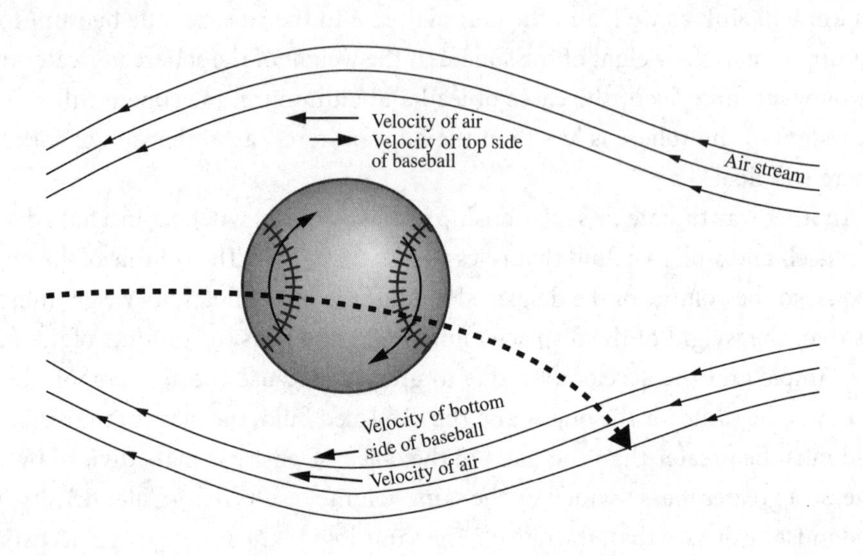

REVIEW QUESTIONS

Questions 1–3 refer to the following pressure amounts.

(A) 9.80×10^6 Pa
(B) 520 kPa
(C) 1.00×10^6 Pa
(D) 690 kPa
(E) 1.10×10^3 kPa

1. A scientist is investigating conditions along the bottom of a lake. The lake is 100.0 meters deep. What is the pressure due to the water at the bottom of the lake? [$\text{Density}_{water} = 1.00 \times 10^3\ \text{kg/m}^3$]

2. What is the total pressure at the bottom of the lake described in question 1? [$\text{Density}_{water} = 1.00 \times 10^3\ \text{kg/m}^3$ $\text{Pressure}_{atmosphere} = 1.00 \times 10^6\ \text{Pa}$]

3. What is the total pressure on a diver 42.0 meters below sea level? [Use $1.00 \times 10^3\ \text{kg/m}^3$ for the density of sea water.]

4. The input piston of a hydraulic system has an area of 0.10 m^2 and the output piston has an area of 0.40 m^2. If the input force is 1500 N, what is the output force?

(A) 2500 N
(B) 3000 N
(C) 3750 N
(D) 4500 N
(E) 6000 N

5. The input piston of a hydraulic device moves a distance of 0.6 m when a force of 1200 N is exerted on it. The result is a force of 3600 N on the output piston. What distance does the output piston move?

(A) 0.2 m
(B) 0.3 m
(C) 0.4 m
(D) 0.6 m
(E) 0.8 m

6. A force of 8 N is exerted on the stopper of a container as shown. If the area of the stopper is 4 cm^2 and the area of the bottom of the jug is 400 cm^2, what is the increase in force on the bottom of the jug?

(A) 100 N
(B) 200 N
(C) 600 N
(D) 800 N
(E) 1600 N

7. Due to buoyant forces, an object experiences an apparent loss of 28.0 grams when submerged in water. What is the volume of the object?

(A) 2.6 cm^3
(B) 14.0 cm^3
(C) 18.2 cm^3
(D) 28.0 cm^3
(E) 37.8 cm^3

8. Which conclusion is reasonable based on the illustration?

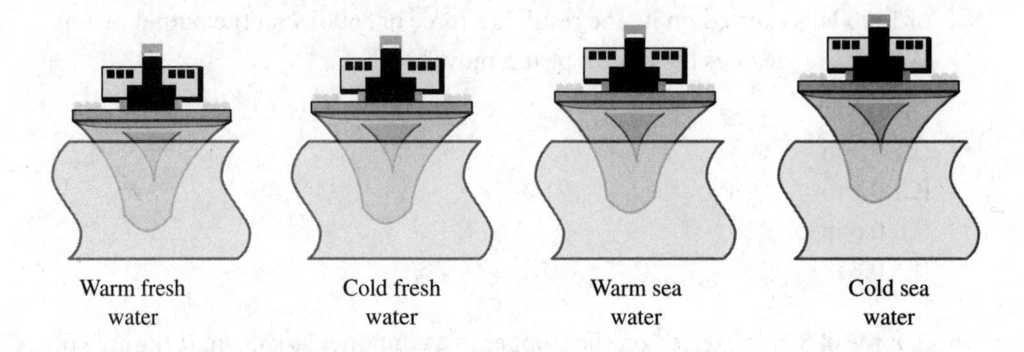

(A) Warm water is denser than cold water.
(B) Cold sea water is denser than warm sea water.
(C) Fresh water is denser than sea water.
(D) A boat is denser in warm sea water than in cold sea water.
(E) A boat is less dense in cold fresh water than in warm fresh water.

Questions 9 and 10 relate to the diagram below, which shows a hydraulic device along with the input force and area of the input and output cylinders.

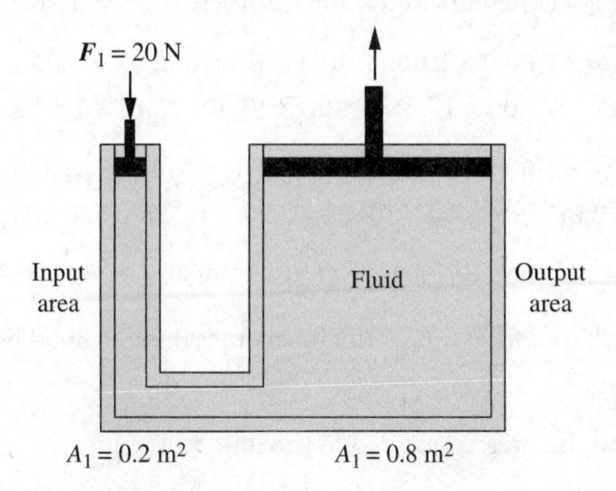

9. What pressure is transmitted throughout the liquid in the device?

(A) 0.1 kPa
(B) 0.2 kPa
(C) 0.4 kPa
(D) 2 kPa
(E) 4 kPa

10. What force is produced on the output piston as a result of the input force?

(A) 60 N
(B) 80 N
(C) 100 N
(D) 120 N
(E) 160 N

ANSWERS AND EXPLANATIONS

1. **(C)** $P_{fluid} = hdg = (100.0\ m)(1.00 \times 10^3\ kg/m^3)(10\ m/s^2) = 1.00 \times 10^6\ Pa$

2. **(E)** $P_{fluid} = hdg = (100.0\ m)(1.00 \times 10^3\ kg/m^3)(10\ m/s^2) = 1.00 \times 10^6 = Pa$; $P_{total} = P_{atmosphere} + P_{fluid} = 1.00 \times 10^6\ Pa + 1.00 \times 10^5\ Pa = 1.10 \times 10^6\ Pa = 1.10 \times 10^3\ kPa$

3. **(B)** $P_{fluid} = hdg = 1.00 \times 10^6 = Pa$; $P_{total} = P_{atmosphere} + P_{fluid} = [(42.0\ m)(1.00 \times 10^3\ kg/m^3)(10\ m/s^2)] + 1.00 \times 10^5\ Pa = 5.20 \times 10^5\ Pa = 520\ kPa$

4. **(E)** $F_2/F_1 = A_2/A_1 \rightarrow F_2 = F_1A_2/A_1 = (1{,}500\ N)(0.40\ m^2)/(0.10\ m^2) = 6000\ N$

5. **(A)** $F_1d_1 = F_2d_2$ so $d_1/d_2 = F_2/F_1$, which becomes $0.6/d_2 = 3600\ N/1200\ N$ so $d_2 = 0.2\ m$.

6. **(D)** First, find the pressure added to the fluid in the jug.

$$P_1 = 8\ N/4\ cm^2 = 2\ N/cm^2$$

Second, find the force that results when the pressure is transmitted to the bottom of the jug. Note that the pressure is transmitted undiminished to all parts of the liquid so the increase in pressure P_2 equals P_1.

$$F_2 = P_2A_2 = (2\ N/cm^2)(400\ cm^2) = 800\ N$$

7. **(D)** The mass of the water displaced by the object must be 28 g. The density of water is 1 g/cm³. Therefore, the volume of water displaced must be 28 cm³. The volume of water displaced by the object must be equal to the volume of the object. The volume of the object is 28 cm³.

8. **(B)** The buoyant force pushing the boat upward increases from left to right in the illustration. Therefore, the weight of the water displaced by the boat must be increasing. The weight depends on the density of the displaced water, which therefore suggests that cold sea water is the densest liquid in which the boat is placed.

9. **(A)** The pressure applied to the input piston is transmitted undiminished throughout the fluid.

$$P_1 = P_2 = F_1/A_1 = 20\ N/0.2\ m^2 = 0.1\ kPa$$

10. **(B)** $F_2/F_1 = A_2/A_1 \rightarrow F_2 = F_1A_2/A_1 = (20\ N)(0.80\ m^2)/(0.20\ m^2) = 80\ N$

CHAPTER 7

Linear Momentum

In everyday usage, the word *momentum* is commonly used to describe objects in motion. You might say that a car has momentum, an athlete has momentum, or even a team has momentum. In other words, the objects are in motion and are difficult to stop. The scientific use of the term is not all that different. For SAT Physics, you will need to understand the scientific definition of the term and how to perform any calculations related to it.

Momentum

The linear momentum, or simply **momentum**, of an object is the product of its mass and velocity. Momentum is generally represent by p, which makes the equation for momentum

$$p = mv$$

Mass is a scalar quantity, and velocity is a vector quantity, which makes momentum a vector quantity. Momentum is in the same direction as the velocity of the object. Common units of momentum are kg · m/s.

An object can have considerable momentum if it has either great mass or velocity. A train can have great momentum because of its great mass, even if it is traveling very slowly. A bullet can have great momentum because of its high velocity, even though its mass is very small. Two objects with different masses can have the same momentum depending on their velocities. An object at rest has no momentum.

Example:

A ball with a mass of 2.0 kg is traveling with a velocity of 4.0 m/s west. What is the ball's momentum? By how much does the momentum change if the velocity is doubled?

$$p_1 = 2.0\text{ kg} \times 4.0\text{ m/s west} = 8.0\text{ kg} \cdot \text{m/s west}$$

$$p_2 = 2.0\text{ kg} \times 8.0\text{ m/s west} = 16\text{ kg} \cdot \text{m/s west}$$

When the velocity is doubled, the momentum is also doubled.

Impulse

Unlike momentum, the everyday use of the word *impulse* is different from its scientific use. The scientific definition of impulse derives from Newton's Second Law. Recall from Chapter 4 that this law states that a net force will cause an object to accelerate, and the amount of acceleration depends on the magnitude of the force and the mass of the object. The equation for Newton's Second Law can be rearranged to relate it to momentum.

$$F_{net} = ma = m\left(\frac{\Delta v}{\Delta t}\right)$$

$$F\Delta t = m\Delta v = mv_f - mv_i$$

The left side of the equation ($F\Delta t$) is defined as the **impulse**. It is equal to the change in momentum ($m\Delta v$). Impulse is a vector quantity. The unit of impulse is the same as the unit of momentum. The equation is sometimes known as the impulse-momentum theorem, which states that the force acting on a mass during a given time interval changes the momentum of the mass.

Example:

Find the applied force that causes a 10 m/s change in the velocity of the box in 5 s if the mass of the box is 5 kg.

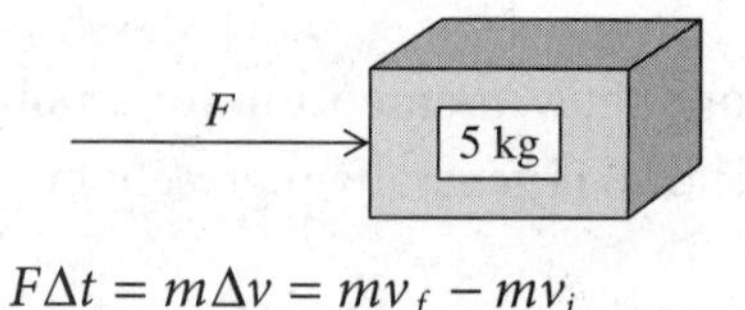

$$F\Delta t = m\Delta v = mv_f - mv_i$$

$$F\Delta t = 5\text{ kg} \times 10\text{ m/s} = 50\text{ kg} \cdot \text{m/s}$$

The impulse is 50 kg · m/s.

$$F = [50\text{ kg} \cdot \text{m/s}]/5\text{ s} = 10\text{ N}$$

SAT Physics often uses graphs to test your understanding of momentum. The impulse caused by a force exerted on an object during a time interval equals the area under the force versus time graph during that time interval.

Example:

Consider the following graph.

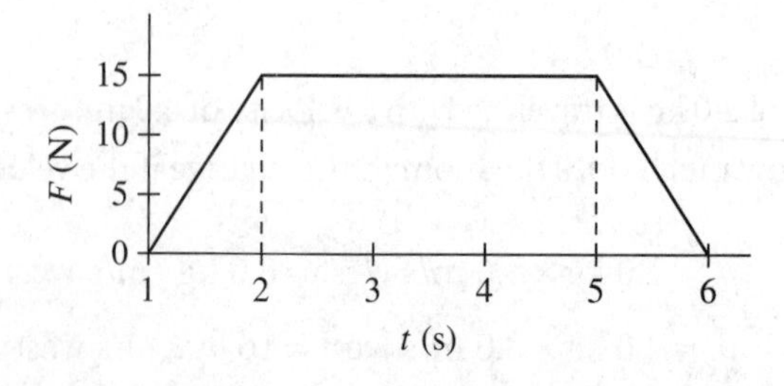

(A) What is the impulse acting on the object between 2.0 s and 5.0 s?
(B) What is the change in momentum between 2.0 s and 5.0 s?

(A) The impulse during the time interval is the area under the graph during that interval.

$$F\Delta t = \text{area} = (15\text{ N})(5.0\text{ s} - 2.0\text{ s}) = 45\text{ N s}$$

(B) The change in momentum equals the impulse, so the change in momentum is also 45 N s.

Conservation of Momentum

If a force acts on an object, the momentum of the object changes. If there is no unbalanced force acting on an object, the total momentum of the system must remain constant. In other words, momentum is conserved. SAT Physics may test your understanding of conservation of momentum in different ways, including inelastic collisions and elastic collisions.

Inelastic Collisions

An inelastic collision occurs when two objects crash and become stuck together, either completely or partially. For example, suppose two train cars on a track collide such that one hooks onto the other. Momentum is conserved, but kinetic energy is not.

Example:

A ball with a mass of 4.0 kg is moving to the right with a velocity of 6.0 m/s collides with a 2.0 kg ball at rest. Upon collision, the balls stick together. What is the velocity immediately after the collision?

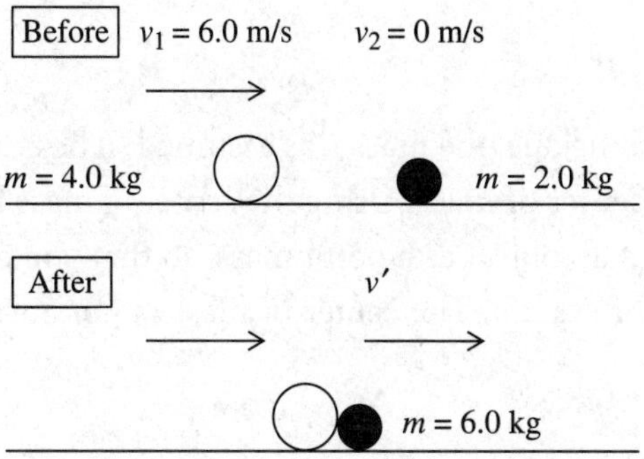

Before $v_1 = 6.0$ m/s $v_2 = 0$ m/s

$m = 4.0$ kg $m = 2.0$ kg

After v'

$m = 6.0$ kg

$$p_{before} = m_1v_1 + m_2v_2$$

$$p_{after} = (m_1 + m_2)v$$

$$v' = \frac{m_1v_1 + m_2v_2}{m_1 + m_2} = \frac{(4.0\text{ kg})(6.0\text{ m/s}) + 0}{6.0\text{ kg}} = 4.0\text{ m/s}$$

Elastic Collisions

An elastic collision occurs when two objects crash but do not stick together. Instead, they bounce off each other. For example, two billiard balls may collide and bounce off in opposite directions. Both momentum and kinetic energy are conserved in this type of collision.

Example:

A white billiard ball with a mass of 0.5 kg is moving at 8 m/s. It collides with a black billiard ball with the same mass that is at rest. After the collision, the white ball continues in the same direction with a velocity of 2.2 m/s. If you ignore friction, what is the velocity of the black ball?

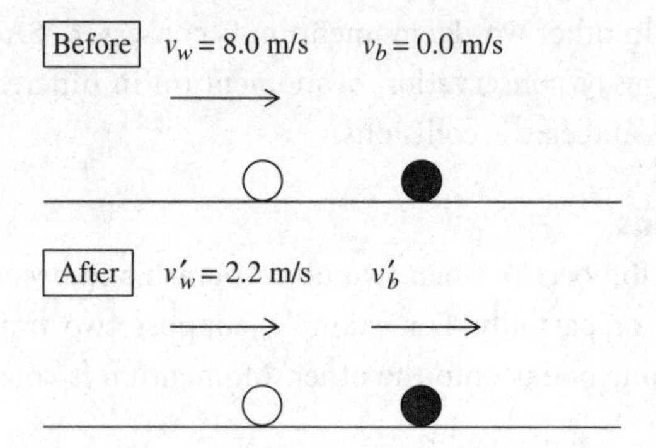

momentum before collision = momentum after collision

$$m_w v_w + m_b v_b = m_w v'_w + m_b v'_b$$

$$(0.5 \text{ kg})(8.0 \text{ m/s}) + 0 = (0.5 \text{ kg})(2.2 \text{ m/s}) + (0.5 \text{ kg})\, v'_b$$

$$v'_b = 5.8 \text{ m/s to the right}$$

Center of Mass

When considering collisions, the mass was assumed to be concentrated in a single point known as the **center of mass**. Using the center of mass is convenient because it enables you to treat an object as a point mass, so that you can ignore the overall shape of the object or system. The center of mass is not always within the object itself.

More specifically, it may be helpful for you to understand how to calculate the center of mass for a collection of particles. For two particles separated by a distance *d*, the center of mass x_{cm} is found as follows.

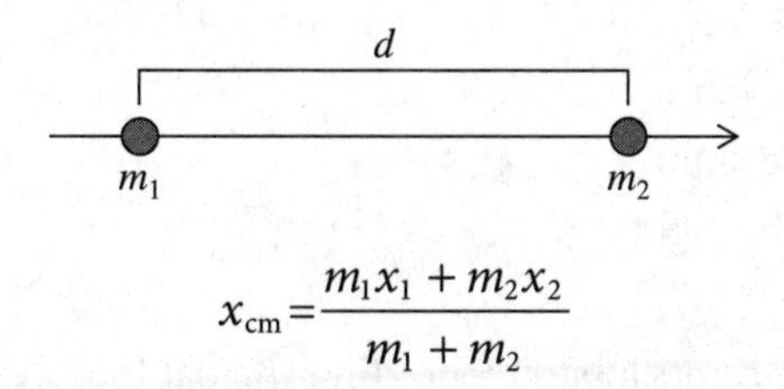

$$x_{cm} = \frac{m_1 x_1 + m_2 x_2}{m_1 + m_2}$$

TEST-TAKING HINT

Pay attention to conventions for directions when it comes to velocity and momentum. By convention, movement to the right is generally described by positive values, whereas movement to the left is generally described by negative values.

REVIEW QUESTIONS

Questions 1–3 refer to the following speeds.

(A) 2.7 m/s
(B) 7.5 m/s
(C) 15 m/s
(D) 140 m/s
(E) 180 m/s

1. Two balls are moving directly toward one another. One has a mass of 4.0 kg and is moving to the right with a velocity of 6.0 m/s. The other has a mass of 2.0 kg and is moving toward the left with a velocity of 4.0 m/s. They collide inelastically. What is the speed of the combined balls immediately after the collision?

2. A force of 10 N acts on a 2-kg object for 3 seconds. What is the change in the speed of the object?

3. A bullet with a mass of 0.020 kg collides with a block of wood with a mass of 3.2 kg. The block of wood is initially at rest. If the bullet and block combination move with a speed of 1.1 m/s after the collision, what was the initial speed of the bullet?

4. A ball with a mass of 0.1 kg is traveling at 5 m/s from right to left. What is the momentum of the ball?

(A) 0.5 kg · m/s to the left
(B) 2.5 kg · m/s to the left
(C) 4.9 kg · m/s to the right
(D) 5.0 kg · m/s to the left
(E) 5.1 kg · m/s to the right

5. A 70-kg runner is traveling from left to right. If the runner has a momentum of 560 kg · m/s to the right, what is the runner's velocity?

(A) 4.9 m/s to the right
(B) 5.6 m/s to the left
(C) 7.5 m/s to the left
(D) 8.0 m/s to the right
(E) 12.6 m/s to the right

6. A baseball player bunts a ball by holding the bat loosely. The momentum of the bat before the collision is 80 units. The momentum of the ball is −40 units before the collision and 10 units after the collision. What is the momentum of the bat after the collision?

(A) 10 units
(B) 20 units
(C) 30 units
(D) 60 units
(E) 70 units

7. In a perfectly elastic collision, if one ball approaches a second ball of the same mass initially at rest, the final speed of the second ball will be

(A) zero
(B) equal to the initial speed of the first ball
(C) one-third the initial speed of the first ball
(D) half the initial speed of the first ball
(E) twice the initial speed of the first ball

8. In a perfectly inelastic collision, if one ball approaches a second ball of the same mass initially at rest, the final speed will be

(A) zero
(B) equal to the initial speed of the first ball
(C) one-third the initial speed of the first ball
(D) half the initial speed of the first ball
(E) twice the initial speed of the first ball

Questions 9 and 10 relate to the diagram below, which shows a force vs time graph for an object.

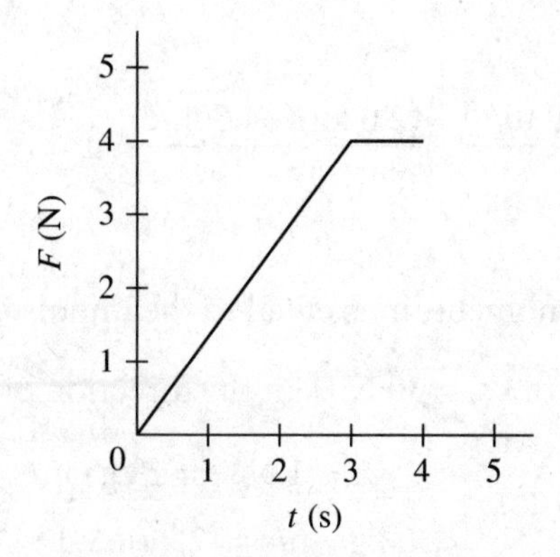

9. What is the impulse delivered by the force graphed in the graph between $t = 0$ and $t = 4$?

(A) 4 kg · m/s
(B) 10 kg · m/s
(C) 12 kg · m/s
(D) 16 kg · m/s
(E) 20 kg · m/s

10. If the mass of the object is 10 kg and it has a velocity of 4 m/s at 3 s, what is its velocity at 4 s?

(A) 3.6 m/s
(B) 4.4 m/s
(C) 6.0 m/s
(D) 9.6 m/s
(E) 14 m/s

ANSWERS AND EXPLANATIONS

1. (A) $v = \dfrac{m_1v_1 + m_2v_2}{m_1 + m_2}$

$= \dfrac{(4.0\text{ kg})(6.0\text{ m/s}) + (2.0\text{ kg})(-4.0\text{ m/s})}{6\text{ kg}}$

$= 2.7\text{ m/s}$

2. (C) The change in momentum is equal to the impulse.

Impulse = (10 N)(3 s) = 30 N · s in the direction of the force

$$1\text{ N}\cdot\text{s} = 1\text{ kg}\cdot\text{m/s}$$

$$30\text{ kg}\cdot\text{m/s} = (2\text{ kg})(\Delta v)$$

$$\Delta v = 15\text{ m/s}$$

3. (E)

$$v_{com} = \frac{m_1v_1 + m_2v_2}{m_1 + m_2}$$

$$(m_1 + m_2)v_{com} = m_1v_1$$

$$v_1 = \frac{m_1 + m_2}{m_1}v_{com}$$

$$= \frac{(0.020\text{ kg}) + (3.20\text{ m/s})}{0.020}1.1\text{ m/s}$$

$$= 180\text{ m/s}$$

4. (A) $p = mv = (0.1\text{ kg})(5\text{ m/s}) = 0.5\text{ kg}\cdot\text{m/s}$ to the left

5. (D) $p = mv$, so $v = p/m$

$v = (560\text{ kg}\cdot\text{m/s})/70\text{ kg} = 8\text{ m/s}$ to the right

6. (C) The total momentum before and after the collision must be the same. The total momentum before the collision is 40 units, so it must be 40 units (30 units + 10 units) after the collision.

7. (B) The original momentum of the system is $p_{initial} = m_1v_1 + 0$. The final momentum of the system will be $p_{final} = 0 + m_2v_2$. Since the masses are the same, the final velocity of ball 2 will equal the initial velocity of ball 1.

8. (D) The original momentum of the system is $p_{initial} = mv_{initial} + 0$. The balls stick together in a perfectly inelastic collision. So $p_{final} = (m_1 + m_2)v_{final} = 2mv_{final}$. By conservation of momentum, $p_{initial} = p_{final}$. So $v_{final} = mv_{initial}/2m = v_{initial}/2$.

9. (B) The impulse equals the area of a triangle of height 4 and base 3 plus the area of the rectangle with height 4 and base 1.

Impulse = (½ × 3 × 4) + (4 × 1) = 10 kg · m/s

10. (B) According to the Impulse-Momentum Theorem,

$$F\Delta t = m(v_f - v_i)$$

The impulse during that time interval is 4 N · s.

$$4\ \text{N}\cdot\text{s} = (10\ \text{kg})(v_f - 4\ \text{m/s})$$

$$v_f = 4.4\ \text{m/s}$$

CHAPTER 8

Circular and Rotational Motion

So far, you have been investigating linear motion, which is motion in a straight line. Many examples of motion, however, are not linear. In this chapter, you will investigate circular and rotational motion. The SAT Physics test may ask both qualitative and quantitative questions involving these topics.

Uniform Circular Motion

An object is said to move in **uniform circular motion** if it moves at a constant speed in a circle. The speed of the object is constant, but the velocity is constantly changing because the direction is constantly changing. In addition to velocity, two other vectors are associated with uniform circular motion—centripetal force and centripetal acceleration.

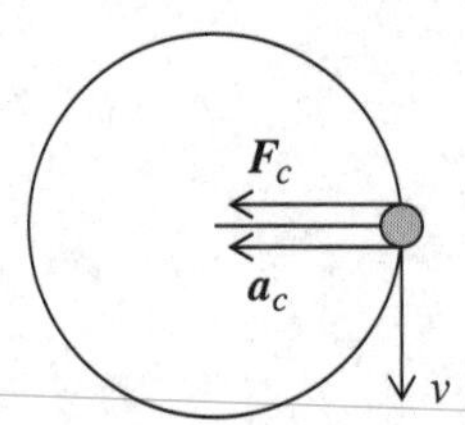

The word *centripetal* means "center seeking." The **centripetal force** (F_c) acts on the object toward the center of the circle. This force does not change the speed of the object. Instead, it constantly accelerates the object toward the center of the circle, which is why it changes direction.

The resulting acceleration, the **centripetal acceleration** (a_c), depends on the velocity of the object and the radius of the circle.

$$a_c = \frac{v^2}{r}$$

The magnitude of centripetal acceleration remains constant, but the direction is constantly changing. The direction is always toward the center of the circle, which must change as the object moves around its circular path. Centripetal acceleration is in the same direction as the centripetal force and perpendicular to the velocity.

According to Newton's Second Law, the net force equals the product of mass and acceleration.

$$F_c = ma = \frac{mv^2}{r}$$

The **period** (T) of an object in uniform circular motion is the time it takes to complete a revolution. Period is measured in a unit of time, such as seconds or minutes. The number of revolutions per unit time is the **frequency** (f) of the object. Frequency is measured in revolutions per unit of time, such as revolutions per second or revolutions per minute. The unit *revolutions per second* is also known as the *hertz.*

Speed equals distance divided by time. For an object in uniform circular motion, the distance is the circumference of the circle and the time is the period.

$$v = \frac{\text{circumference}}{\text{period}} = \frac{2\pi r}{T}$$

Example:

A ball with a mass of 3.0 kg is swung in a circular path at the end of a rope with length 2.0 m. If the speed of the ball is 1.0 m/s, what is the tension on the rope? Assume the rope to be massless.

The tension on the rope is equal to the centripetal force.

$$T = \frac{mv^2}{r} = \frac{(3.0\text{ kg})(1.0\text{ m/s})^2}{2.0\text{ m}} = 1.5\text{ N}$$

Rotational Motion

If you think of a circle as a straight line that has been rolled up, it makes sense that the quantities you apply to straight-line motion can be used to analyze rotational motion. The rotational components are described as angular—**angular displacement** (θ), **angular velocity** (ω), and **angular acceleration** (α).

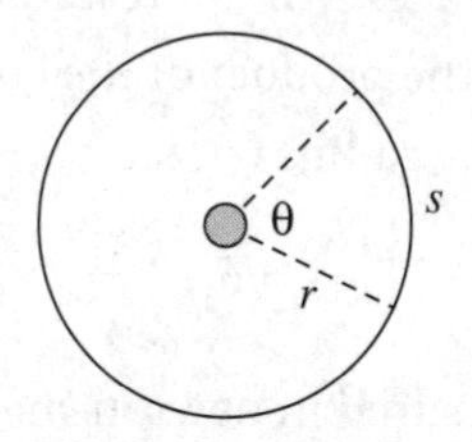

If the disk in the diagram rotates such that the arc length (s) equals the radius (r), the subtended angle, which is the angular displacement θ, will equal 1 radian.

$$s = r\theta$$

$$\theta = s/r$$

If the disk rotates one complete revolution, s becomes equal to the circumference. This means that $\theta = 2\pi$ radians and 2π radians $= 360°$.

Although you will not be required to do so on the SAT Physics exam, it will help you to understand how to derive the next relationships. By differentiating the equation $s = r\theta$, you find the relationship for angular velocity.

$$s = r\theta$$

$$ds/dt = r(d\theta/dt)$$

$$v = r\omega$$

Thus, angular velocity $\omega = v/r$. The unit of angular velocity is radians per second.

Differentiating again gives the angular acceleration.

$$v = r\omega$$

$$dv/dt = r(d\omega/dt)$$

$$a = r\alpha$$

Thus, angular acceleration $\alpha = a/r$. The unit of angular acceleration is radians per second squared.

You can see that each angular component is related to r, the distance to the center of the circle. When velocity is constant, $s = vt$ becomes $\theta = \omega t$.

Example:

What is the angular velocity of the second hand of a clock?

$$\omega = 1 \text{ rev/min} = (2\pi/60 \text{ radians/s}) = 0.105 \text{ rad/s}$$

Torque

With the basics of rotational kinematics mastered, you can comfortably approach questions involving rotational dynamics, such as torque. Essentially, **torque** is a measure of how much a force acting on an object causes that object to rotate. The force must be applied to any point other than the center of mass. A net force applied to an object's center of mass will not result in rotation.

More formally, torque (τ) is the product of the component of the force perpendicular to the radius ($F_\perp$) and the radius (r).

$$\tau = F_\perp r$$

Torque is a vector quantity measured in newton-meters (N · m).

Consider the massless lever in the following diagram.

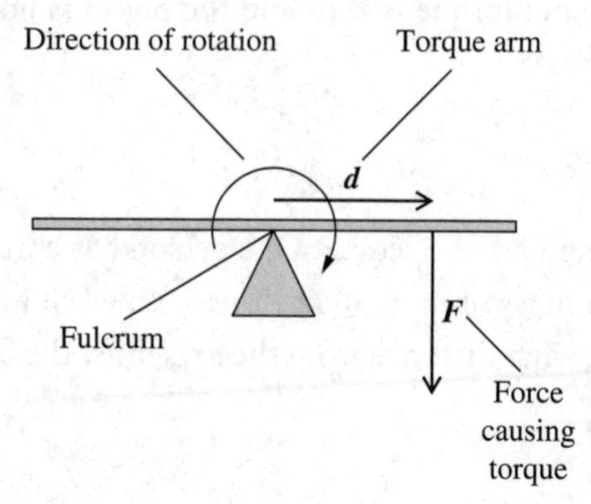

The force exerted on one end of the board causes the board to rotate in the direction indicated by the curved arrow. The distance between the point of rotation and the fulcrum, identified by d, is known as the torque arm. If the torque arm is 3.0 m long and the force is 4.0 N, the magnitude of the torque is calculated as follows.

$$\tau = (3.0\ \text{m})(4.0\ \text{N}) = 12\ \text{N} \cdot \text{m}$$

Torque associated with clockwise rotation is considered to be directed into the page, whereas counterclockwise torque is directed out of the page. You can use the right-hand rule to identify direction. Wrap the fingers of your right hand in the direction of rotation caused by the force. Your thumb will point in the direction of the torque vector.

In the preceding example, the force was exerted at a right angle to the torque arm. This made the calculation ordinary multiplication. If the angle is not a right angle, the calculation must account for the angle made between the force vector and the lever. Consider the lever attached to the wall as shown in the following diagram.

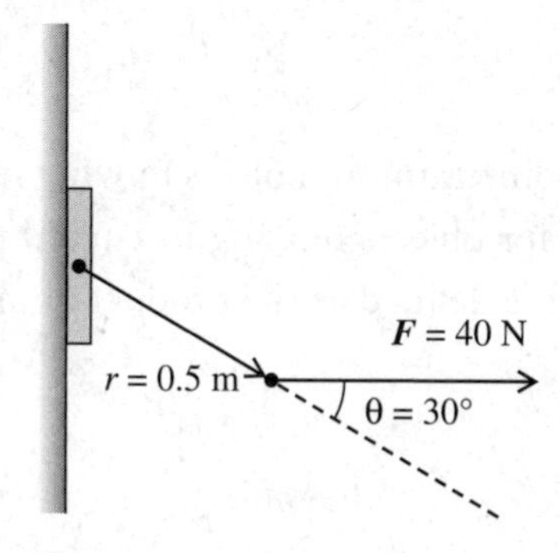

Suppose a force of 40 N is exerted on the lever 0.5 m from its axis of rotation. The force forms an angle of 30° with the lever arm. What is the torque on the lever arm?

$$\begin{aligned}\tau &= Fr \sin \theta \\ &= (40\ \text{N})(0.5\ \text{m}) \sin 30^\circ \\ &= 10.0\ \text{N} \cdot \text{m pointing out of the page}\end{aligned}$$

If a system is at equilibrium, both the sum of the forces and the sum of the torques must equal zero. If the net torque is zero and the object is not rotating, the object will not begin rotating.

Example:

Two people are on either end of a seesaw. One person has a mass of $m_1 = 60$ kg and the other person has a mass of $m_2 = 30$ kg. The seesaw is 8 m long and pivoted on an axis at its center. At what distance from the axis must the first person sit to keep the seesaw horizontal?

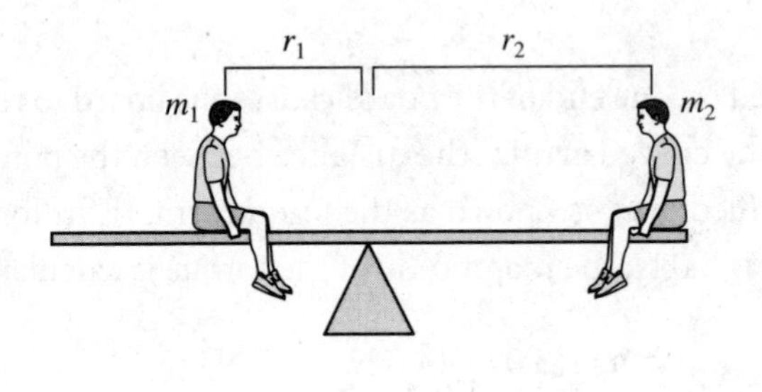

$$\tau_{left} = \tau_{right}$$
$$F_1 r_1 = F_2 r_2$$

The force is applied perpendicular to the seesaw by gravity.

$$(m_1 g)r_1 = (m_2 g)r_2$$
$$m_1 r_1 = m_2 r_2$$
$$r_1 = m_2 r_2 / m_1 = (30 \text{ kg})(4.0 \text{ m})/60 \text{ kg} = 2.0 \text{ m to the left of the axis}$$

Angular Momentum

Just as you found linear momentum for objects moving in a straight line, you can find **angular momentum** for objects moving in curved paths. Angular momentum (L) for a circular orbit is defined as the product of mass, velocity, and radius of orbit.

$$L = mvr$$

You will revisit the topic of angular momentum in Chapter 9, but for now it would be helpful for you to become familiar with the general concept. Recognize that angular momentum is conserved if there are no external torques on an object. Therefore, if the radius of an orbit is decreased, the speed will decrease. This explains why ice skaters spin faster when they tuck their bodies in more tightly.

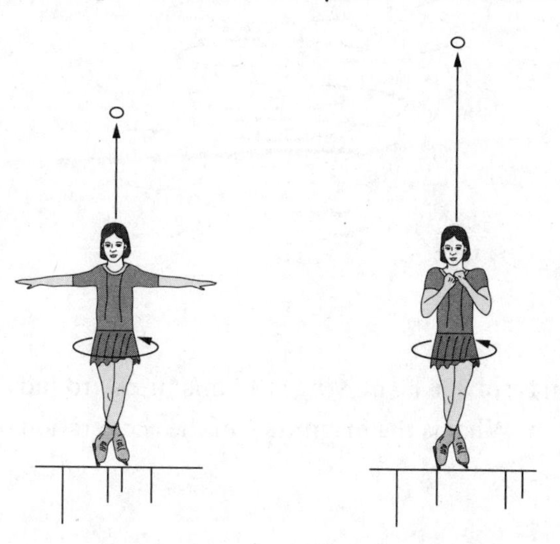

TEST-TAKING HINT

Many linear and rotational equations are related. Make a list relating equations so you can study them together. If you become confused by a test question, you may be able to recall the equation you need by remembering the related equation.

REVIEW QUESTIONS

Questions 1–2 refer to the following directions.

(A) To the left
(B) To the right
(C) Straight up
(D) Straight down
(E) Down and to the left

1. A ball is swung around at the end of a string in a horizontal circle rotating counterclockwise as shown. The string breaks when the ball is in the position shown. In which direction will the ball move?

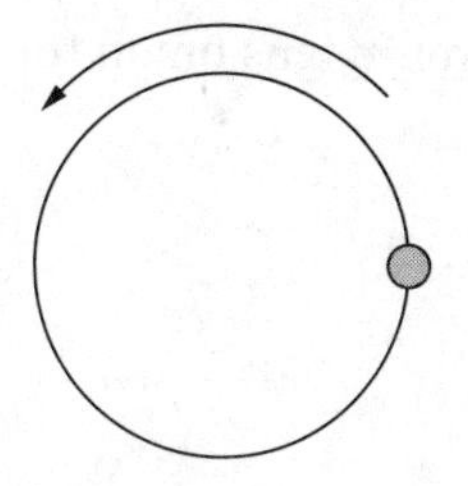

2. A meteorologist is studying a tornado, which is spinning counterclockwise to the plane of the ground. In which direction does the angular velocity vector point?

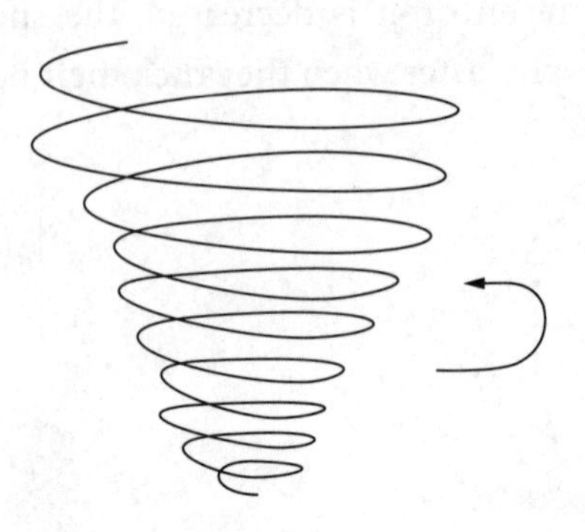

3. A car with a mass of 800 kg moving at 12 m/s turns around a circle with a radius of 30.0 m. What is the magnitude of the acceleration of the car?

(A) 4.0 m/s²
(B) 4.8 m/s²
(C) 5.5 m/s²
(D) 9.0 m/s²
(E) 11.3 m/s²

4. A 60-kg student goes on a ride at an amusement park that spins quickly. Then the floor drops out. If the ride has a radius of 3.0 m and makes 10 revolutions in 30 seconds, what is the centripetal force in newtons acting on the student?

(A) 60π
(B) 80π
(C) $40\pi^2$
(D) $60\pi^2$
(E) $80\pi^2$

5. A 2-kg ball is spun horizontally on the end of a string in a circle with a radius of 8.0 m. If the maximum tension on the string is 36 N, what is the maximum speed of the ball?

(A) 2 m/s
(B) 4 m/s
(C) 12 m/s
(D) 16 m/s
(E) 20 m/s

6. A record player turns with constant angular speed. As a ladybug crawls from the rim of the record toward the center, what happens to the magnitude of total centripetal force exerted on the ladybug?

(A) It is always zero.
(B) It increases.
(C) It decreases.
(D) It is not zero but remains the same.
(E) It increases or decreases, depending on the direction of rotation.

7. The diagram below shows two masses balanced on a scale. What is m_2 in terms of m_1 if the bar of the scale is horizontal and massless?

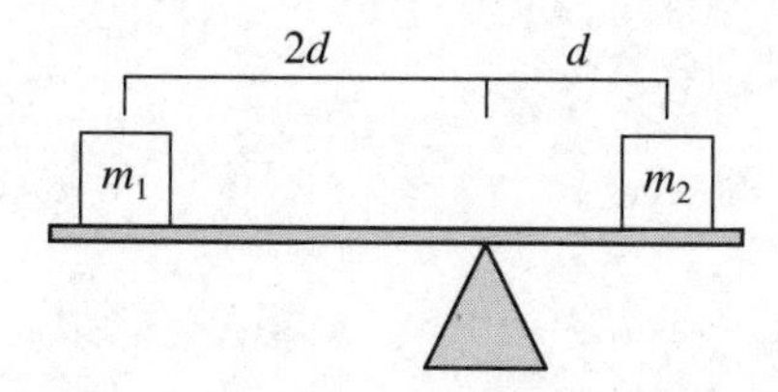

(A) $1/2m_1$
(B) $2m_1$
(C) m_1
(D) m_1^2
(E) $4m_1$

8. The angular velocity of the tires on a car increase from 30 rad/s to 50 rad/s in 2 s. What is the angular acceleration of the tires?

(A) 6 rad/s^2
(B) 10 rad/s^2
(C) 12 rad/s^2
(D) 18 rad/s^2
(E) 20 rad/s^2

Questions 9 and 10 relate to the diagram below, which shows a mass swinging at the end of a string at 2.0 m/s.

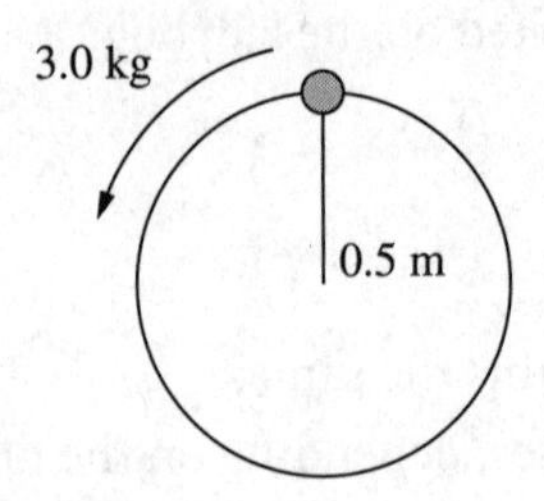

9. What is the centripetal acceleration of the object?

(A) 2.0 m/s²
(B) 2.5 m/s²
(C) 4.0 m/s²
(D) 8.0 m/s²
(E) 10.0 m/s²

10. What is the centripetal force on the object?

(A) 8.0 J
(B) 16 J
(C) 24.0 J
(D) 28 J
(E) 32 J

ANSWERS AND EXPLANATIONS

1. **(C)** The ball is moving to the top of the page when the string breaks. It would continue to move in that direction.

2. **(C)** Use the right-hand rule to bend your fingers in the direction of the tornado's spin. Your thumb, which points upward, shows the direction of the angular velocity vector.

3. **(B)** Use the equation for centripetal acceleration: $\boldsymbol{a}_c = v^2/r = (12 \text{ m/s})^2/30.0 \text{ m} = 4.8 \text{ m/s}^2$.

4. **(E)** Because it takes 30 seconds to complete 10 revolutions, $T = 3.0$ s. Find the angular speed as $\omega = 2\pi/T = 2\pi/3$. Find the acceleration as $a_c = v^2/r = \omega^2 r = (2\pi/3)^2 \times 3 = 4\pi^2/3$. Finally, find the force as $F = ma_c = 60 \text{ kg} \times 4\pi^2/3 = 80\pi^2$.

5. **(C)** Tension $= F_c = mv^2/r$, so $v = \sqrt{\dfrac{rF_c}{m}} = \sqrt{\dfrac{(8)(36)}{2}} = 12$ m/s.

6. **(C)** It decreases. $F_c = mv^2/r = m\omega^2 r$. Since ω is constant and r is decreasing, F_c must also be decreasing.

7. **(B)** The torque on each side of the axis must be equal.

$$\tau_{\text{left}} = \tau_{\text{right}}$$
$$F_1 d = F_2 2d$$
$$(m_1 g)2d = (m_2 g)d$$
$$m_1 2d = m_2 d$$
$$m_2 = 2m_1$$

8. **(B)** Angular acceleration α equals the change in angular velocity divided by time.

$$(50 \text{ rad/s} - 30 \text{ rad/s})/2\text{s} = 10 \text{ rad/s}^2$$

9. **(D)** Use the equation for centripetal acceleration: $\boldsymbol{a}_c = v^2/r = (2.0 \text{ m/s})^2/0.5 \text{ m} = 8.0 \text{ m/s}^2$.

10. **(C)** Use the force equation: $\boldsymbol{F} = \boldsymbol{ma} = mv^2/r = (3.0 \text{ kg})(2.0 \text{ m/s})^2/0.5 \text{ m} = 24.0$ J.

CHAPTER 9

Gravity

The reason you are able to sit in a chair as you read these words is because of gravity. This force not only holds you and other objects on Earth's surface but also holds the solar system together. The SAT Physics exam will present you with several types of questions involving gravity. While some questions may involve direct calculations, the majority will test your understanding of relationships among quantities and the effects of changes on one or more of those quantities.

Newton's Law of Universal Gravitation

According to Newton's Law of Universal Gravitation, a force of attraction exists between every pair of masses in the universe. The magnitude of the force is directly proportional to the product of the masses and inversely proportional to the square of the distance between their centers. This relationship is described by the following equation, in which F_g is the gravitational force, m_1 and m_2 are the masses, and r is the distance between their centers. The constant G, known as the **gravitational constant**, equals 6.67×10^{-11} N · m²/kg².

$$F_g = \frac{Gm_1m_2}{r^2}$$

The SAT Physics test will generally not test your recall of the gravitational constant, but instead ask questions about your understanding about the relationships between the other variables. For example, you should recognize that the gravitational force is a vector quantity. The gravitational force exerted by m_1 pulls m_2 toward its center, whereas the gravitational force exerted by m_2 pulls m_1 toward its center. The forces exerted by the masses are equal in magnitude but opposite in direction.

Example:

Two masses m are separated by a distance r are attracted to each other by a gravitational force F.

(A) If the mass of one of the objects is doubled, what will be the new gravitational force in terms of F?

(B) If the mass of both objects is doubled, what will be the new gravitational force in terms of F?

(C) If the distance between the objects is doubled, what will be the new gravitational force in terms of F?

(A) $F = \frac{Gmm}{r^2}$ and one of the masses becomes $2m$.

$$F_2 = \frac{G(2m)m}{r^2} = \frac{2Gmm}{r^2} \text{ or } 2F.$$

(B) $F = \frac{Gmm}{r^2}$ and both masses become $2m$.

$$F_2 = \frac{G(2m)(2m)}{r^2} = \frac{4Gmm}{r^2} \text{ or } 4F.$$

(C) $F = \frac{Gmm}{r^2}$ and r becomes $2r$.

$$F_2 = \frac{Gmm}{(2r)^2} = \frac{Gmm}{4r^2} \text{ or } 1/4F.$$

Acceleration Due to Gravity

You often use the value of 10 m/s^2 to represent the acceleration due to gravity near Earth's surface. This value can be derived from Newton's Law of Universal Gravitation. Begin with Newton's Second Law, $F = ma$. Then substitute the equation for the gravitational force.

$$F = m_{\text{object}}a = \frac{Gm_{\text{object}}m_{\text{Earth}}}{r^2},$$

which becomes

$$a = \frac{Gm_{\text{Earth}}}{r^2}$$

$$a = \frac{(6.67 \times 10^{-11}\ \text{N} \cdot \text{m}^2/\text{kg}^2)(5.98 \times 10^{24}\ \text{kg})}{(6.37 \times 10^6\ \text{m})^2} = 9.8\ \text{m/s}^2$$

This value is true at or near Earth's surface. However, if the distance from Earth's surface increases, the radius will also change. Like the gravitational force, the acceleration due to gravity is also inversely proportional to the distance between the centers of the masses. A distance two Earth radii from the center of Earth would result in an acceleration that is one-fourth the value at the surface.

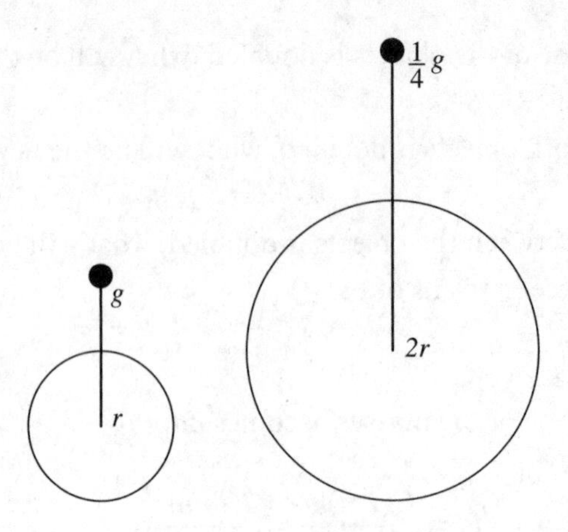

Notice that the mass of the object being pulled by Earth cancels out of the equation. This explains why objects fall to Earth with the same acceleration regardless of mass.

Kepler's Laws

In the early 1600s, Johannes Kepler formulated three laws of planetary motion. Developed before Newton proposed his laws of motion and universal gravitation, Kepler's laws were based on observations made by the astronomer Tycho Brahe.

Kepler's First Law (Law of Ellipses)

The shape of each planet's orbit is an ellipse with the sun at one focus. Although planetary orbits are often approximated as circles, they are actually elliptical in shape.

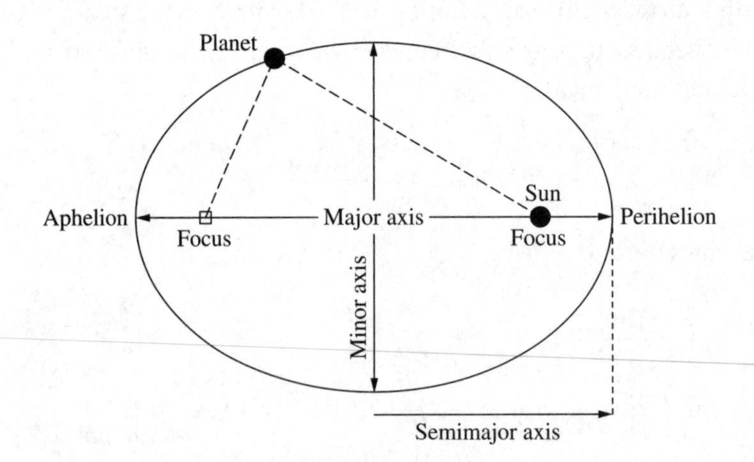

Kepler's Second Law (Law of Equal Areas)

Each planet sweeps out equal areas in equal times. The speed at which a planet moves changes throughout its orbit. The speed increases as the planet moves closer to the sun and decreases as it moves away from the sun. If an imaginary line were drawn between a planet and the sun, the area swept out by the line during any time period would be the same. The following diagram shows areas swept out during equal periods of time.

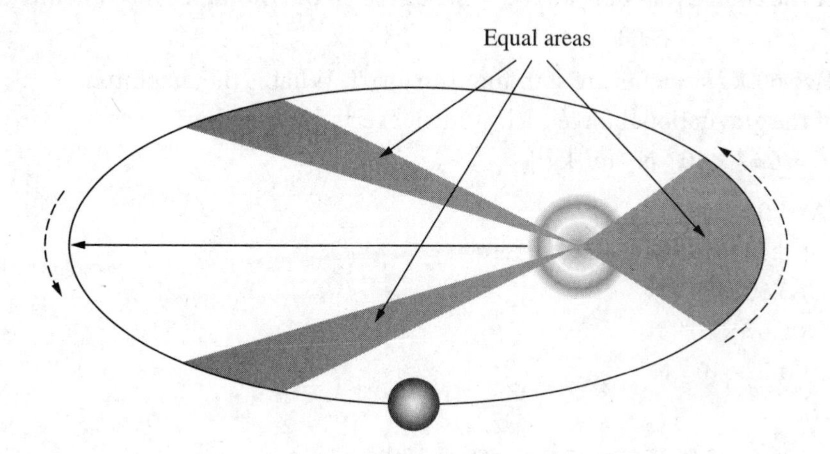

Kepler's Third Law (Law of Harmonies)

The time it takes for a planet to orbit the sun, its orbital period, is related to its distance from the sun. The ratio of the square of the period to the cube of the semimajor axis, T^2/a^3, is the same for all planets.

Orbital Speed

To determine the orbital speed of a planet, you need to approximate its path as a circle. The centripetal force acting on the planet is equal to the force of gravity exerting by the sun. You can therefore set the equation for universal gravitation equal to the equation for centripetal force.

$$\frac{Gm_{\text{sun}}m_{\text{planet}}}{r^2} = \frac{m_{\text{planet}}v^2}{r}$$

$$v = \sqrt{\frac{Gm_{\text{sun}}}{r}}$$

Therefore, the orbital speed of a planet depends on the gravitational constant, the mass of the sun, and the planet's distance to the sun. Now you can see why a planet moves faster when it is closer to the sun (r smaller) and slower when it is farther away (r larger).

TEST-TAKING HINT

Make sure you are comfortable with different types of relationships. For example, quantities that are directly related change together. Both increase or decrease together. Quantities that are indirectly related change in opposite directions. If one increases, the other decreases.

REVIEW QUESTIONS

Select the choice that best answers the question or completes the statement.

1. Two 60-kg students are standing 1 m apart. What is the magnitude of the gravitational force each student exerts on the other? [$G = 6.67 \times 10^{-7}$ N · m²/kg²]

   (A) 4.00×10^{-9} N
   (B) 2.40×10^{-7} N
   (C) 3.60×10^{-7} N
   (D) 6.67×10^{-4} N
   (E) 4.02×10^{-3} N

2. What is the relationship between *G* and *g*?

   (A) *G* describes gravity on Earth, whereas *g* describes gravity in space.
   (B) *G* is a universal constant, whereas *g* is acceleration due to the gravitational force.
   (C) *G* is the gravitational force, whereas *g* is gravitational acceleration.
   (D) *G* is the speed of an object due to gravity, whereas *g* is the acceleration due to gravity.
   (E) *G* describes the gravitational force exerted by an object, whereas *g* describes the gravitational force exerted on an object.

3. The mass of Mars is 6.42×10^{23} kg and its radius is 3.37×10^{6} m. What is the ratio of its gravitational acceleration on the surface of Mars to that on the surface of Earth? [$G = 6.67 \times 10^{-7}$ N · m²/kg²]

   (A) 0.38
   (B) 0.65
   (C) 0.87
   (D) 1.03
   (E) 2.60

4. A satellite is orbiting Earth at an altitude that is equal to the radius of the planet. The gravitational force exerted on the satellite by the planet is *F*. The satellite is then brought down to the Earth's surface. What is the new gravitational force exerted on the satellite?

   (A) $F/4$
   (B) $F/2$
   (C) $F + 4$
   (D) $2F$
   (E) $4F$

5. A satellite orbits the sun as shown in the diagram below. If the satellite travels from A to B during the same period of time as it travels from C to D, which of the following conclusions can be reached?

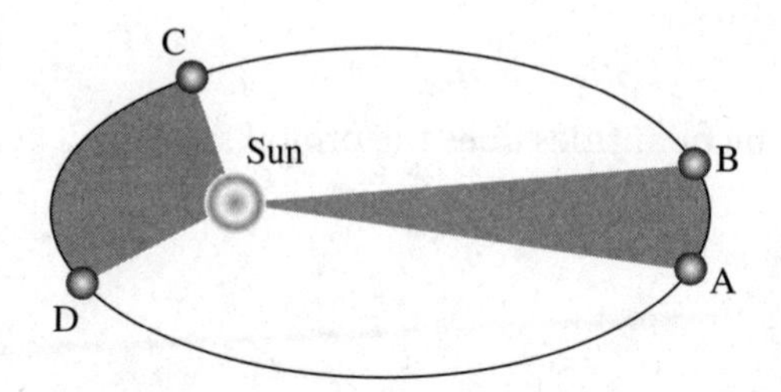

(A) The satellite travels at a greater speed from C to D than from A to B.
(B) The triangle containing A and B covers less area than the triangle containing C and D.
(C) The gravitational force on the satellite is greater between A and B than between C and D.
(D) The mass of the satellite is greater between C and D than between A and B.
(E) The ratio of the period to the average distance to the sun is greater between A and B than between C and D.

6. Earth has a period of 1 year and an average distance to the sun of 1 AU. If the average distance between Saturn and the sun is 9.54 AU, what is Saturn's period of revolution?

(A) 5.4 years
(B) 9.1 years
(C) 19.1 years
(D) 29.5 years
(E) 57.2 years

7. Two planets A and B orbit a star in elliptical paths. The semimajor axis of planet A has a length l and the semimajor axis of planet B has a length $16l$. If planet A orbits with period T, what is the period of planet B's orbit?

(A) $16T$
(B) $24T$
(C) $64T$
(D) $T/4$
(E) $T/64$

8. Each of the following quantities affects Earth's motion through space.

I. mass of Earth
II. mass of sun
III. distance to sun

Upon which of the quantities does the orbital speed v depend?

(A) I only
(B) I and II only
(C) I and III only
(D) II and III only
(E) I, II, and III

Questions 9 and 10 relate to the diagram below, which shows that two identical masses are separated by a distance r.

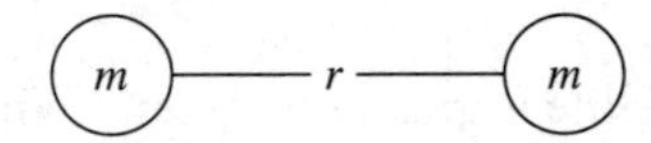

9. How will the gravitational force between the masses be affected if the distance between them is changed to $r/2$?

(A) It decreases by a factor of 4.
(B) It decreases by a factor of 2.
(C) It increases by a factor of 2.
(D) It increases by a factor of 4.
(E) It increases by a factor of 8.

10. Which of the following will have the same effect on gravitational force as doubling the distance between the masses?

(A) halving both masses
(B) halving one mass
(C) doubling one mass and halving the other
(D) doubling both masses
(E) doubling both masses and dividing the distance between them

ANSWERS AND EXPLANATIONS

1. **(B)** $F = \dfrac{Gmm}{r^2}$

$$= \frac{(6.67 \times 10^{-11}\ \text{N} \cdot \text{m}^2/\text{kg}^2)(60\ \text{kg})(60\ \text{kg})}{(1\ \text{m})^2}$$

$$= 2.40 \times 10^{-7}\ \text{N}$$

2. **(B)** The values G and g are often confused. The quantity represented by G is a universal constant found within the equation describing universal gravitation. The quantity represented by g is the acceleration due to gravity and varies with the magnitude of the gravitational force in a particular location.

3. **(A)** Use the acceleration equation to find g:

$$g = \frac{Gm_{\text{Mars}}}{r^2}$$

$$= \frac{(6.67 \times 10^{-11}\ \text{N} \cdot \text{m}^2/\text{kg}^2)(6.42 \times 10^{23}\ \text{kg})}{(3.37 \times 10^6\ \text{m})^2}$$

$$= 3.77\ \text{m/s}^2$$

Then compare with g on Earth, which is 9.81 m/s^2. 3.77/9.81 = 0.38.

4. **(E)** The original force is found as $F = \dfrac{Gm_{\text{Earth}}\, m_{\text{satellite}}}{r^2}$. The radius is then halved.

$$F_2 = \frac{Gm_{\text{Earth}}\, m_{\text{satellite}}}{\left(\frac{r}{2}\right)^2} = \frac{4Gm_{\text{Earth}}\, m_{\text{satellite}}}{r^2} = 4F$$

5. **(A)** Kepler's Second Law states that a line connecting a planet to the sun sweeps out equal areas in equal times. The planet therefore must move faster when it is closer to the sun than when it is farther from the sun.

6. **(D)** The period is T and the average distance is a as described by Kepler's Third Law, $\dfrac{T^2}{a^3}$. This ratio is the same for all planets.

$$\frac{1}{1} = \frac{T^2}{(9.54)^3}$$

$$T^2 = (9.54)^3$$

$$T = 29.5\ \text{years}$$

7. **(C)** According to Kepler's Third Law, the ratio T^2/a^3 is constant. Set the period of planet B equal to xT. Then

$$\frac{T^2}{a^3} = \frac{x^2T^2}{16^3a^3}$$

$$\frac{16^3a^3T^2}{a^3T^2} = x^2$$

$$16^3 = x^2$$

$$(4^2)^3 = x^2$$

$$(4^3)^2 = x^2$$

$$4^3 = x$$

$$64 = x$$

8. **(D)** The orbital speed $v = \sqrt{\frac{Gm_{sun}}{r}}$ depends on the gravitational constant, the mass of the sun, and the planet's distance to the sun. The equation does not include the mass of the planet.

9. **(D)** $F = \frac{Gm_1m_2}{r^2}$, so if r becomes $r/2$, $F = \frac{Gmm}{\left(\frac{r}{2}\right)^2} = \frac{4Gmm}{r^2}$. Therefore, the force is multiplied by 4.

10. **(A)** $F = \frac{Gmm}{r^2}$, so if r is doubled, $F_2 = \frac{Gmm}{(2r)^2} = \frac{Gmm}{4r^2}$, or ¼ F.

    Another way to achieve the same result is to reduce each mass by ½.

    $F = \frac{Gmm}{r^2}$, so if both masses are halved, $F = \frac{G\left(\frac{m}{2}\right)\left(\frac{m}{2}\right)}{r^2} = \frac{G\frac{mm}{4}}{r^2}$, or ¼ F.

CHAPTER 10

Heat and Temperature

Energy can be used to describe motion on macroscopic scales in terms of motion and work. It can also be used on a microscopic level to describe the properties of matter. The SAT Physics test will ask several different kinds of questions involving the motion and energy of particles of matter. In this chapter, you will review some of those topics, including heat, temperature, and phase changes.

Kinetic Theory

Matter is anything that has mass and takes up space. The **kinetic theory** describes the physical properties of matter in terms of the motion of its particles, which can be molecules, atoms, or ions. According to the kinetic theory, all matter is made up of tiny particles that are in constant motion. In a gas, the particles move rapidly in all directions. In a liquid, the particles are held closer together and can only slide past one another. In a solid, the particles are arranged tightly together in a regular pattern and only vibrate in place. The kinetic theory, therefore, can be used to explain the differences among the states of matter. More on this topic will be covered later in this chapter as well as in the next chapter.

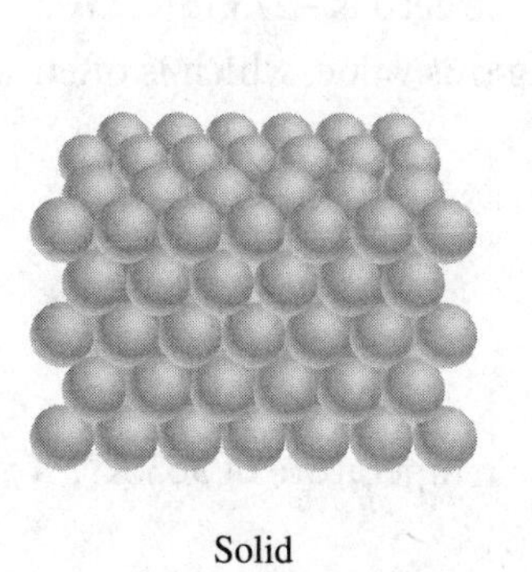
Solid

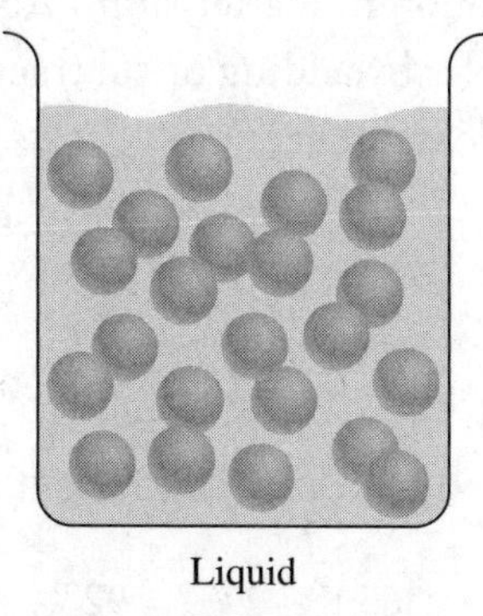
Liquid

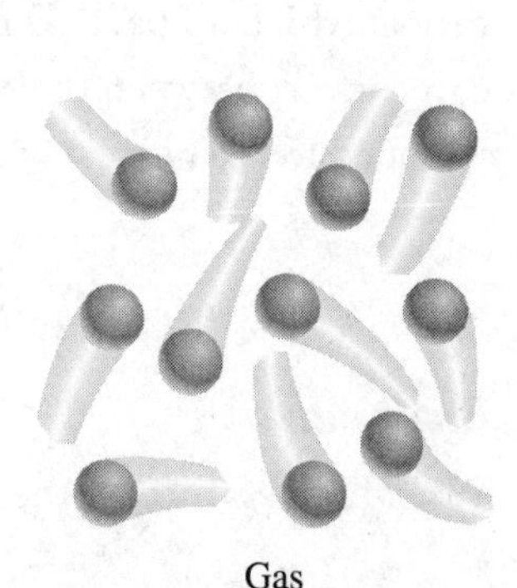
Gas

Thermal Energy

Objects have kinetic energy as a result of their motion. Particles of matter, therefore, have kinetic energy because they are in motion. Like any other object, the kinetic energy of a particle of matter depends on the mass and velocity of the particle.

$$KE = \frac{1}{2} mv^2$$

The total kinetic energy of all the particles in a sample of matter is its **thermal energy**. Like other forms of energy, thermal energy is measured in joules.

Thermal energy flows from warmer substances to cooler substances. The transfer of thermal energy from a warmer substance to a cooler substance is known as **heat**. When a substance absorbs heat, its particles move faster and farther apart. When a substance loses heat, its particles slow down and remain closer together.

Keep in mind that you may come across a related term, **internal energy**. Thermal energy is a portion of internal energy. In a sample of matter, internal energy is the total amount of energy of the particles and includes potential energy in addition to kinetic energy.

Temperature

The average kinetic energy of the molecules in a sample of matter is its **temperature**. When a sample is heated, therefore, its temperature can rise because the motion of its particles increases. When a sample cools, the particles slow down thereby decreasing the average kinetic energy and temperature. Simply put, temperature is a measure of how hot or cold a substance is.

Note that temperature describes an average, whereas thermal energy describes a total. Therefore, two samples, such as a pot of hot tea and a cup of hot tea, can have the same temperature but different amounts of thermal energy.

Temperature is measured with a thermometer. The SI unit of temperature is the kelvin (K), measured on the Kelvin scale. However, temperature is more commonly measured on the Celsius scale (°C). The two scales are closely related. At standard pressure, water freezes at 0°C and boils at 100°C. The Kelvin scale increases by the same increments as the Celsius scale. The Kelvin scale starts at 0 K, which is **absolute zero**. This temperature is defined as the theoretical temperature at which all particle motion in matter stops. Absolute zero is –273.15°C. You can convert between the scales by adding or subtracting this value, which is often rounded to the nearest whole number.

$$K = °C + 273$$

Example:

What temperature on the Celsius scale is equivalent to a temperature of 365 K?

$$365\text{ K} - 273° = 92°\text{C}$$

Specific Heat

The **specific heat** of a substance is the amount of heat required to raise the temperature of one mass unit by one degree Celsius. The following equation shows the relationship among the heat added to a substance Q, the specific heat c, the mass m, and the temperature T.

$$Q = mc\Delta T$$

The equation holds true both for substances that absorb energy from their surroundings and those that transfer energy to their surroundings. When energy is absorbed, Q and ΔT are positive. When energy is transferred out of a substance, Q and ΔT are negative.

Water has one of the highest specific heats of any common substance at 4.186 J/g°C. This means that it takes 4.186 J of energy to raise the temperature of 1 gram of water by 1°C. It is this property that makes water extremely important in temperature regulation. The specific heat of copper is 0.385 J/g°C. A lot less heat is required to raise the temperature of one gram of copper by 1°C than it is to raise the temperature of water by the same amount. The table shows the specific heats of some common substances.

Specific Heat

SUBSTANCE	C (J/g°C)
Air	1.01
Aluminum	0.902
Gold	0.129
Iron	0.450
Mercury	0.140
NaCl	0.864
Ice	2.03
Water	4.18

The term *heat capacity* is often used to describe a sample as a whole rather than as a unit mass. The heat capacity of a sample is the product of the mass of the sample and the specific heat of the material.

Example:

What is the specific heat of copper if 204.75 J of energy raises the temperature of 15 g of copper from 18°C to 53°C?

$$Q = mc\Delta T$$

$$204.75 = 15 \times c \times 35$$

$$c = 0.39 \text{ J/g°C}$$

Phase Changes

It was stated earlier the temperature of a substance *can* rise if it absorbs heat and therefore increases its thermal energy. The reason for choosing the word *can* is that an increase in thermal energy does not always result in a temperature increase. During a phase change, the increased energy goes into overcoming the attractive forces between particles and, therefore, does not change the temperature. The heat involved in phase changes can be represented by the following equation, where L is the heat of transformation.

$$Q = mL$$

The value of L depends on the substance as well as the process. The heating curve for water is useful for reviewing the different processes.

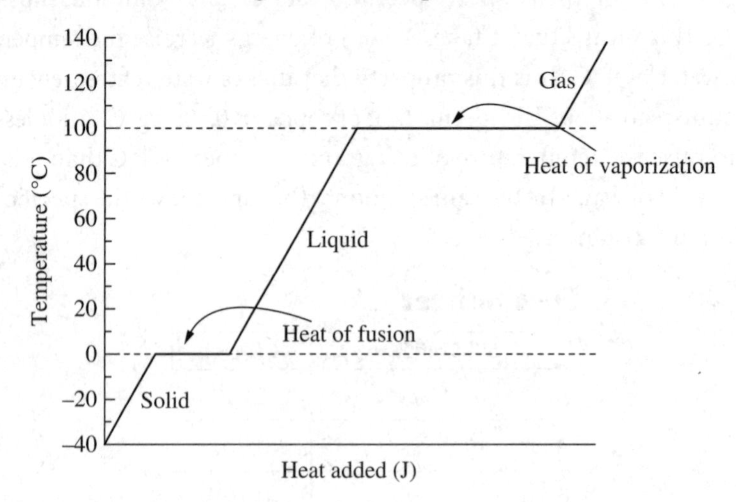

Beginning toward the left of the graph, you find water in the solid state at −40°C. As heat is added, the temperature increases until it reaches 0°C, which is the melting point of water. At this temperature, any added heat causes the solid ice to change into liquid water. The flat portion of the graph represents the phase change. The amount of heat required for this change is known as the **heat of fusion**. For water, the heat of fusion is 334 J/g.

Once all of the water in the sample is liquid, adding heat will again raise the temperature. The temperature continues to rise until it reaches 100°C, which is the boiling point of water. At this temperature, any added heat causes the liquid water to change to water vapor (steam). The flat portion of the graph again represents the change of phase. The amount of heat required for this change is known as the **heat of vaporization**. For water, the heat of vaporization is 2260 J/g.

If you review the graph in reverse, you find the opposite processes. When heat is released at 100°C, the gas condenses into a liquid. When heat is released at 0°C, the liquid freezes into a solid.

Thermal Expansion in Solids

Now that you have reviewed general information about all phases, it would be helpful to consider some specific properties of certain phases of matter. For example, you may know that when a metal bridge heats up, it expands. This is the reason expansion joints are included in bridge construction. Most solids expand when heated and contract when cooled. (Water is an exception to the rule.)

The change in length L, known as linear expansion, is proportional to the original length L_o and the change in temperature T.

$$\Delta L = \alpha L_o \Delta T$$

In this equation, α is the coefficient of linear expansion and depends on the material from which an object is made. When the same lengths of different materials experience the same temperature change, the thermal expansion will be different. For example, the coefficient of linear expansion for brass is $19 \times 10^{-6}/°C$, which means that each centimeter of brass will increase in length by 19×10^{-6} cm for each increase of 1°C. The coefficient of linear expansion for steel is $11 \times 10^{-6}/°C$. For the same temperature increase of 1°C, each centimeter of steel will increase in length by only 11×10^{-6} cm.

As you can see, the changes in length are quite small. The result is more dramatic, however, when two different kinds of metal are fused together. This is the principle behind a bimetallic strip, or compound bar. The strip is made by joining two different metals together, such as brass and steel. When heated, the brass expands more than the steel so the strip curves. The curving can cause the strip to either open or close an electric circuit, such as in a heating system. When cooled, the strip will bend in the opposite direction.

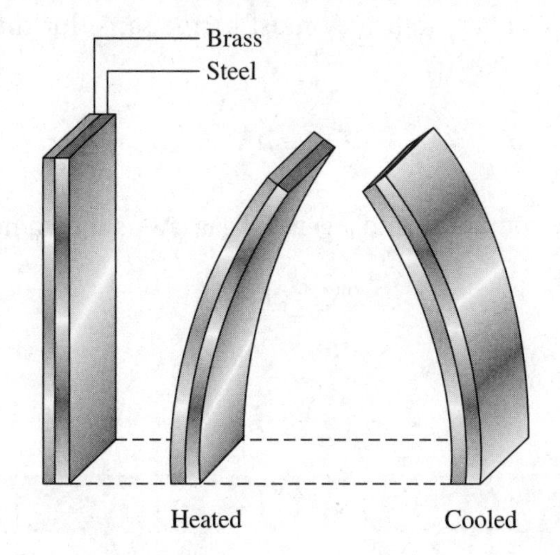

Example:

An engineer is working with a 2-m brass rod and a 1-m aluminum rod. The rods are each fixed at one end and have a gap of 2.0×10^{-3} m between them when the temperature is 25°C. What must the temperature be for the two rods to touch?

α_b coefficient of linear expansion (brass) $= 19 \times 10^{-6}/°C$

α_a coefficient of linear expansion (aluminum) $= 23 \times 10^{-6}/°C$

The rods will touch when they expand a length equal to the gap.

$$\alpha_b L_b \Delta T + \alpha_a L_a \Delta T = 2.0 \times 10^{-3}$$

$$\Delta T = 2.0 \times 10^{-3} / \alpha L_b + \alpha L_a$$

$$= 2.0 \times 10^{-3} / (19 \times 10^{-6})(2.0) + (23 \times 10^{-6})(1.0)$$

$$= 32.8°\text{C}$$

The original temperature was 25°C, so the rods will touch when the temperature rises to 57.8°C.

The Gas Laws

Several laws describe the relationships among the pressure, volume, and temperature of gases. According to **Charles' Law,** volume is directly proportional to temperature if pressure is held constant. The relationship is summarized in the following equation. Note that the volume can be expressed in any units as long as the same units are used for V_1 and V_2.

$$\frac{V_1}{V_2} = \frac{T_1}{T_2}$$

TEST-TAKING HINT

You will rarely be asked to memorize specific values for SAT Physics. However, some values come up often and may be helpful to know, such as absolute zero and the melting and boiling points of water.

According to **Boyles' Law,** volume is inversely related to pressure if temperature is held constant. The relationship is summarized in the following equation. Again, note that the units can vary, but they must be the same for the initial and final values.

$$P_1 V_1 = P_2 V_2$$

The two laws can be combined into a general gas law as shown here.

$$\frac{P_1 V_1}{T_1} = \frac{P_2 V_2}{T_2}$$

REVIEW QUESTIONS

Questions 1–3 refer to the following amounts of energy.

(A) 3.3 J
(B) 36.9 J
(C) 2.7×10^3 J
(D) 3.9×10^3 J
(E) 1.0×10^4 J

1. A scientist adds 36 J of heat to 12 g of aluminum. By how much will the temperature increase? The specific heat of aluminum is 0.902 J/g°C.

2. How much heat is lost when a sample of aluminum with a mass of 16.231 g cools from 320.0°C to 52.0°C? The specific heat of aluminum is 0.902 J/g°C.

3. How much energy is needed to raise the temperature of 71.0 g of water from 24.0 to 59.0°C? The specific heat of water = 4.184 J/g°C.

4. A temperature of 90°C is equivalent to which of the following Kelvin temperatures?

   (A) −183 K
   (B) 17 K
   (C) 190 K
   (D) 363 K
   (E) 413 K

5. What is absolute zero?

   (A) the temperature on the Celsius scale at which water freezes
   (B) the temperature at which the Celsius and Kelvin scales intersect
   (C) the temperature at which heat energy is transferred from a substance
   (D) the temperature at which molecular motion ceases
   (E) the temperature on the Kelvin scale at which water boils

6. The temperature of a sample of matter is most closely associated with the

   (A) average kinetic energy of the molecules
   (B) internal energy of the sample
   (C) total kinetic energy of the molecules
   (D) average potential energy of the sample
   (E) specific heat of the molecules

7. Engineers constructed a steel bridge with a length of 2000 m. If the coefficient of linear expansion for steel is 1.0×10^{-5}/°C, how much expansion should the engineers expect if the temperature rises from 10°C to 30°C?

   (A) 0.4 m
   (B) 0.6 m
   (C) 1.0 m
   (D) 1.2 m
   (E) 1.4 m

8. The volume of an ideal gas is doubled without changing the temperature. What will happen to the pressure of the gas?

(A) It will be reduced by ¼.
(B) It will be halved.
(C) It will stay the same.
(D) It will be doubled.
(E) It will be quadrupled.

Questions 9 and 10 relate to the diagram below, which shows the heating curve for water.

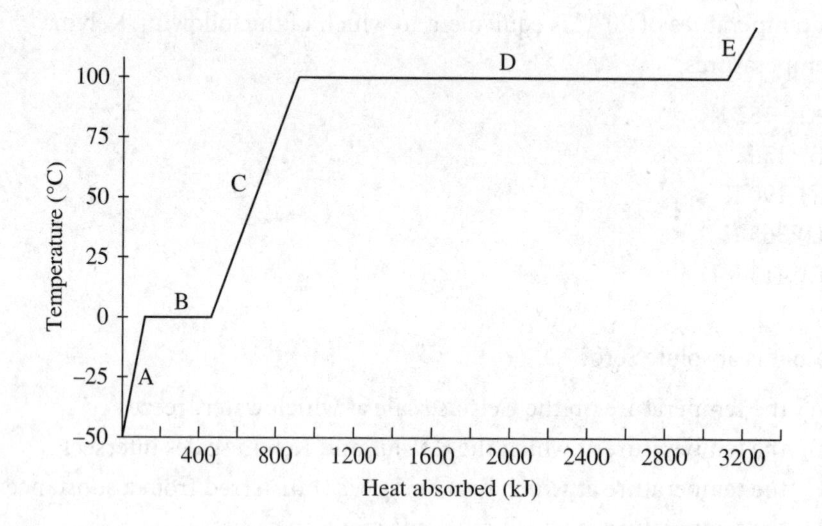

9. Which part of the graph represents the heat of vaporization?

(A) A
(B) B
(C) C
(D) D
(E) E

10. Why is section B of the graph shorter than section D?

(A) The volume of the solid is less than the volume of the liquid.
(B) The pressure at which water melts is less than the pressure at which it boils.
(C) Less water is destroyed during melting than during boiling.
(D) Melting occurs more quickly than boiling.
(E) Less heat is required to melt the solid than to boil the liquid.

ANSWERS AND EXPLANATIONS

1. **(A)** Estimate 1.0 J/g°C as the specific heat of aluminum. $Q = mc\Delta T$, so $\Delta T = Q/mc$, so $\Delta T = 36$ J/(12 g)(1 J/g°C) = 3 J. The closest answer is A.

2. **(D)** Estimate the mass as 16 g and the specific heat of aluminum as 1.0 J/g°C. $Q = mc\Delta T = (161\ \text{J/g°C})(268°\text{C}) = 4.3 \times 10^3$ J. The closest answer is D.

3. **(E)** Estimate the mass as 70 g and the specific heat of water as 4.0 J/g°C. $Q = mc\Delta T = (70\ \text{g})(4\ \text{J/g°C})(35°\text{C}) = 9.8 \times 10^3$ J. The closest answer is E.

4. **(D)** $T = °\text{C} + 273 = 90 + 273 = 363$ K

5. **(D)** Absolute zero, which is the starting point of the Kelvin scale, is a theoretical temperature at which molecular motion would stop.

6. **(A)** The temperature of a sample is the average kinetic energy of the molecules in the sample.

7. **(A)** $\Delta L = \alpha L_0 \Delta T = (1.0 \times 10^{-5}/°\text{C})(2000\ \text{m})(20°\text{C}) = 0.4$ m

8. **(B)** $P_1V_1 = P_2V_2$ and $V_2 = 2V_1$, so $P_1V_1 = P_2 2V_1$. Therefore, $P_2 = P_1/2$.

9. **(D)** The heat of vaporization is the amount of heat required to change the liquid to the gas. It is represented by the flat portion of the graph at the boiling point of water.

10. **(E)** The lines on the graph relate the amount of heat to the temperature. The flat portions indicate changes of phase, during which temperature remains constant. The length of the flat portions is proportional to the amount of heat required to change the phase of a sample of water.

CHAPTER 11

Thermodynamics

Thermodynamics is the study of how heat, energy, and work affect a system. Unlike the kinetic theory, which you reviewed in the previous chapter, thermodynamics deals with large-scale changes in a system. There are four basic laws of thermodynamics, but SAT Physics will primarily deal with only the first two and their related processes.

First Law of Thermodynamics

The first law of thermodynamics relates the law of conservation of energy to processes involving heat. According to this law, the change in the internal energy of a system ΔU equals the heat added to the system Q minus the work the system does W.

$$\Delta U = Q - W$$

If work is done on a system, the value of W is negative. Therefore, you might see the equation as $\Delta U = Q + W$. If you look at the equation, you can see that it describes that energy cannot be created or destroyed. If the energy of the system changes, the energy of the environment outside the system must also change.

Recall that work is done when a force is exerted to move a body some distance. In a thermodynamic system, work is often done on or by a gas in a cylinder confined by a piston. When the piston is pushed down, it does work on the gas. When the gas is heated, it will expand and push the piston upward, thereby doing work on the piston.

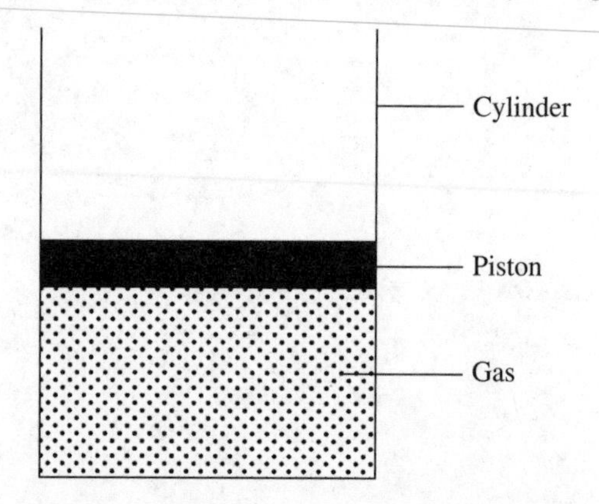

Example:

A system does 30 J of work when 80 J of heat are added to it. What is the change in the internal energy of the system?

$$\Delta U = Q - W$$

$$\Delta U = 80\text{ J} - 30\text{ J} = 50\text{ J}$$

Special Thermodynamic Processes

The first law of thermodynamics explains that a system's internal energy can be changed by the transfer of energy either as heat, work, or some combination of the two. Several processes are special conditions of the first law.

Adiabatic

A process is said to be *adiabatic* if it is thermally insulated from its environment. In other words, no energy is transferred as heat. If a change is adiabatic, $Q = 0$, so $\Delta U = -W$. Consider a gas in a cylinder confined by a piston. The temperature of the gas will change as the internal energy changes. Therefore, the gas will warm up as the gas is compressed and cool down as the gas expands. Adiabatic processes are considered theoretically only because perfect thermal insulation does not exist. For compression or expansion that occurs rapidly, however, the process approaches adiabatic conditions because heat takes time to flow.

Isothermal

In an *isothermal* process, the internal energy remains constant, and there is no change in temperature. As a result, $\Delta U = 0$, so $Q = W$. The heat flow and the work done exactly balance each other. Any energy added to a system as heat is removed as work done by the system. Any energy added to the system by work done on it is removed as heat.

Isovolumetric

No work is done in an *isovolumetric* process. As a result, $W = 0$, so $\Delta U = Q$. Any energy added to the system as heat increases the internal energy. Any energy removed from the system as heat decreases the internal energy.

Isolated System

A system is said to be *isolated* if it experiences no heat or work interactions with the surrounding environment. There is no change in the internal energy for a closed system.

Refrigeration and Heat Engines

In a cyclic system, the change in internal energy is zero, and the final and initial values of internal energy are the same.

$$\Delta U_{net} = 0, \text{ so } Q_{net} = W_{net}$$

The process is somewhat similar to an isothermal process. However, the process repeats itself.

A refrigerator takes advantage of a cyclic process. Heat is transferred from the interior of the refrigerator to an evaporating refrigerant (Q_c). In another part of the system, energy is transferred as heat from a hot condensing refrigerant to the air outside the refrigerator (Q_h).

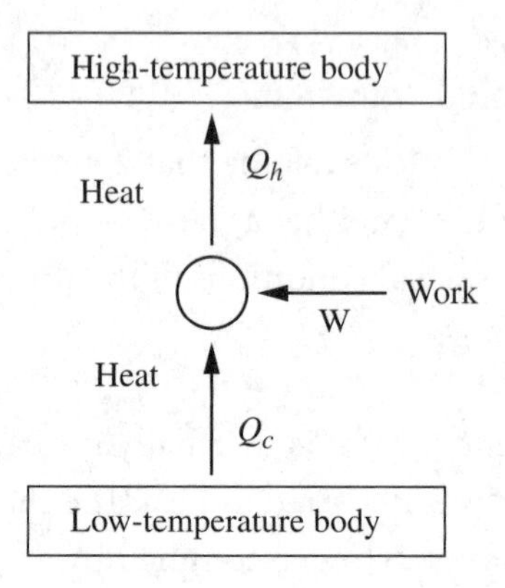

The difference between the two values is the net heat, and, therefore, the net work done during each cycle of the process.

$$W_{net} = Q_h - Q_c$$

A **heat engine** is a device that converts heat to mechanical energy by doing work. It uses a process that is opposite that of a refrigerator. The basic concept behind a heat engine is that mechanical energy can be obtained from thermal energy when heat is allowed to flow from a high temperature to a low temperature. During this process, some energy is transformed to mechanical work as summarized in the following diagram.

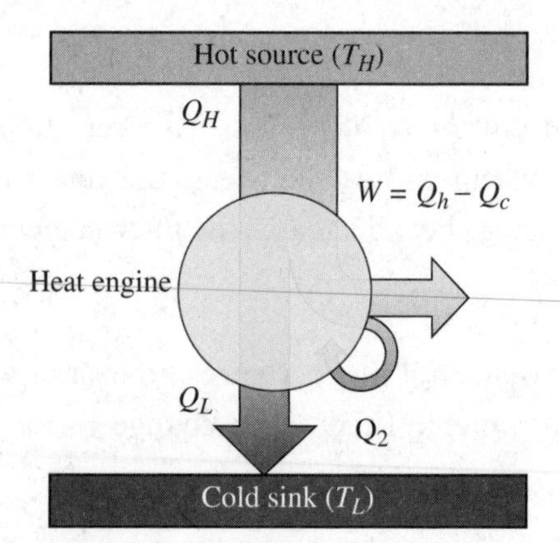

The amount of work the heat engine can do depends on the difference between the amount of energy transferred as heat into the engine and out of the engine.

No cyclic process can convert all of the energy transferred as heat into work. In addition, it cannot transfer energy as heat from a cooler body to a warmer one without work being done. The efficiency of an engine (*eff*) is a measure of the useful energy taken out of a process compared with the total energy put into the process

$$eff = \frac{W_{net}}{Q_h}$$

which can be rewritten as

$$eff = \frac{Q_h - Q_c}{Q_h} = 1 - \frac{Q_c}{Q_h}$$

Another way to look at this equation is as follows:

eff = 1 – energy removed as heat/energy added as heat

Multiply the resulting decimal from each equation by 100% to find the efficiency as a percentage. Now you can see that a heat engine would only be 100% efficient if no heat were removed. Because this is not possible, all heat engines have an efficiency that is less than 100%.

Example:

A heat engine receives 160 J of energy from a hot reservoir, does work, and exhausts 50 J of energy into a cold reservoir. How much work is done by the engine? What is its efficiency?

$$W_{net} = Q_h - Q_c = 160\text{ J} - 50\text{ J} = 110\text{ J}$$

The percent efficiency is $W_{net}/Q_h \times 100\%$.

$$W_{net}/Q_h \times 100\% = 110\text{ J}/160\text{ J} = 69\%$$

Second Law of Thermodynamics

The second law of thermodynamics deals with **entropy**, which is a measure of the disorder of a system. According to the second law, spontaneous processes that proceed in an isolated system lead to an increase in entropy. What that means is that an isolated system will naturally change toward a state of greater disorder. Unlike other quantities in physics, entropy is not conserved in natural processes. Entropy in the universe is increasing.

When dealing with entropy S, it is the change rather than the absolute amount that is important. The change in entropy ΔS can be described by the following equation, in which Q is the amount of heat added to the system and T is the kelvin temperature.

$$\Delta S = Q/T$$

Heat Transfer

Throughout this chapter, you have been considering heat transfer into and out of systems. Heat is transferred in three ways.

Conduction

If you put a metal spoon in a bowl of hot soup, the spoon will eventually become hot. The method through which heat is transferred to the spoon is by **conduction**, which is the transfer of heat between materials that are in contact with one another. Materials through which heat passes easily, such as metals, are known as conductors. Materials that resist the transfer of heat are known as insulators. Wood, rubber, and air are good insulators.

Convection

When a fluid is heated, the particles move faster and farther apart. As a result, the density of a fluid decreases. A heated fluid therefore rises and the cooler fluid above it sinks into its place. The cooler fluid then becomes heated and the process repeats to produce a convection current. Eventually, heat is transferred throughout the fluid through this method, which is known as **convection**.

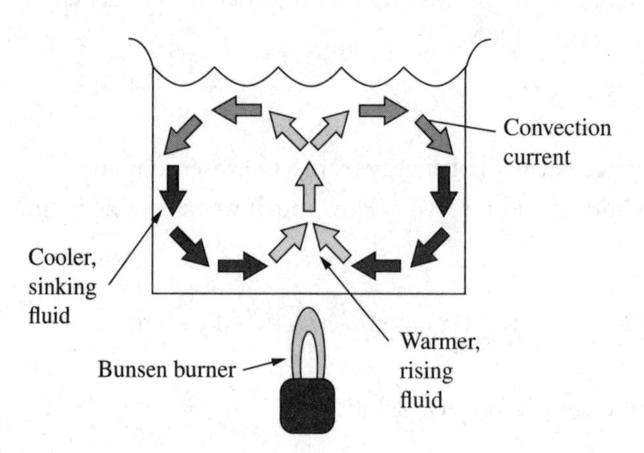

TEST-TAKING HINT

If you are unsure of an answer, try to rule out any answers you know to be incorrect. For example, if you know that internal energy increases or decreases, you can eliminate some answers accordingly. Once your possible choices are reduced, you can try to figure out the correct one.

Radiation

Heat from the sun is transferred to Earth through the vacuum of space. **Radiation** is a method of heat transfer that can occur through a vacuum. This method of heat transfer involves electromagnetic waves, which will be discussed in more detail in Chapter 16. The warmth you feel from a campfire results from radiation.

REVIEW QUESTIONS

Questions 1–3 refer to the following types of heat transfer.

(A) Conduction
(B) Friction
(C) Convection
(D) Radiation
(E) Entropy

1. A person holds a pot of water over a campfire. Through which process is heat transferred from the fire to the person's hand?

2. When a person enters a shower stall, closes the curtain, and turns on the water, the shower curtain blows against her legs. Which process is at work?

3. When a person sips coffee that is too hot, it burns the roof of his mouth. Through which process is heat transferred from the coffee to the mouth?

4. A total of 110 J of work is done on a gaseous refrigerant as it is compressed. If the internal energy increases by 85 J during the process, how much energy is transferred as heat from the refrigerant?

(A) −25 J
(B) −10 J
(C) 0 J
(D) −85 J
(E) −195 J

5. What is the efficiency of a heat engine that receives 200 J of energy from combustion and loses 130 J as heat to exhaust?

(A) 0.159
(B) 0.350
(C) 0.550
(D) 0.650
(E) 0.800

6. What happens to a gas that is compressed rapidly by a piston? [Assume friction is negligible.]

(A) Its temperature decreases.
(B) Its temperature increases.
(C) Its temperature remains the same.
(D) It does positive work on the piston.
(E) It becomes a solid.

7. The first law of thermodynamics is an application of which other law?

(A) the law of universal gravitation
(B) the law of harmonies
(C) the law of entropy
(D) the ideal gas law
(E) the law of conservation of energy

8. A scientist finds that $\Delta U = -W$. Which type of process is the scientist observing?

(A) isothermal
(B) isolated
(C) adiabatic
(D) isovolumetric
(E) reverse

9. Fifty joules of heat flow into a system. The system does 70 joules of work. What happens to the internal energy of the system?

(A) It remains constant.
(B) It increases by 120 J.
(C) It decreases by 120 J.
(D) It increases by 20 J
(E) It decreases by 20 J.

10. Convection currents can form in liquid water, but not solid metal, because

(A) fluids expand when heated
(B) electromagnetic waves carry energy
(C) work is done when heat is transferred
(D) the specific heat capacity of water is high
(E) the arrangement of water molecules resists the transfer of heat

ANSWERS AND EXPLANATIONS

1. **(D)** The hand is not touching the fire and currents do not carry heat to the hand. The process through which heat is transferred from the fire is radiation.

2. **(C)** The moving water creates convection currents in the enclosed air of the shower, which blows the shower curtain.

3. **(A)** Heat is transferred directly from the hot coffee to the roof of the person's mouth by conduction.

4. **(A)** Work is done on the gas so the value of W is negative. $W = -110$ J. The internal energy increases, so ΔU has a positive value. $\Delta U = 85$ J. Rearrange the first law $\Delta U = Q - W$ to solve for $Q = \Delta U + W = 85 \text{ J} + (-110 \text{ J}) = -25$ J.

5. **(B)** $Q_h = 200$ J and $Q_c = 130$ J

   $eff = 1 - Q_c/Q_h = 1 - 130 \text{ J}/200 \text{ J} = 0.350$

6. **(B)** The piston does work on the gas, causing the gas to become warmer. If the process is rapid, the heat cannot escape the system immediately and the temperature of the gas rises.

7. **(E)** The first law of thermodynamics is a form of the law of conservation of energy. It states that energy cannot be destroyed and that if the internal energy of a system changes, the energy of the surrounding environment must also change.

8. **(C)** If a change is adiabatic, $Q = 0$. Therefore, $\Delta U = Q - W$ becomes $\Delta U = 0 - W$, so $\Delta U = -W$.

9. **(E)** $\Delta U = Q - W = 50 \text{ J} - 70 \text{ J} = -20$ J. The negative sign indicates a decrease.

10. **(A)** The water at the bottom of the pot is heated, causing it to expand. As the heated water expands, its density decreases. The less dense water rises as the denser water above it sinks, forming convection currents.

CHAPTER 12

Static Electricity

A bright flash of lightning and the shock you receive when touching a doorknob have something in common. They both involve electric charges. In this chapter, you will review the buildup of electric charges, or static electricity. In the next chapter, you will review the flow of electric charges as current through circuits.

Electric Charge

Matter is made up of atoms, which in turn are made up of particles. One of those particles, the electron (*e*), carries a negative electric charge. The symbol for charge is *q*, and the SI unit of charge is the Coulomb (C). The charge of an electron is -1.6×10^{-19} C. A proton carries a charge that is equal in magnitude to that of an electron, but has a positive value of 1.6×10^{-19} C. In addition to electrons and protons, neutrons are also found within atoms. Neutrons are electrically neutral.

electron	-1.6×10^{-19} C
proton	1.6×10^{-19} C

Objects become charged when electrons are transferred from one to another. Objects described as negatively charged have a surplus of electrons, whereas objects described as positively charged have a deficiency of electrons. Charges are conserved whenever they are transferred, which means that electric charges are neither created nor destroyed. For example, when a rubber rod is stroked with a piece of wool, electrons are transferred from the wool to the rubber. As a result, the rubber rod becomes negatively charged because it gains electrons and the wool becomes positively charged because it loses electrons.

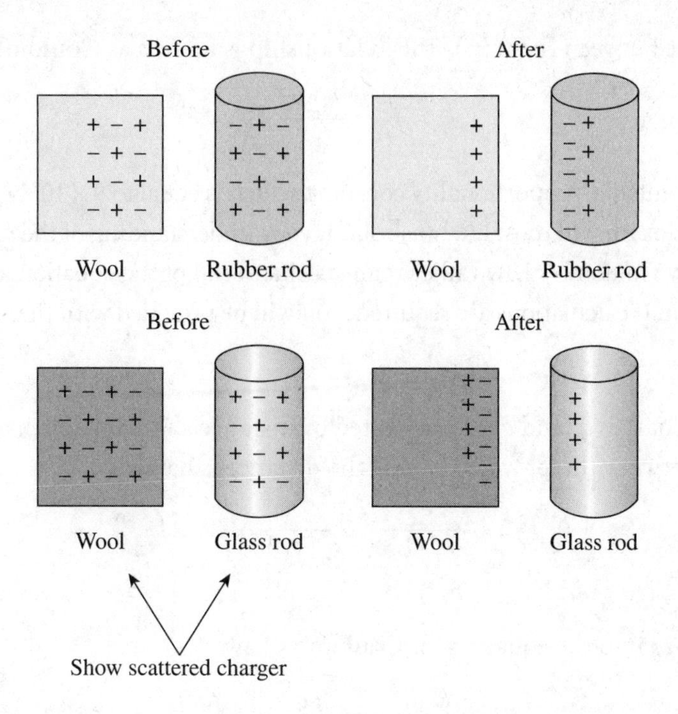

Conversely, if a glass rod is rubbed with a piece of silk, electrons are transferred from the glass to the silk. In this case, the glass rod becomes positively charged because it loses electrons, and the silk becomes negatively charged because it gains electrons. In each case, the net loss of electrons by one object equals the net gain of electrons by the other object. Charge is conserved.

The **law of charges** states that unlike charges attract one another, and like charges repel one another. The same is true for charged objects such as the rods and cloths described earlier. The law of charges is easily demonstrated with the use of pith balls suspended from strings thin enough to be considered massless. Two positively charged pith balls or two negatively charged pith balls will repel one another. A positively charged pith ball and a negatively charged pith ball will attract one another.

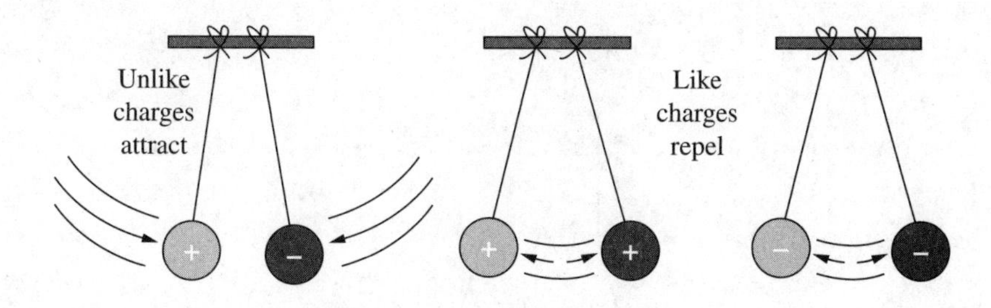

Coulomb's Law

The attraction or repulsion between charged particles constitutes an electric force. The magnitude of the force between charged particles F_E is proportional to the product of the two charges (q_1 and q_2) and varies inversely as the square of the distance between them (r^2). This relationship is known as **Coulomb's Law.**

$$F_E = \frac{kq_1q_2}{r^2}$$

The k represents the proportionality constant, which is equal to $9 \times 10^9\ \text{N}\cdot\text{m}^2/\text{C}^2$. The SAT Physics exam will most likely ask about your understanding of the relationships described by Coulomb's Law rather than specific recall of the equation and calculations using it. If calculations are required, you will be provided with the value of k.

Example:

Two point charges q_1 and q_2 are separated by distance r. What happens to the electric force between these two charges if the distance is halved?

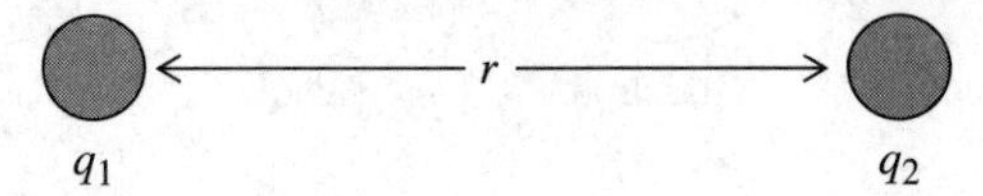

Substitute ½ r into the equation for Coulomb's Law.

$$F_E = \frac{kq_1q_2}{\left(\frac{1}{2}r\right)^2} = \frac{kq_1q_2}{\frac{1}{2}(r)^2} = 4\frac{kq_1q_2}{r^2}$$

The force is quadrupled.

Electroscope

An **electroscope** is an instrument used to determine the presence of small electric charges. It consists of two thin metal leaves suspended from a metal knob. The metal leaves hang down when not charged. When a negatively charged object is brought near the metal knob at the top, the electrons in the knob are repelled into the leaves. The knob becomes positively charged and both leaves become negatively charged. Because the same charge is transferred to each leaf, they repel one another and separate.

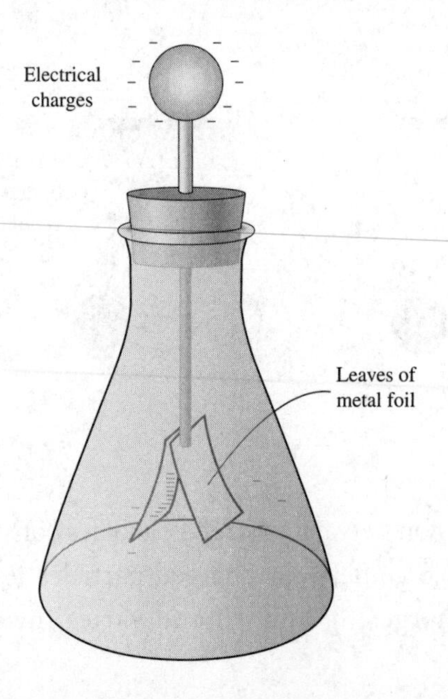

If, instead, a positively charged object is brought near the metal knob at the top, electrons are drawn into knob. The knob becomes negatively charged, and the leaves become positively charged. Again, the leaves have the same charge, so they repel one another and separate.

Electric Field

An **electric field** exists in the space around an electric charge where another charge will experience a force. An electric field can be visualized by drawing arrows known as electric field lines. Each line points in the direction in which a positive test charge would experience a force. A positive test charge would be repelled by a positive charge, so the electric field lines point outward from a positive charge. A positive test charge would be attracted toward a negative charge, so the electric field lines point inward toward a negative charge.

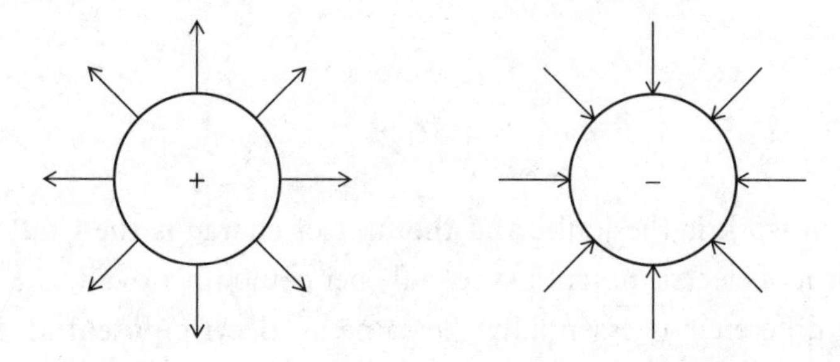

When two charges are brought near one another, the lines become curved as shown in the following diagram. The lines curve away from like charges and toward unlike charges.

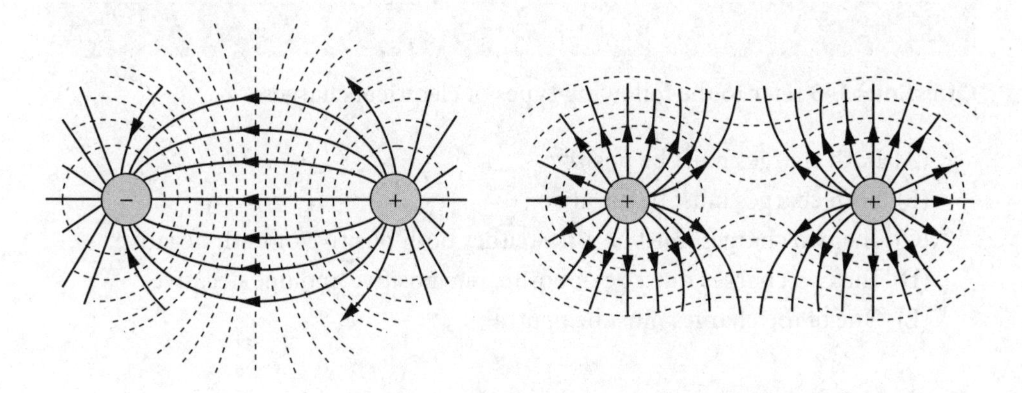

The magnitude of the electric field is known as the **electric field intensity** *E*. Another way to describe this quantity is as the force per unit charge in a region of space. The relationship is described by the following equation.

$$E = \frac{F}{q}$$

The SI units of electric field intensity are newtons per Coulomb (N/C).

According to Coulomb's Law, the electric force between two point charges is given by $F_E = \frac{kq_1q_2}{r^2}$. Therefore, you can rewrite electric field intensity as shown here.

$$E = \frac{F}{q} = \frac{kq_1q_2}{q_1r^2} = k\frac{q_2}{r^2}$$

Viewing the equation in this format shows that the electric field at a given point depends on the charge of the object establishing the field and the distance to a point in space.

Electric Potential

Moving a charge against an electric field requires work. The work required to move charge q between two points in an electric field is known as potential difference V.

$$\Delta V = \frac{\text{work}}{q}$$

The SI unit of work is the joule, and the unit of charge is the Coulomb, which makes the unit of electric potential the joule per coulomb, or volt.

Potential difference is essentially the same as **electric potential**. For electric potential, however, the work is defined as the amount required to bring a positive charge from infinity to some point. You will read more about electric potential in your review of electric circuits.

TEST-TAKING HINT

When answering questions, make sure you read through all the answer choices before deciding. Two answer choices may be very similar, so you could overlook the difference if you rush to choose one.

REVIEW QUESTIONS

Questions 1–3 refer to the following types of electrical charges.

(A) Both charges must be positive.
(B) Both charges must be negative.
(C) The two charges must be alike, either both positive or both negative.
(D) The two charges must be opposite, one positive and one negative.
(E) One of the charges must be neutral.

1. The electric force between two charges is negative. Which statement must be true?

2. The diagram below represents the electric field between two point charges.

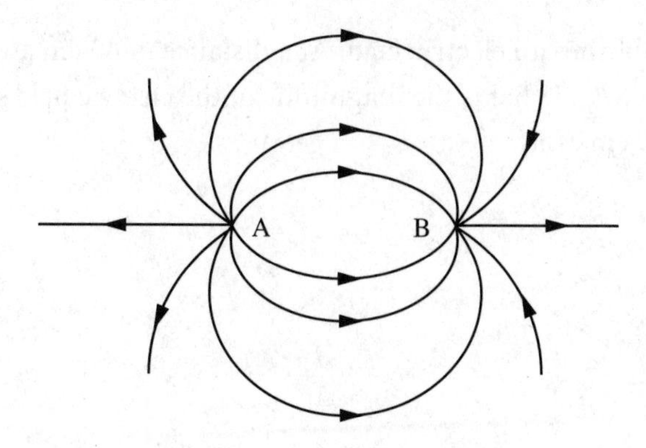

What must be true about the charges?

3. A rod is brought near the knob of an electroscope. The metal leaves of the electroscope separate. What can you conclude about the charges of the objects?

4. How does the charge of an electron compare to the charge of a proton?

(A) The charge of a proton is double the magnitude of the charge of an electron.
(B) The charge of a proton is the same as the charge of an electron.
(C) The charge of a proton is equal and opposite to the charge of an electron.
(D) The charge of a proton is half the magnitude of the charge of an electron.
(E) The charge of a proton is four times the charge of an electron.

5. Two point charges q_1 and q_2 are separated by distance r. What happens to the electric force if the charges are doubled?

(A) It is quartered.
(B) It is halved.
(C) It remains the same.
(D) It is doubled.
(E) It is quadrupled.

6. Particle A with charge q exerts a force F to the right on a particle B with a charge of $-2q$. What is the magnitude and direction of the force exerted by particle B on particle A?

(A) F to the right
(B) F to the left
(C) ½ F to the right
(D) ½ F to the left
(E) $2F$ to the right

7. Charge q establishes an electric field. At a distance of 40 cm away, the strength is 60 N/C. What is the magnitude of the electric field strength at a distance of 80 cm away?

(A) 6 N/C
(B) 15 N/C
(C) 20 N/C
(D) 30 N/C
(E) 60 N/C

8. Which of the following is measured by electrical potential?

(A) force per unit charge
(B) distance between charges
(C) work per unit charge
(D) force between charges
(E) electric proportionality constant

Questions 9 and 10 relate to the diagram below, which shows two charges located along the x-axis.

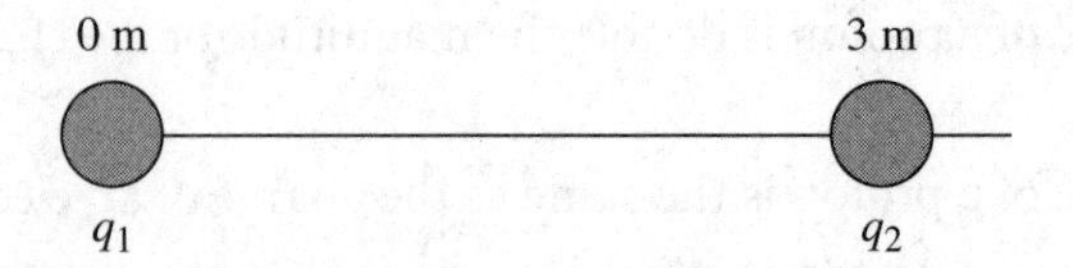

9. What is the magnitude of the force F between the charges in terms of q_1 and q_2?

(A) $\left(\frac{1}{9}\text{m}\right)kq_1q_2$

(B) $\left(\frac{1}{3}\text{m}\right)kq_1q_2$

(C) kq_1q_2

(D) (3 m) kq_1q_2

(E) (9 m) kq_1q_2

10. Suppose $q_1 = 12 \times 10^{-6}$ C and $q_2 = 9.0 \times 10^{-6}$ C. Where along the x-axis must a negative charge q_3 be placed such that the resulting force on it is zero?

(A) 1.00 m
(B) 1.15 m
(C) 1.60 m
(D) 2.15 m
(E) 3.45 m

ANSWERS AND EXPLANATIONS

1. **(D)** According to Coulomb's Law, like charges will result in a positive force so the charges must be unlike for the force to be negative.

2. **(D)** Electric field lines represent the direction in which a positive test charge will experience a force. Because like charges repel one another, the test charge would be forced away from a positive charge and toward a negative charge. Thus, A must be positive and B must be negative.

3. **(C)** Like charges repel one another. For the leaves to separate, they had to be charged with like charges. The charges could be either positive or negative. The nature of each charge cannot be identified from the fact that the leaves of the electroscope separate.

4. **(C)** Electrons and protons are charged particles within atoms. Despite the difference in mass, the magnitude of the charge on the particles is the same. However, the charge of an electron is considered negative, whereas the charge on a proton is positive.

5. **(E)** Substitute $2q_1$ and $2q_2$ into the equation for Coulomb's Law.

$$F_E = \frac{kq_1q_2}{r^2} = \frac{k2q_1 2q_2}{r^2} = 4\frac{kq_1q_2}{r^2}$$

6. **(B)** According to Coulomb's Law, the forces exerted by each particle on the other are the same. Because the particles have opposite charges, an attractive force is established between them. Therefore, the magnitudes of the forces are the same, but the direction is opposite.

7. **(B)** The electric field strength is inversely related to the square of the distance. Therefore, if the distance is doubled, the original value E is divided by 4.

8. **(C)** Electric potential is defined as the amount of work required to bring a positive charge from infinity to some point.

9. **(A)** $F = \frac{kq_1q_2}{r^2} = \frac{kq_1q_2}{3^2} = \left(\frac{1}{9}\text{m}\right)kq_1q_2$

10. **(C)** The force exerted on q_3 by q_1 must equal the force exerted on q_3 by q_2 for the forces to cancel.

$$\frac{kq_1q_3}{r^2} = \frac{kq_2q_3}{r^2}$$

The constant k and q_3 drop out of the equation, making it $\frac{q_1}{r^2} = \frac{q_2}{r^2}$. Substitute in the given values to get

$$\frac{12 \times 10^{-6}\ \text{C}}{x^2} = \frac{9.0 \times 10^{-6}\ \text{C}}{(3.0\ \text{m} - x)^2}$$

which becomes

$$F = \frac{12}{x^2} = \frac{9.0}{(3.0\text{ m} - x)^2}$$

Rearrange to solve for x: $9.0\,x^2 = 12\,(3.0\text{ m} - x)^2$, so $x\sqrt{9} = (3.0\text{ m} - x)\sqrt{12}$,

$x = (3.0\text{ m} - x)\sqrt{\frac{12}{9.0}}$. Then $x = 1.15(3.0\text{ m} - x)$, so $2.15x = 3.45$ and $x = 1.60$ m.

CHAPTER 13

Current and Circuits

In the previous chapter, you reviewed static electricity, or the buildup of charge. Under the right conditions, charges can be made to flow as current electricity. In this chapter, you will review current electricity and types of circuits.

Voltage

In the previous chapter, you briefly reviewed potential difference, or **voltage**. A battery is a device that has a voltage across its terminals. When a battery is connected to a bulb by wires, a charge begins to flow from one terminal through the wires and bulb and back to the other terminal of the battery. As you learned, the charges that "move" are actually electrons.

One terminal, the positive terminal, has a deficiency of electrons. The other terminal, the negative terminal, has an excess of electrons. This difference was set up by work done against the force of repulsion. Recall that work per unit charge is potential difference (V).

Current

The amount of charge moving through a conductor per second is known as **current**. The unit of current is the coulomb per second, or ampere. The symbol I is used to represent current. Even though it is electrons that are considered to be the carriers of charge, the convention for a long time has been to take the direction of the current as if it were positive charges that flow. This is, therefore, known as conventional current and is generally used to describe current flow on SAT Physics.

Current does not flow unimpeded by conductors. The opposition to the flow of charges offered by a material is known as its **resistance**. The unit of resistance is the ohm (Ω). It may help to think of resistance as friction through conductors.

Resistance depends on several factors, including the type of material, length, temperature, and cross-sectional area. Resistance is lower in conductors and higher in insulators. Resistance is directly proportional to length, meaning that resistance increases as length increases. Generally, resistance increases with temperature. It is inversely proportional to the cross-sectional area, which means that resistance increases as a wire becomes thinner.

The relationship among current, resistance, and voltage is described by **Ohm's Law**, which states that resistance is directly proportional to voltage and indirectly proportional to current.

$$R = \frac{V}{I}$$

You may also see or need to use the relationship in different formats.

$$I = \frac{V}{R} \quad \text{and} \quad V = IR$$

Circuits

Current requires a complete path known as a **circuit** through which to travel. The resistors in a circuit can be arranged in series, in parallel, or some combination of the two.

Series Circuit

A **series circuit** is one in which two or more resistors are arranged in such a way that the same current passes through each of them.

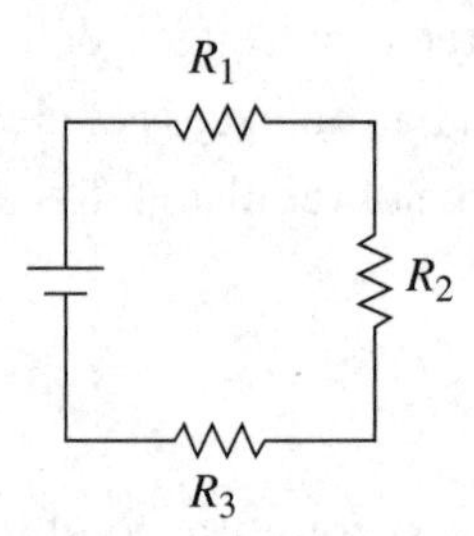

- The total resistance is the sum of the individual resistances.

$$R_T = R_1 + R_2 + R_3$$

- The same current passes through all the resistors.

$$I_T = I_1 = I_2 = I_3$$

The total current is found by dividing the total voltage by the total resistance.

$$I_T = \frac{V_T}{R_T}$$

- The voltage is divided among the resistances.

$$V_1 = I_1R_1 \qquad V_2 = I_2R_2 \qquad V_3 = I_3R_3$$

The total voltage is then the sum of the individual voltages.

$$V_T = V_1 + V_2 + V_3$$

Example:
Three resistors of 10 Ω, 20 Ω, and 30 Ω are connected in series with a 12-V battery.
(A) What is the total resistance?
(B) What current flows through the circuit?

(A) $R_T = R_1 + R_2 + R_3 = 10\ \Omega + 20\ \Omega + 30\ \Omega = 60\ \Omega$

(B) $I_T = \dfrac{V_T}{R_T} = \dfrac{9\ \text{V}}{60\ \Omega} = 0.15\ \text{amp}$

Parallel Circuit
A **parallel circuit** is one in which two or more resistors are arranged in such a way that current has more than one path through which to flow and each resistor has the same potential difference across it.

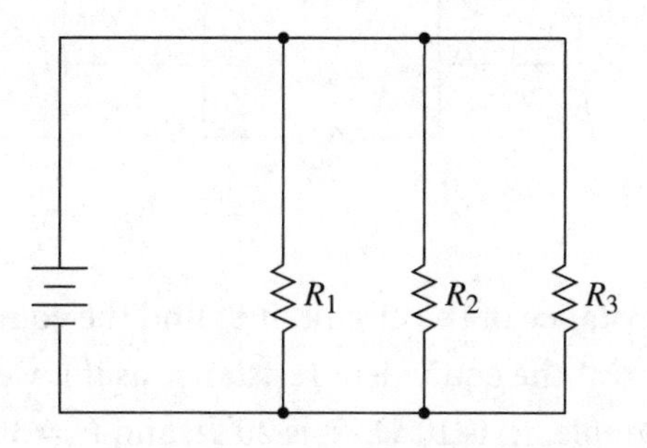

- The total resistance is the inverse sum of the inverse resistances.

$$\frac{1}{R_T} = \frac{1}{R_1} + \frac{1}{R_2} + \frac{1}{R_3}$$

- The total current is the sum of the currents through each resistor.

$$I_T = I_1 + I_2 + I_3$$

- The voltage across each resistance is the same.

$$V_T = V_1 = V_2 = V_3$$

Example:

Three resistors of 3 kΩ, 6 kΩ, and 12 kΩ are connected in parallel with a 12-V battery.

(A) What is the total resistance?

(B) What is the total current?

(A) $\frac{1}{R_T} = \frac{1}{3\text{ k}\Omega} + \frac{1}{6\text{ k}\Omega} + \frac{1}{12\text{ k}\Omega} = \frac{4}{12\text{ k}\Omega} + \frac{2}{12\text{ k}\Omega} + \frac{1}{12\text{ k}\Omega} = \frac{7}{12\text{ k}\Omega}$

$R_T = 12/7\text{ k}\Omega$

(B) $I_1 = 12\text{ V}/12\text{ k}\Omega = 1\text{ mA}$

$I_2 = 12\text{ V}/6\text{ k}\Omega = 2\text{ mA}$

$I_3 = 12\text{ V}/3\text{ k}\Omega = 4\text{ mA}$

$I_T = 1\text{ mA} + 2\text{ mA} + 4\text{ mA} = 7\text{ mA}$

Combination Circuits

A combination circuit is one that includes elements of both series circuits and parallel circuits.

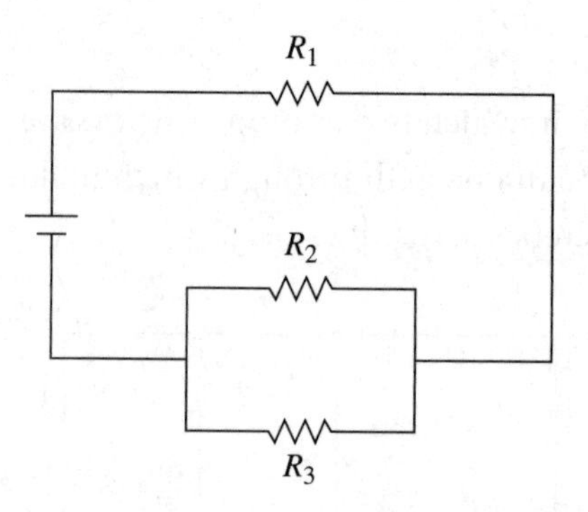

- To find the total resistance of the circuit, first find the equivalent resistance of R_2 and R_3. Then treat the equivalent resistance as if it were in series with R_1. Suppose, for example, R_1 is 10 Ω, R_2 is 20 Ω, and R_3 = 30 Ω. The equivalent resistance of R_2 and R_3 is as follows

$$\frac{1}{R_{23}} = \frac{1}{20\ \Omega} + \frac{1}{30\ \Omega} = \frac{3}{60\ \Omega} + \frac{2}{60\ \Omega} = \frac{5}{60\ \Omega}$$

so $R_{23} = 12\ \Omega$.

$$R_T = R_{23} + R_1 = 12\ \Omega + 10\ \Omega = 22\ \Omega$$

- The total current is found by Ohm's Law:

$$I_T = \frac{V_T}{R_T}$$

Suppose the voltage in this circuit is 12 V.

$$I_T = \frac{12\ V}{22\ \Omega} = 0.55\ A$$

- The voltage across each resistor is found separately because the voltage supplied by the battery is divided proportionally between the parallel combination and R_1.

$$V_1 = I_1R_1 = (0.55\ \text{A})(10\ \Omega) = 5.5\ \text{V}$$

The voltage across the parallel portion of the circuit must therefore be the difference between this voltage and the voltage supplied by the battery, or 6.5 V.

Circuit Measurements

You have been calculating measurements of current through a resistor and voltage across a resistor. To actually measure the values, you can use devices placed within the circuit. An **ammeter** is a device used to measure current. It must be placed in a circuit in series with a resistor so that the same current passing through the resistor passes through the ammeter. An ammeter has a low resistance so that it does not significantly change the total resistance of the circuit.

A **voltmeter** is a device used to measure voltage. It is placed in parallel with a resistor so that the voltage across the resistor is the same as the voltage across the voltmeter. A voltmeter has a high resistance so that current will not bypass the resistor by traveling through it.

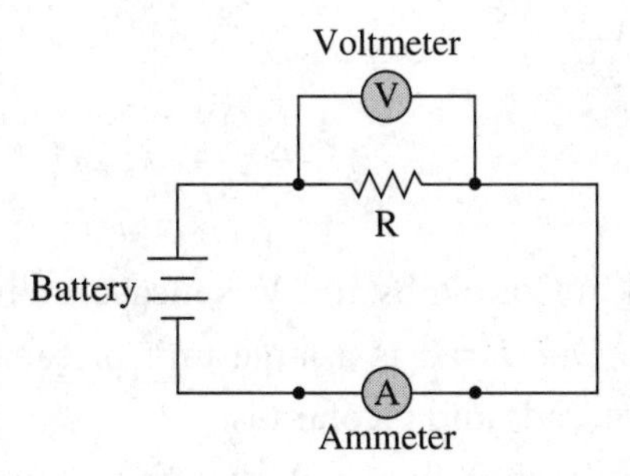

Heating and Power

Whenever energy flows through a resistor, some heat is produced. The amount of heat depends on the current, the resistance, and the time it takes for current to flow through the resistor. These quantities are related in the following equation, in which the heat produced H in joules is directly proportional to the square of the current I in amperes, the resistance R in ohms, and the time in seconds. This relationship is known as **Joule's Law of Heating.**

$$H = I^2Rt$$

Recall that power is the rate at which work is done or energy is transferred. Power in an electric circuit can be described by the following equations, in which P is power, I is the current through a resistor, V is the voltage across a resistor, and R is the resistance.

$$P = IV \qquad P = I^2R \qquad P = V^2/R$$

Capacitors

If two conducting plates are connected to a battery, charge flows from the battery to the plates. One plate becomes positively charged and the other becomes negatively charged. The plates form a **capacitor**, which stores charge. A capacitor consists of two conductors separated by an insulator. That insulator could be air, oil, paper, or other materials known as dielectrics.

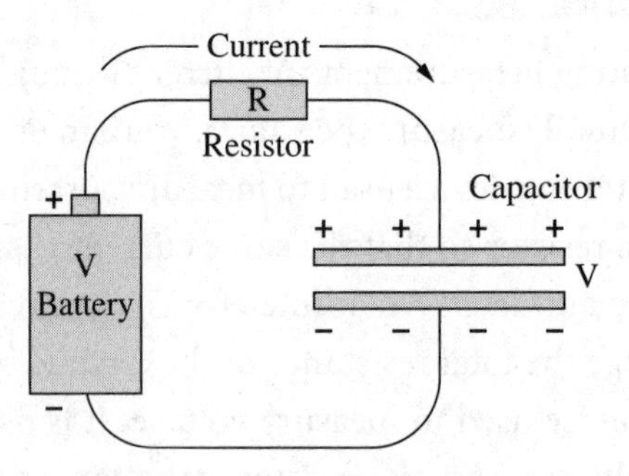

The capacitance C is the charge per unit voltage of the plates and is represented by the following equation, in which q is the charge of one of the plates and V is the voltage across the plates.

$$C = \frac{Q}{V}$$

When Q is measured in Coulombs and V is measured in volts, the unit of capacitance is the farad F. One farad is a large unit of capacitance. Typical units of capacitance are microfarads and picofarads.

The charge on a capacitor does not change instantaneously. The change in charge occurs over time interval Δt.

$$\Delta Q = I\Delta t$$

Charge flowing into a capacitor builds up rather than passing through it. That charge cannot build up indefinitely. Instead, the charge builds up until the voltage balances the external voltage pushing charge on the capacitor.

TEST-TAKING HINT

Practice solving problems involving different types of circuits. Begin with the basic series and parallel circuits, and then work with combination circuits. Once you become familiar with the processes for finding current, voltage, and resistance, you can apply it to any circuit you encounter.

REVIEW QUESTIONS

Questions 1–3 refer to the following resistance values.

(A) 1.3 Ω
(B) 8.0 Ω
(C) 20.0 Ω
(D) 250.0 Ω
(E) 400.0 Ω

1. What is the equivalent resistance of the circuit?

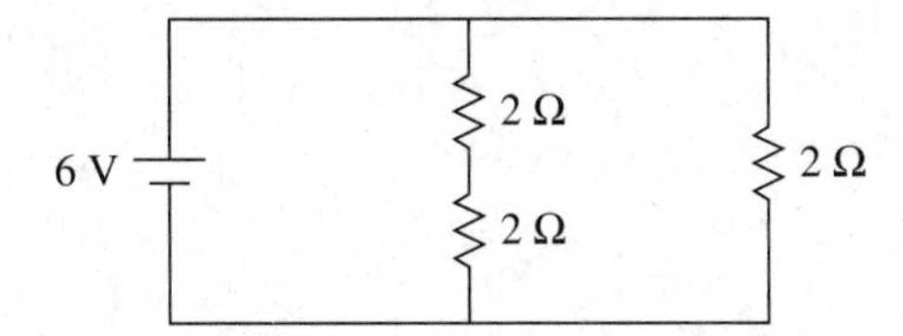

2. What is the value of a resistor that would limit the current in a circuit to 0.02 A when connected to an 8-V power supply?

3. What is the equivalent resistance of the circuit below?

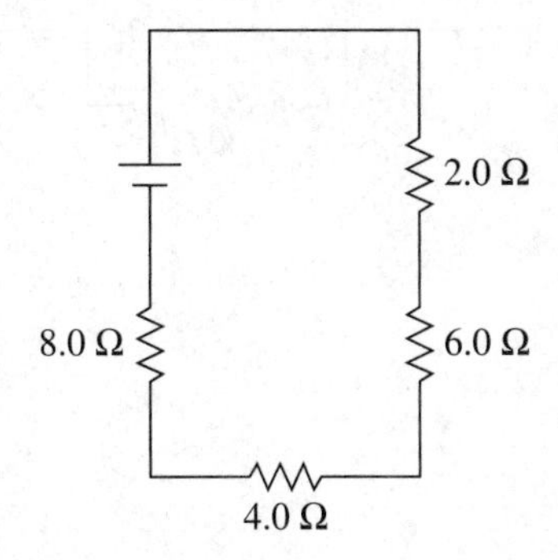

4. Two resistors are connected in series in a circuit. The same two resistors are connected in parallel in another circuit. Which statement about the circuits is true?

(A) The equivalent resistance of the series circuit is greater than in the parallel circuit.
(B) The resistor in series will have the same potential difference across them.
(C) The equivalent resistance in parallel must be greater than the resistance in series.
(D) The total current in parallel will be the same through each resistor.
(E) The equivalent resistance of both circuits will be identical.

5. What is the voltage across the 6.0-Ω resistor?

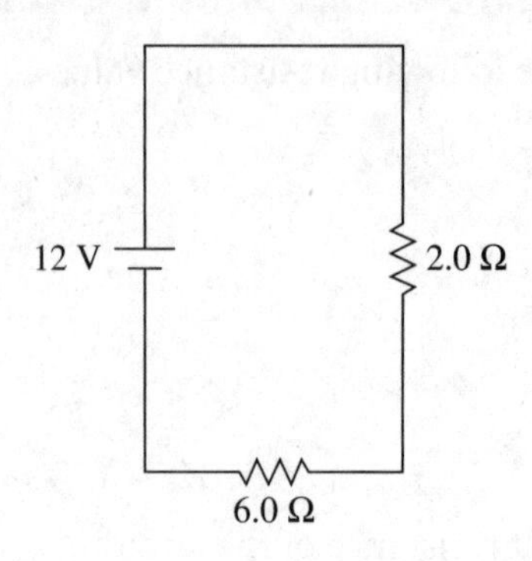

(A) 4.5 V
(B) 8.0 V
(C) 9.0 V
(D) 12.5 V
(E) 16.0 V

6. Which current will flow through the 250-Ω resistor in the circuit below?

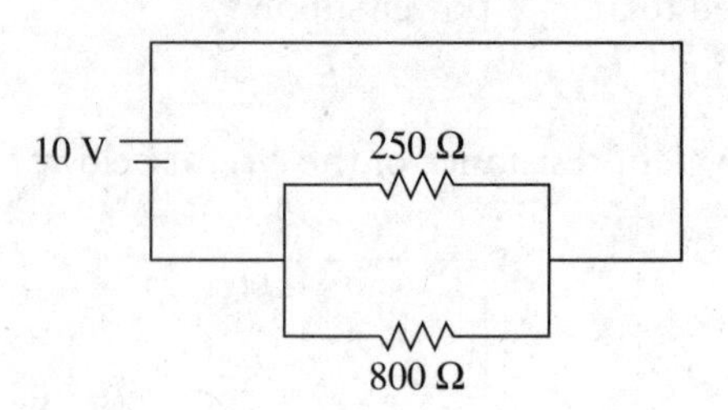

(A) 10 mA
(B) 40 mA
(C) 50 mA
(D) 125 mA
(E) 250 mA

7. What is the power dissipated in the resistor of the circuit shown below?

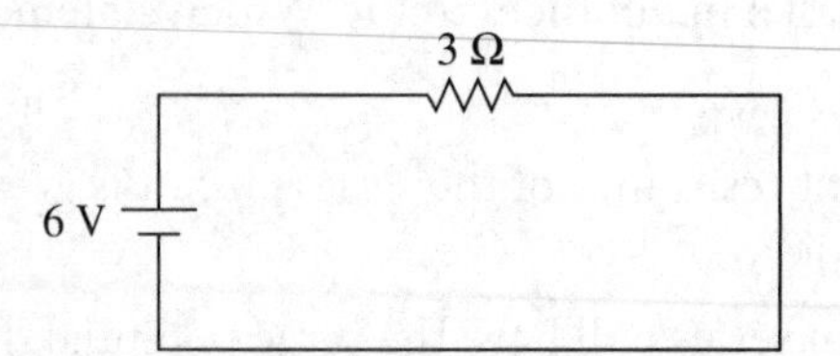

(A) 2 W
(B) 3 W
(C) 6 W
(D) 9 W
(E) 12 W

8. A radio with a resistance of 40 Ω has a current of 0.1 A flowing through it. What is the voltage?

(A) 0.4 V
(B) 2.5 V
(C) 1.0 V
(D) 4.0 V
(E) 4.3 V

Questions 9 and 10 relate to the diagram below, which shows an electric circuit. A current of 6 A passes through R_1.

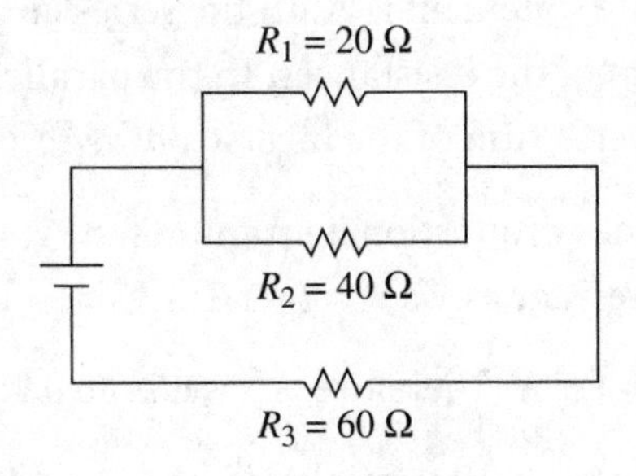

9. What is the potential difference across R_1?

(A) 30 V
(B) 60 V
(C) 80 V
(D) 120 V
(E) 160 V

10. What is the rate at which R_1 uses electrical energy?

(A) 120 W
(B) 144 W
(C) 360 W
(D) 600 W
(E) 720 W

ANSWERS AND EXPLANATIONS

1. **(A)** Find the resistance of the path with two resistors: 2.0 Ω + 2.0 Ω = 4.0 Ω. Then find the total resistance as $\frac{1}{R} = \frac{1}{4\ \Omega} + \frac{1}{2\ \Omega} = \frac{1}{4\ \Omega} + \frac{2}{4\ \Omega} = \frac{3}{4\ \Omega}$, so $R = 1.3\ \Omega$.

2. **(E)** The problem can be solved using Ohm's Law: $R = \frac{V}{I} = \frac{8\ \text{V}}{0.02\ \text{A}} = 400\ \Omega$.

3. **(C)** The total resistance for a series circuit is the sum of the individual resistances. $RT = 2.0\ \Omega + 6.0\ \Omega + 4.0\ \Omega + 8.0\ \Omega = 20.0\ \Omega$.

4. **(A)** The only accurate statement is A. In the series circuit, the equivalent resistance is the sum of the resistances. In the parallel circuit, the equivalent resistance is the inverse sum of the inverse of the resistances.

5. **(C)** The voltage divides proportionately among the resistances. Set the voltage through the 6.0-Ω resistor as V_1. $V_1 = I_1R_1$. The value I_1 is equal to I_T in a series circuit. $I_T = \frac{12\ \text{V}}{8\ \Omega} = 1.5$ A. Therefore, $V_1 = (1.5\text{A})(6.0\ \Omega) = 9.0$ V.

6. **(B)** Despite the setup of the circuit, this is a basic parallel circuit. Therefore, the current through each resistor is the quotient of the voltage and resistance.

$$I = \frac{V}{R} = \frac{10\ \text{V}}{250\ \Omega} = 40\ \text{mA}$$

7. **(E)** First find the current: $I = \frac{V}{R} = \frac{6\ \text{V}}{3\ \Omega} = 2$ A. Then find the power: $P = IV = (2\ \text{A})(6\ \text{V}) = 12$ W.

8. **(D)** The problem can be solved using Ohm's Law: $V = IR = (0.1\ \text{A})(40\ \Omega) =$ 4.0 A.

9. **(D)** $V_1 = I_1R_1 = (6\ \text{A})(20\ \Omega) = 120$ V

10. **(E)** $P_1 = I_1^2R_1 = (6\ \text{A})^2(20\ \Omega) = 720$ W

CHAPTER 14

Magnetism

When you think of magnets, you may consider locker magnets, refrigerator magnets, and children's toys. However, the applications of magnets extend far beyond these simple devices. Magnets are the basic elements of devices from doorbells to power stations. Understanding the principles of magnetism will enable you to answer a variety of questions on the SAT Physics test.

Magnets

The effects of a magnet are strongest in two regions, known as the poles, called the north pole and the south pole. The names arise from the fact that a magnet suspended by its midpoint and allowed to swing freely will become oriented such that one pole faces the north and the other faces the south. Three common types of magnets are shown here.

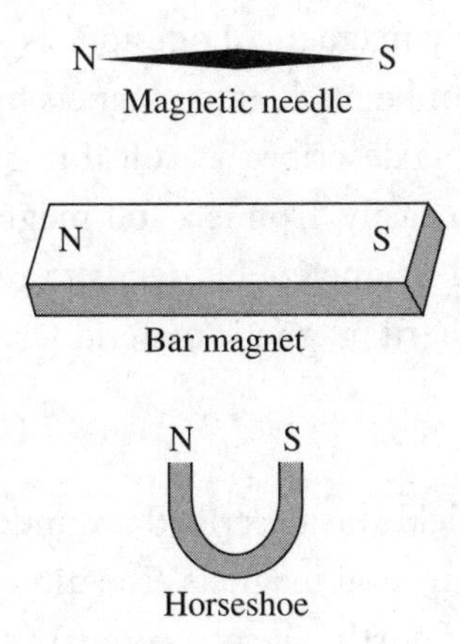

Just as the law of charges described the interaction of like and unlike charges, the law of magnets describes the interaction of magnetic poles. Like magnetic poles repel one another, and unlike magnetic poles attract one another.

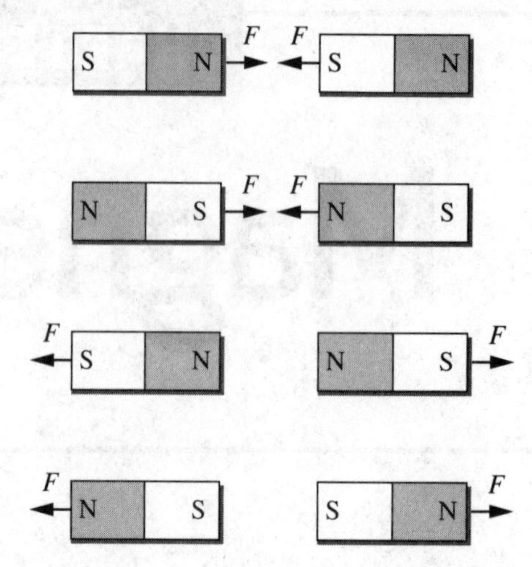

A major difference between magnetic poles and electric charges is that magnetic poles cannot be isolated. Magnetic poles always occur in pairs. If a magnet were cut in half, each portion would have both a north pole and a south pole.

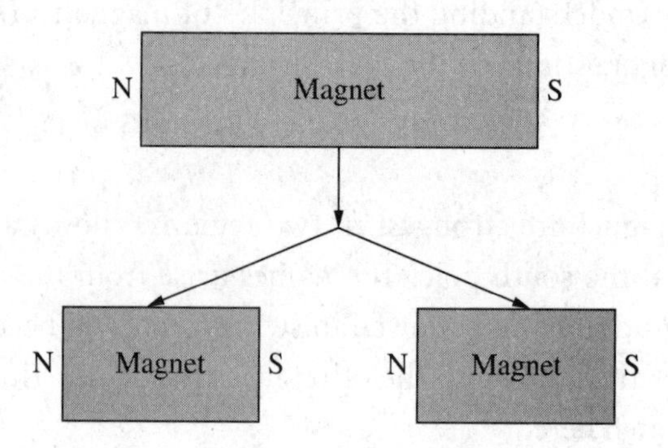

Some materials are naturally magnetic. Lodestone is a rock that acts as a natural magnet. Other materials can be made into magnets by stroking them with a permanent magnet. A magnet is described as soft if it is easily magnetized but also tends to lose its magnetism easily. Iron is a soft magnetic material. A hard magnetic material is difficult to magnetize but tends to retain its magnetism. Cobalt and nickel are examples of hard magnetic materials.

Magnetic Field

The concept of an electric field was described around electric charge. Similarly, a magnetic field is described around magnets. The direction of the magnetic field $\boldsymbol{B}$ is the direction in which the north pole of a compass needle would point if placed in a particular location. The magnitude of the magnetic field is greatest near the poles and decreases as the distance from the poles increases. The magnetic field can be described by magnetic field lines, as shown in the following diagram, as well as iron filings sprinkled around a magnet. As you can see, magnetic field lines do not overlap. Instead, they form closed loops from the north pole to the south pole outside the magnet. Iron filings are convenient for demonstrating the magnetic field around a magnet. The diagram on the left shows how to draw the field, and the diagram on the right shows how iron filings become aligned in the presence of a bar magnet.

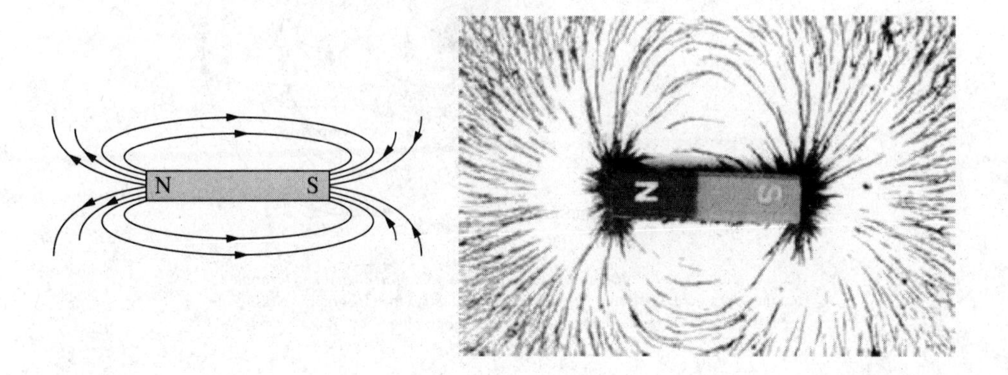

As in the case of electric charges, magnetic fields interact due to the attraction and repulsion between magnetic poles. The following diagram shows how magnetic fields are altered.

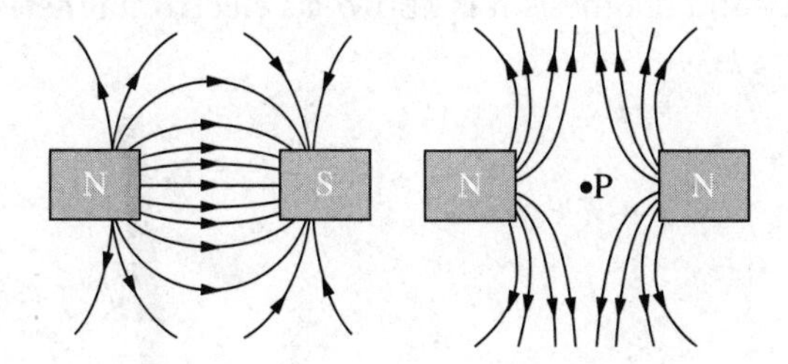

In diagrams and questions, an X is used to show that ***B*** is directed into the page, a dot is used to show that *B* is directed out of the page, and an arrow is used to show that ***B*** is along the plane of the page.

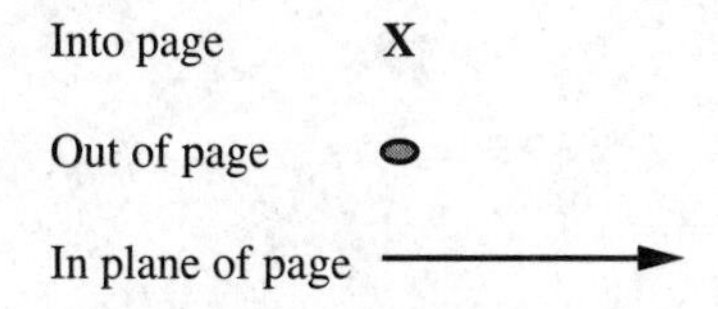

Compasses are useful because Earth behaves as if it had a giant bar magnet through its center. It has a magnetic field extending from one pole to the other. However, the magnetic poles do not line up exactly with the geographic poles. The difference between the geographic pole and the direction toward which a compass points is described as magnetic declination. The location of magnetic north varies over time, and the magnetic declination varies with location.

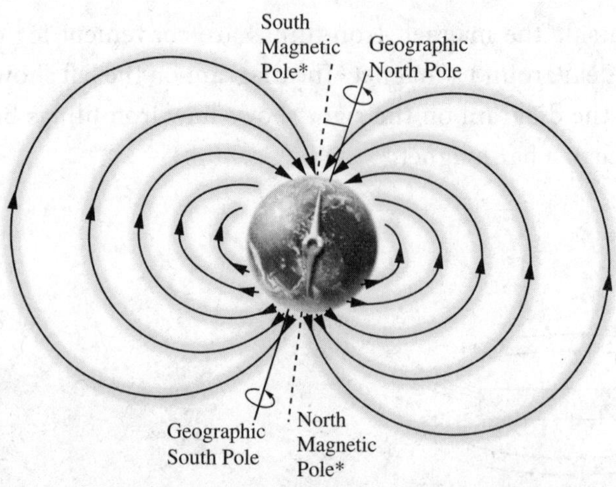

*Since magnetic field lines originate at the north pole of a magnet and terminate at the south pole, the South Magnetic Pole is close to the Geographic North Pole, and the North Magnetic Pole is close to the Geographic South Pole.

Electromagnetism

If a wire carrying current is passed through a sheet of plastic with iron filings on it, the filings become arranged in circles around the wire. The reason for this is that an electric current establishes a magnetic field around it. This relationship between electricity and magnetism is known as **electromagnetism**.

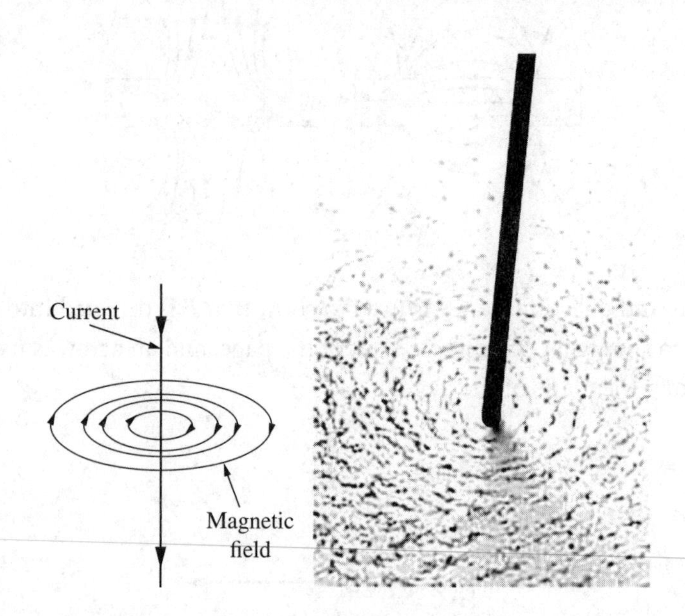

If, instead of iron filings, compasses were placed around the wire, the direction of the magnetic field could be investigated. You could easily observe that when the direction of the current is reversed, the directions of the compass needles are also reversed. The direction of the magnetic field can be determined using the right-hand rule. If the wire were held with the right hand such that the thumb is in the direction of the current, the fingers curl in the direction of $\boldsymbol{B}$. The magnitude of $\boldsymbol{B}$ is proportional to the current in the wire and inversely proportional to the distance from the wire.

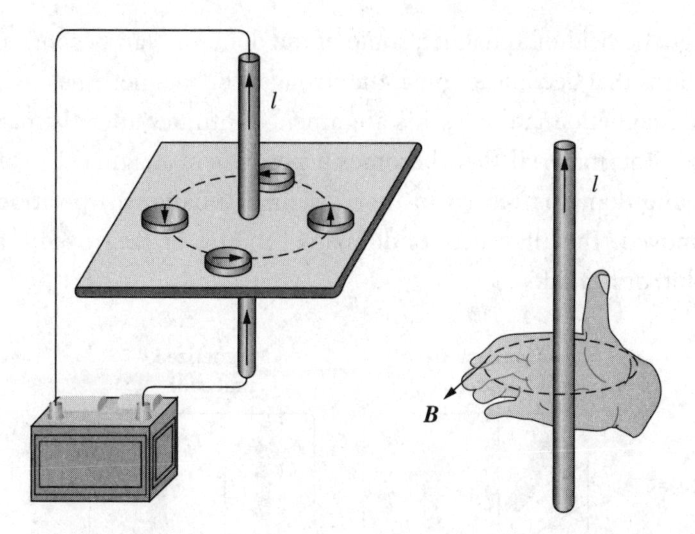

If the wire is wound into a coil, the magnetic field resembles the field around a bar magnet. A series of closely spaced coils of wire is known as a **solenoid**. The magnetic field of a solenoid depends on the current in the wire as well as the number of coils. As the current and or the number of coils increases, the strength of the solenoid increases. The strength can also be increased by inserting an iron rod into the center. A solenoid with an iron core is known as an **electromagnet**. Electromagnets have many useful applications because the strength can be controlled and the magnet can be turned on and off using a switch.

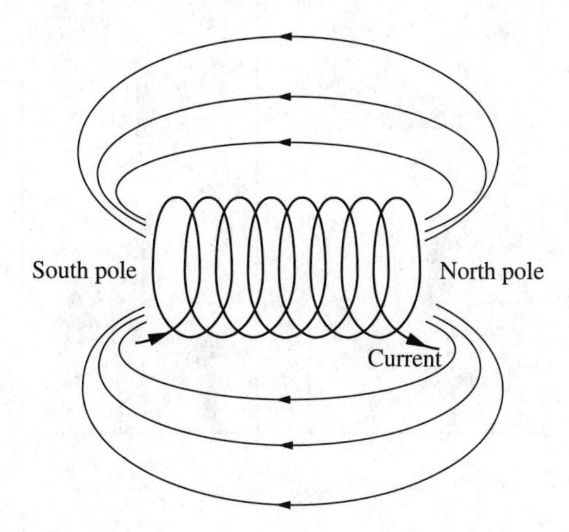

Magnetic Domains

A model of domains is often used to explain the magnetic properties of materials. The electrons of atoms within the material are described as spinning. (The actual motion of electrons will be better explained by quantum mechanics.) The spinning motion produces a tiny magnetic field. The fields of electrons with opposite spins cancel out. For nonmagnetic materials, most or all of the fields cancel out. For magnetic materials, however, the fields do not cancel out completely. Large groups of atoms with net spins that align form **domains**. If the material is placed in the magnetic field of a magnet, some of the domains can become aligned. The more domains that become aligned, the stronger the magnetic field will be.

In hard magnetic materials, this alignment continues after the magnetic field is removed. The material then becomes a permanent magnet. In soft magnetic materials, the domains return to their original (unaligned) positions once the field is removed. The alignment of domains can also be destroyed if a magnet is dropped, hit, or heated.

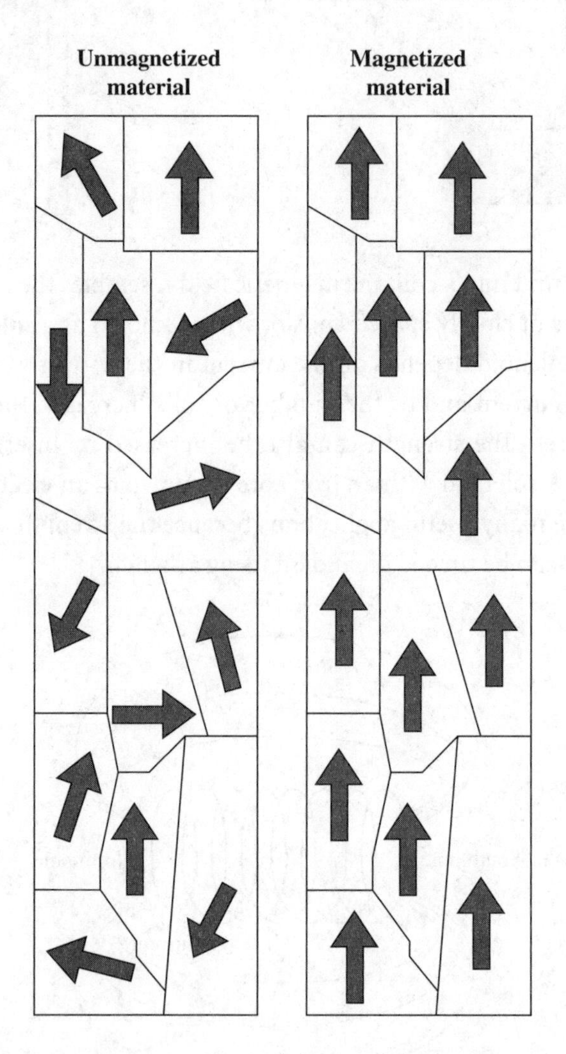

Magnetic Force

A charged particle moving through a magnetic field experiences a magnetic force F_m. No force is exerted if the charged particle is stationary. The magnitude of the force is maximum when the charge is moving perpendicular to the magnetic field and zero when the particle moves along magnetic field lines.

The magnetic force exerted on a positive charge q moving at velocity v perpendicular to the field is described by the following equation.

$$\vec{F}_m = q\vec{v} \times \vec{B} \quad \text{so} \quad B = \frac{F_m}{qv}$$

When force is measured in newtons, charge in Coulombs, and velocity in meters per second, the unit of the magnetic field strength is the tesla T. The strength of the magnetic field is often referred to as magnetic flux Φ.

The right-hand rule is again used to find the direction of the magnetic force on a positive charge. This time, extend your fingers in the direction of $\boldsymbol{B}$ and your thumb pointing in the direction of v. The direction of the magnetic force will extend out of the palm of your hand. This rule is sometimes described as the second right-hand rule.

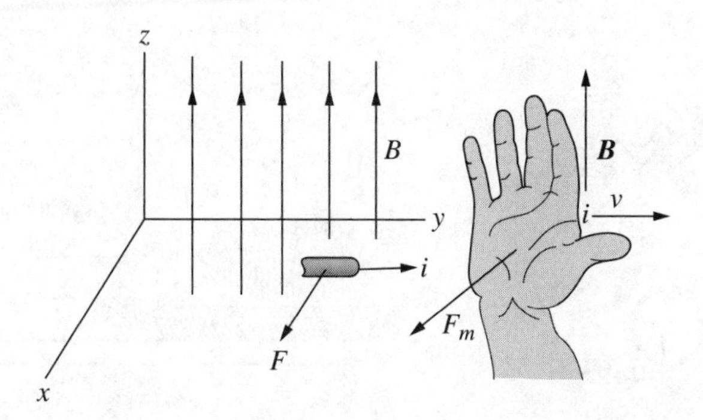

An electric current is made up of a stream of moving charges. Therefore, a current-carrying wire experiences a magnetic force just as a single charge does. The resultant force is the sum of the individual forces on the particles. The magnitude of the total magnetic force for a segment of straight wire L carrying current I in a uniform magnetic field B is described by the following equation.

$$\vec{F}_m = I\vec{L} \times \vec{B}$$

Note that the equation assumes the current is at an angle of 90° to the magnetic field. If not, the equation becomes the following, where θ is the angle between the wire and the magnetic field.

$$F_m = BIL \sin\theta$$

Example:

A current of 30 A travels through a wire with a length of 0.20 m between the poles of a magnet at an angle of 45°. If the magnetic field has a uniform value of 0.2 T, what is the magnitude of the force on the wire?

$$F_m = BIL \sin\theta = (0.2\ \text{T})(30\ \text{A})(0.20\ \text{m})(0.7) = 0.84\ \text{N}$$

Electromagnetic Induction

Just as an electric current can establish a magnetic field, a magnetic field can be made to induce an electric current. The key is that the magnetic field must be changing. Either the magnet must be moving with respect to a current-carrying wire or the wire must be moving with respect to the magnet. The process through which an electric current can be induced by a changing magnetic field is known as **electromagnetic induction**. The phenomenon of electromagnetic induction was first noticed and investigated by English scientist Michael Faraday in 1831. **Faraday's law of induction** is its quantitative expression. Faraday's law is a fundamental relationship derived from Maxwell's equations.

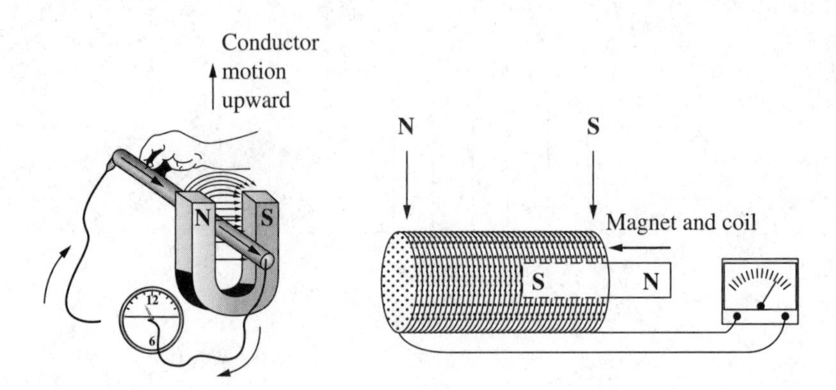

Faraday's law says that the induced voltage in a coil is proportional to the number of loops at the rate at which the magnetic field changes within those loops. It is expressed by emf $= N\dfrac{\Delta(BA)}{\Delta t}$, where N is the number of loops in the wire, B is the strength of the magnetic field, A is the area of one loop, and Δt is the change in time. For simple problems, Faraday's law can be reduced to emf $= NBLv$.

As you can see from the equation, you can increase the magnitude of the emf by increasing the length of the wire moving through the field, the relative speed, or the strength of the magnetic field.

Lenz's law is a common way to understand how electromagnetic circuits obey Newton's third law of motion and the law of conservation of energy. Lenz's law says that an induced electromotive force (emf) always gives rise to a current whose magnetic field opposes the change in original magnetic flux.

Generator

A loop of wire turning within a magnetic field induces a current because of the relative motion between the wire and the field. The mechanical energy used to turn the loop of wire is therefore transformed into electrical energy. A device capable of this type of transformation is known as a **generator**. In electric power plants, the mechanical energy might be supplied from steam produced by burning coal or from nuclear reactions. It might also be supplied by falling water, wind, or even tidal changes.

It will be helpful to be familiar with basic technology related to generators. The armature is the power-producing component, which is the wire through which the current is induced. The armature is mounted on a rotate shaft. A commutator is a device that maintains the direction of the current. Carbon brushes are devices that transfer current between the stationary parts of a generator and the moving parts. A split ring is a metal ring that connects parts together.

DC Generator

When the armature is vertical, such that the area of the loop is perpendicular to the magnetic field, the wire is moving parallel to the field, and, therefore, no current is induced in the wire. As it rotates, the segments of wire move relative to the magnetic field, so a current is induced. The current is maximum when segments of the loop have rotated 90° and move perpendicularly relative to the magnetic field. The loop continues to rotate as the current decreases down to zero. The current always flows in the same direction and is known as direct current, or DC. This type of generator is therefore known as a DC generator.

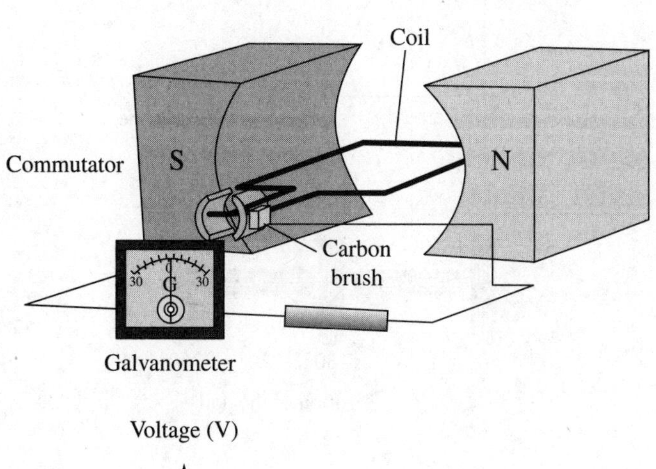

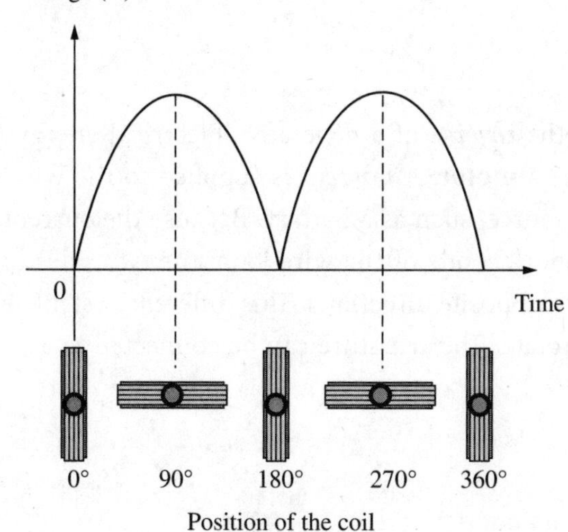

AC Generator

If the commutator is replaced by two separate slip rings, the direction of the current keeps changing. The current is known as alternating current, or AC, and the generator is known as an AC generator.

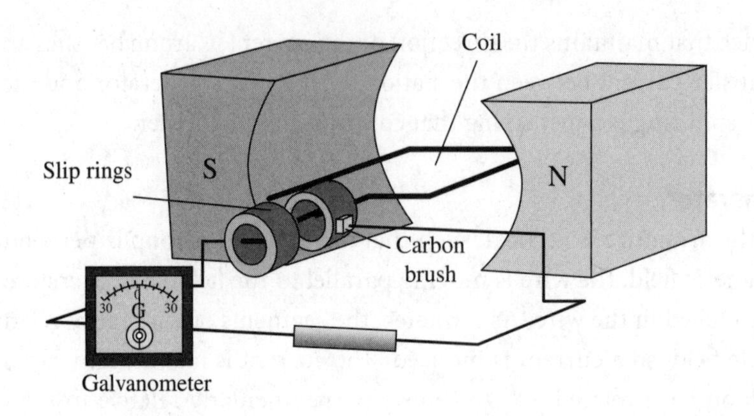

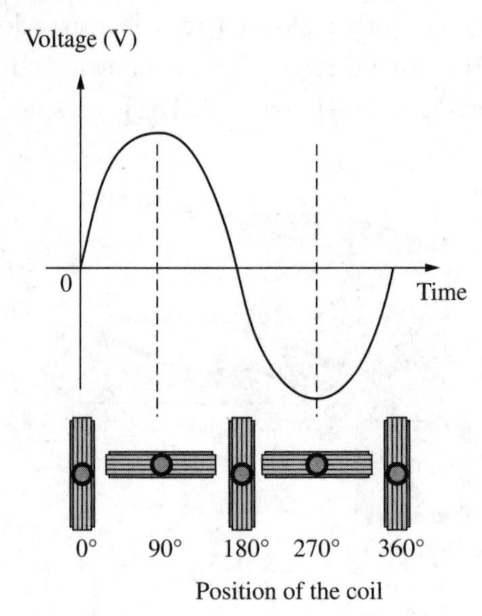

Motors

A motor is the reverse of a generator. Electrical energy converted to mechanical energy is a **motor**. A current is supplied to the wire in a magnetic field by an external source, such as a battery. Because the current is flowing in opposite directions on each side of the wire loop, the two sides of the wire loop experience forces in opposite directions. This difference establishes a torque that causes the loop to rotate. The armature can be connected to a shaft that can be used to do work.

TEST-TAKING HINT

Be aware when using the right-hand rules that you are using conventional current. It developed from Benjamin Franklin, who theorized that electric current arose from the flow of positive charges moving from the positive terminal of a battery to the negative terminal.

REVIEW QUESTIONS

Questions 1–3 refer to the following answer options.

(A) into the page
(B) out of the page
(C) to the left
(D) to the right
(E) counterclockwise

1. A negative charge is placed in the magnetic field shown, which is directed out of the page. In which direction will the magnetic force on the charge be exerted?

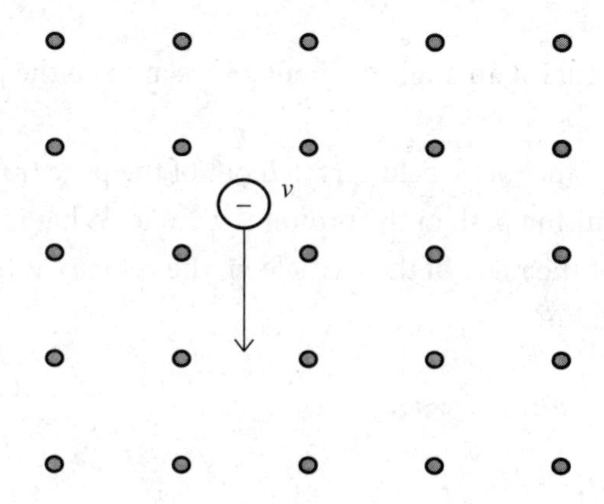

2. A wire carries current coming directly out of the page. Which of the answer choices shown above describes the magnetic field produced by the wire?

3. In which direction will the force on the wire be directed?

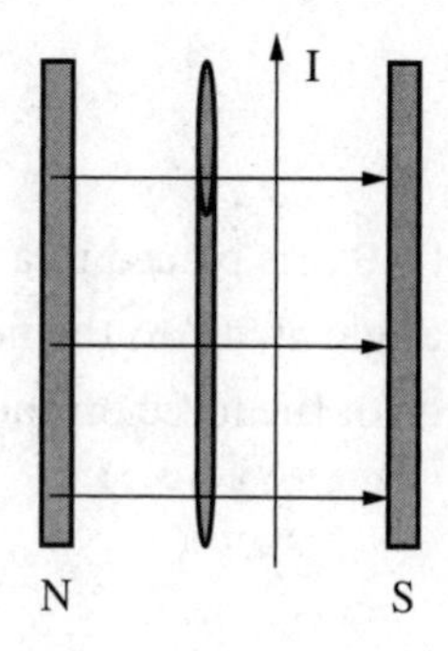

4. What would be the result of cutting a magnet in half?

(A) Both magnetic poles would be destroyed.
(B) Each half would have two identical poles.
(C) One end of each half would not have a magnetic pole.
(D) Each half would have a north pole and a south pole.
(E) Each half would have two magnetic poles on each end.

5. What is the cause of magnetic declination?

(A) Earth constantly rotates on its axis.
(B) Earth's magnetic poles are different from the geographic poles.
(C) Earth's distance from the sun varies throughout its elliptical orbit.
(D) Earth's circumference is greatest at the equator and smallest near the poles.
(E) Earth's axis tilts at an angle of about 23° relative to the plane of its orbit.

6. A proton enters a magnetic field directed out of the page traveling from left to right. The resulting path of the proton is a circle. What is the radius of the circle in terms of the mass of the particle m, the velocity v, the charge q, and the magnetic field B?

(A) $mvBq$

(B) $\frac{qv}{mB}$

(C) $\frac{qB}{mv}$

(D) $\frac{mB}{qv}$

(E) $\frac{mv}{qB}$

7. A wire carrying a current of 6.0 A is placed in a magnetic field of 0.3 T such that the current flows at an angle of 90° to the field. The length of the wire is 0.2 m. What is the magnitude of the force on the wire?

(A) 0.36 N
(B) 0.60 N
(C) 1.2 N
(D) 1.8 N
(E) 6.5 N

8. A proton moving east experiences a force of 7.2×10^{-19} N upward as a result of Earth's magnetic field. If the speed of the proton is 2.0×10^5 m/s, what is the magnitude of the magnetic field at this location? (Assume that Earth's magnetic field is essentially true north in this location.)

(A) 1.4×10^{-6} T
(B) 2.3×10^{-5} T
(C) 3.6×10^{-5} T
(D) 4.5×10^{-5} T
(E) 5.2×10^{-4} T

Questions 9 and 10 relate to the diagram below, which shows an electric generator.

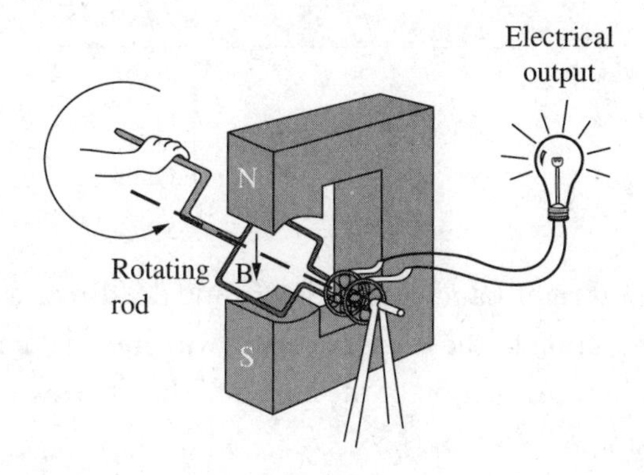

9. Which transformation takes place in the generator shown?

(A) chemical energy to electrical energy
(B) heat energy to mechanical energy
(C) electrical energy to mechanical energy
(D) electrical energy to heat energy
(E) mechanical energy to electrical energy

10. How can the emf of a generator be increased?

(A) reverse the poles of the permanent magnet
(B) increase the rate at which the wire moves through the magnetic field
(C) shorten the length of the wire moving through the field
(D) reduce the strength of the magnetic field through which the wire moves
(E) maintain the wire in a stationary position

ANSWERS AND EXPLANATIONS

1. **(D)** Use the second right-hand rule to determine the direction. Extend your fingers in the direction of the magnetic field, which is to the right. Point your thumb upward in the direction of the current. The force extends from the palm of your hand into the page.

2. **(E)** The direction of the magnetic field can be determined using the right-hand rule. If the wire were held with the right hand such that the thumb is in the direction of the current, the fingers curl in the direction of ***B***.

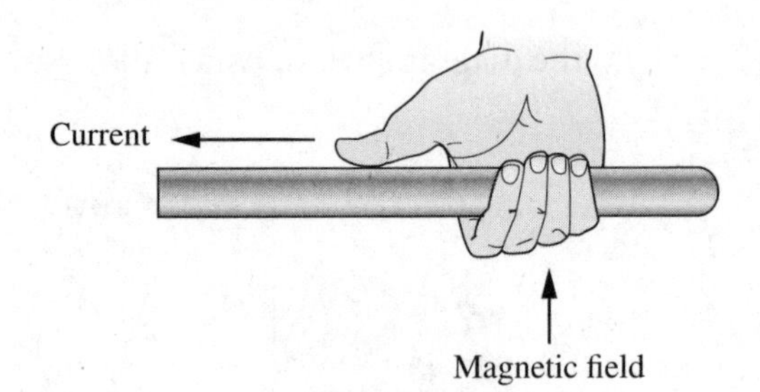

3. **(D)** Use the second right-hand rule to determine the direction. Extend your fingers in the direction of the magnetic field, which is to the right. Point your thumb upward in the direction of the current. The force extends from the palm of your hand.

4. **(D)** Magnetic poles cannot exist independently. Every magnet has both a north pole and a south pole. Therefore, cutting a magnet in half would result in two complete smaller magnets, each with a north and south pole.

5. **(B)** Magnetic declination describes the difference between the direction a compass points and geographic north. It arises from the fact that the geographic poles are different from the magnetic poles, and it varies with location and time.

6. **(E)** If the proton moves in a circle, the magnetic force (qvB) must equal the centripetal force ($\frac{mv^2}{r}$). Therefore, $qvB = \frac{mv^2}{r}$. Rearranging to solve for r shows that $r = \frac{mv}{qB}$.

7. **(A)** $F_m = BIL = (0.3\ \text{T})(6\ \text{A})(0.2\ \text{m}) = 0.36\ \text{N}$

8. **(B)** $B = \frac{F_m}{qv} = \frac{7.2 \times 10^{-19}\ \text{N}}{(1.60 \times 10^{-19}\ \text{C})(2.0 \times 10^{5}\ \text{m/s})} = 2.3 \times 10^{-5}\ \text{T}$

9. **(E)** An electric generator operates when the armature is turned. The motion of the armature requires mechanical energy. As the wire moves through the magnetic field, an electric current is induced. Therefore, electrical energy is produced. The mechanical energy is transformed into electrical energy.

10. **(B)** The induced emf is directly proportional to the length of the wire, the relative speed, and the magnitude of the magnitude field according to the equation emf = LvB. The only change listed that would increase emf would be to increase the rate, or speed, at which the wire moves relative to the field.

CHAPTER 15

Waves

You may already be familiar with how to form water waves or waves on a rope, but being comfortable describing the properties of waves and the interactions between them will help you answer several questions on the SAT Physics exam. In this chapter, you will focus on waves that disturb materials, such as ropes, water, and air. In the following chapter, you will review light waves and similar types of waves.

Wave Motion

Drop a pebble into a lake, and you will observe ripples extending outward in all directions. The ripples are produced because a disturbance is transmitted away from the source. The disturbance is the **wave**, and it carries energy from one location to another.

A water wave is an example of a **mechanical wave**, which is a traveling disturbance that requires a medium through which to travel. A medium is a material through which a wave moves. In the case of a water wave, the medium is water. As the wave travels through the medium, the particles of the medium are disturbed and vibrate back and forth. Once the wave passes a given point, the particles return to their normal resting positions.

Not all waves require a medium. **Electromagnetic waves** are made up of electric and magnetic fields and do not require a medium. Light, microwaves, and ultraviolet radiation are electromagnetic waves. These kinds of waves will be discussed in the next chapter.

Transverse and Longitudinal Waves

Waves can be further classified as transverse or longitudinal. A **transverse wave** causes the particles of the medium to vibrate perpendicular to the direction of motion of the wave. A **longitudinal wave** causes the particles of the medium to vibrate parallel to the direction of motion of the wave.

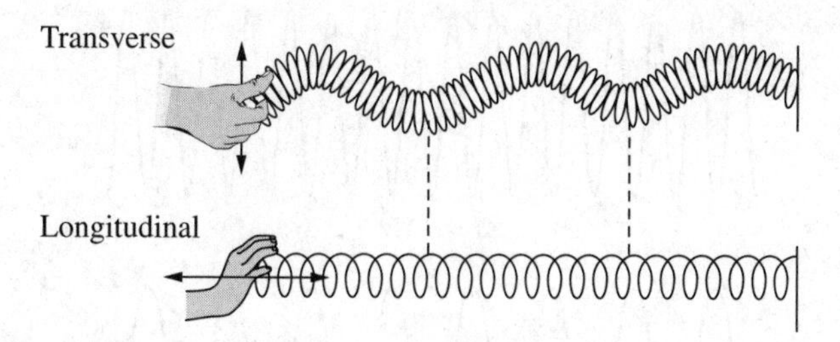

Both kinds of waves can be described by several basic properties. It is often convenient to represent a wave as a graph similar to the following diagram. As you can see, the maximum displacement of the particles of the medium leads to high points and low points. For a transverse wave, the high points are also known as crests and the low points as troughs. For a longitudinal wave, the high points relate to regions where particles of the medium are most crowded and the low points to regions where the particles are most spread out. The crowded regions are known as compressions and the expanded regions as rarefactions.

The maximum displacement of the particles of a medium describes the **amplitude** of a wave. The distance between similar points on consecutive waves is the **wavelength** of the wave, λ. For example, one wavelength might be measured from crest to crest, trough to trough, or two other similar points.

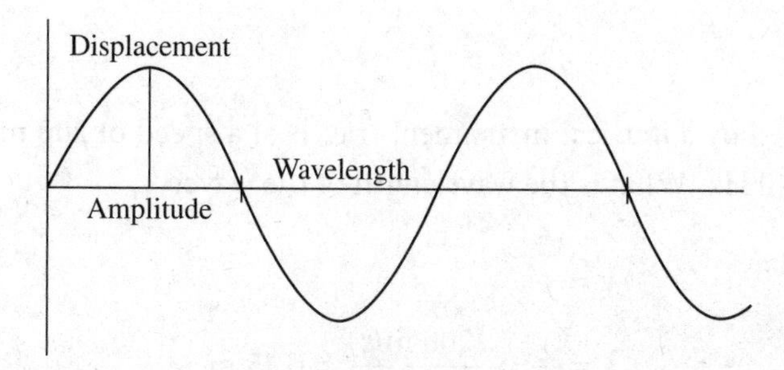

The **frequency** of a wave, *f*, describes the number of waves that pass a given point per second. Notice that the waves in the following diagram all have the same amplitude, but different frequencies and wavelengths. At a given speed, frequency and wavelength are inversely proportional, so the wave with the greatest wavelength has the smallest frequency and the wave with the smallest wavelength has the greatest frequency.

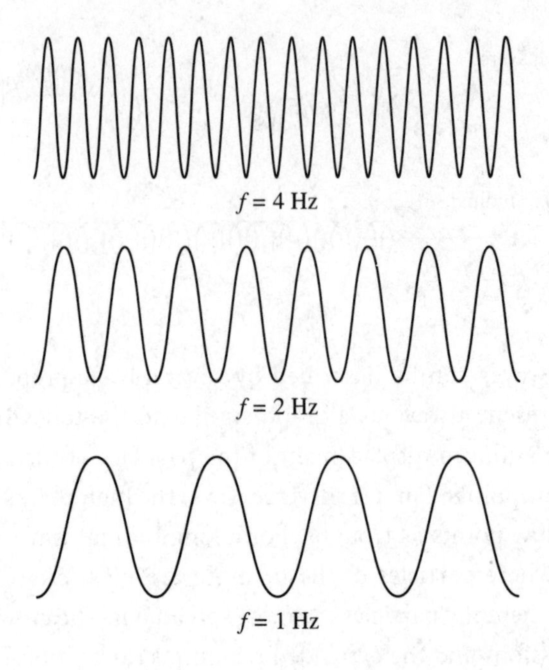

The speed of a wave equals the product of the frequency of a wave and its wavelength. This relationship is summarized by the following equation.

$$v = f\lambda$$

Example:

A wave produced by a musical instrument travels at a speed of 200 m/s and has a frequency of 440 Hz. What is the wavelength of the wave?

$$v = f\lambda$$

$$\lambda = \frac{v}{f} = \frac{(200 \text{ m/s})}{(440 \text{ Hz})} = 0.45 \text{ m}$$

Resonance

All objects have a **natural frequency** or group of frequencies at which they vibrate. The natural frequency depends on the physical properties of the object, including its mass, shape, and elasticity. However, an object may be set into vibration if an external force is applied to it. The resulting vibration is known as a **forced vibration**. The amplitude of the forced vibration depends on the difference between the applied frequency f and the natural frequency f_o. It is at a maximum when the applied frequency equals the natural frequency, $f = f_o$. The effect of the maximum is known as **resonance**, and the natural frequency is therefore known as the **resonant frequency**. In other words, if the frequency of forced vibrations matches the natural frequency of an object, there is a dramatic increase in the amplitude of the result wave.

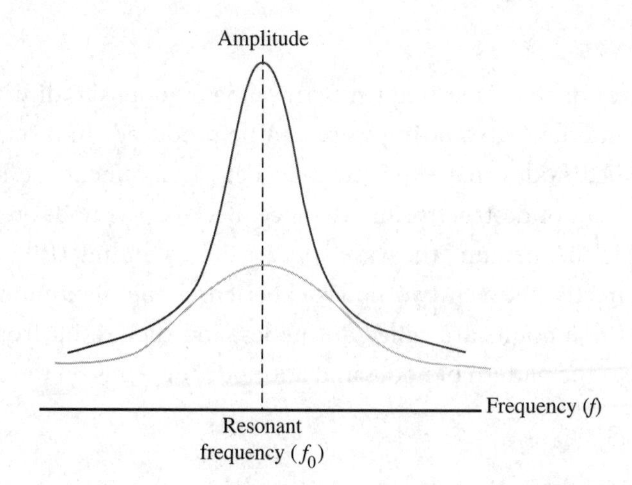

A common example used to describe this phenomenon is a simple child's swing. The swing, like other objects, has a natural frequency. If you exert a force on a swing by pushing, your timing determines the resulting effect. Even a small force can result in a large amplitude if delivered along with the natural frequency of the swing.

Interference

Two waves traveling in the same medium at the same time experience interference. **Constructive interference** occurs when two waves combine in such a way that the amplitude of the resulting wave is greater than either of the two individual waves. Waves that interfere constructively are said to be *in phase.* Once the waves pass one another, they retain their original properties. **Destructive interference** occurs when two waves combine in such a way that the amplitude of the resulting wave is less than either of the two individual waves. Again, the waves retain their original properties after they pass one another. The rule that describes the interference of waves is known as the law of superposition.

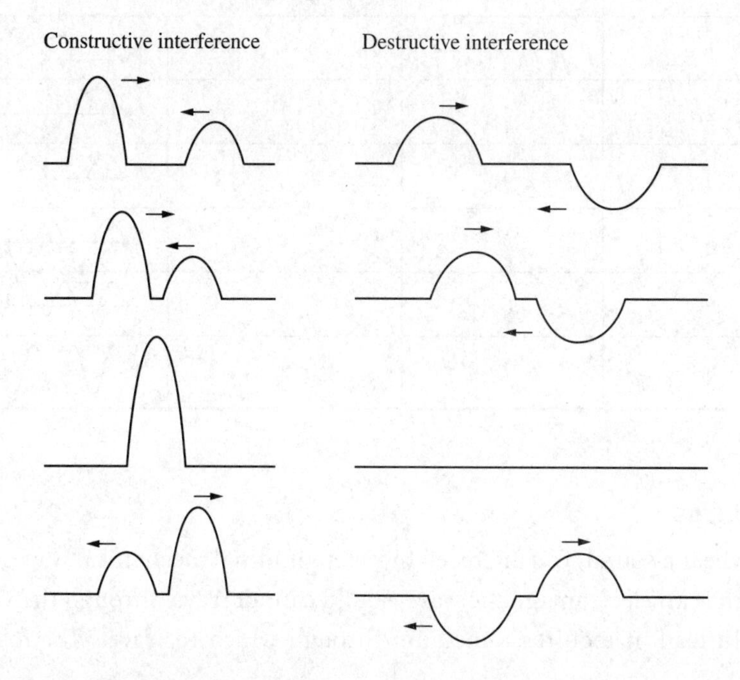

Standing Waves

When two waves of the same frequency traveling in opposite directions in the same medium interfere, a **standing wave** can be produced. In a standing wave, fixed points called nodes that experience no net displacement are formed. The nodes are the result of destructive interference. Because the nodes remain in the same positions in the medium, the wave appears to be standing still.

The point directly between two nodes experiences the maximum amount of displacement. These points are called antinodes, and they result from constructive interference. The pattern of nodes and antinodes depends on the frequency of the waves.

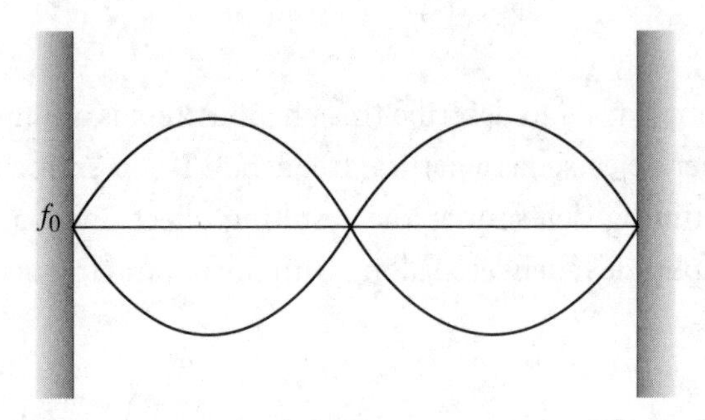

The natural frequencies are often described as the *harmonics* of an object or system. The first harmonic is the standing wave with two nodes and one antinode. It has the longest wavelength and the lowest frequency. The lowest frequency is also known as the fundamental frequency. All other harmonics are whole number multiples of the fundamental frequency. Each additional harmonic adds one node and antinode.

Harmonic	Number of Nodes	Number of Antinodes	Pattern
1	2	1	
2	3	2	
3	4	3	
4	5	4	
5	6	5	
6	7	6	

Sound Waves

When you hear a sound, you are receiving a longitudinal mechanical wave. Unlike light, which is an electromagnetic wave, sound cannot travel through the vacuum of space. Instead, it requires a medium through which to travel. The following diagram relates to the compressions and rarefactions of the medium with the properties of the wave.

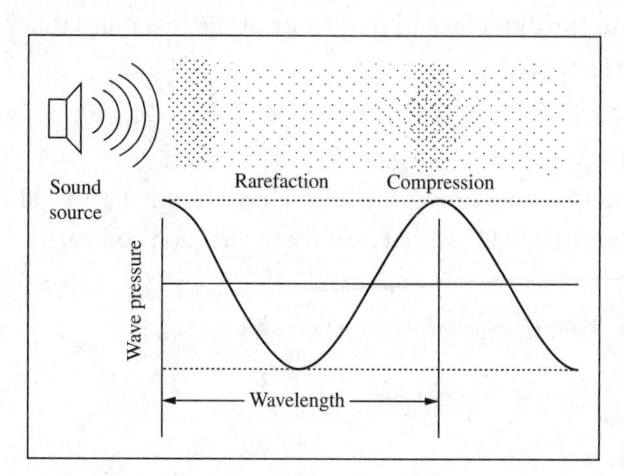

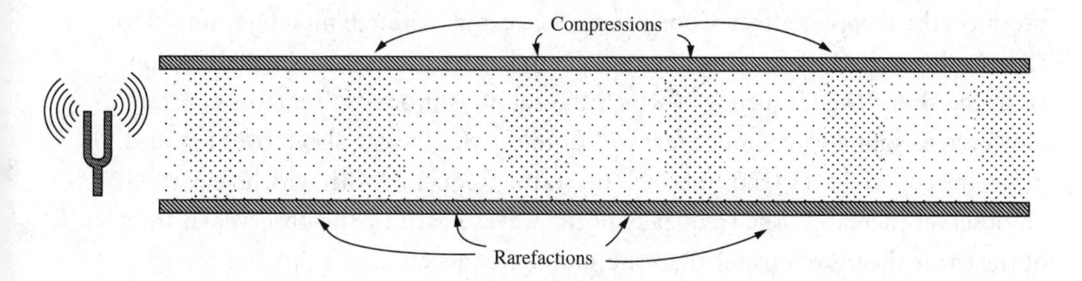

A sound is produced when an object vibrates. Any vibrating object produces a sound, even if you cannot hear it. The frequencies of sound waves are classified as *audible* if they are within the range of human hearing. This range is described as between 20 and 20,000 Hz. Sound waves are considered *infrasonic* if they are below 20 Hz and *ultrasonic* if they are above 20,000 Hz.

The frequency of a sound wave determines a property known as pitch. The **pitch** of a sound describes how high or how low the sound is. A sound with a high frequency, such as that produced by a flute, has a high pitch. A sound with a low frequency, such as that produced by a tuba, has a low pitch.

The frequencies of sound can be used to distinguish music from noise. While noise generally consists of a mixture of frequencies with no discernable relationship, music consists of frequencies with an obvious mathematical relationship. Two waves with a whole number ratio between their frequencies interfere to form a wave with a repeating pattern described as music. For example, suppose two waves—for which their frequencies are in a ratio of 2:1—interfere. The resulting wave has a regular, repeating pattern known as an octave.

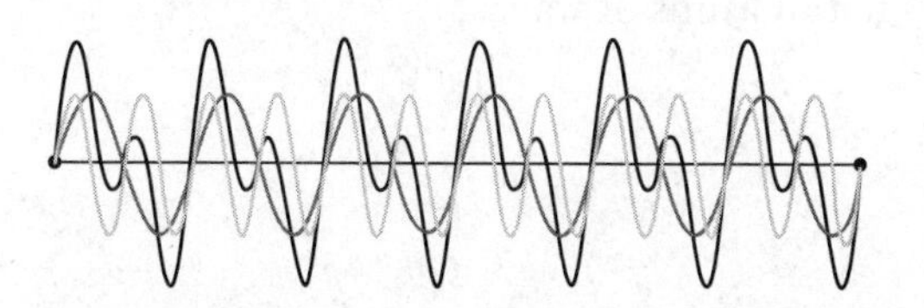

Periodic, repeating fluctuations in the intensity of a sound that result when two waves with slightly different frequencies interfere are known as **beats**. The beat frequency is equal to the difference in frequency of the two notes that produce the beats through interference.

Example:

Two sounds occur at the same time. One sound has a frequency of 248 Hz, and the other has a frequency of 250 Hz. What beat frequency do you hear?

$$\text{beat frequency} = 250\text{ Hz} - 248\text{ Hz} = 2\text{ Hz}$$

Doppler Effect

A relative change in frequency due to motion of a sound of source or its observer produces the **Doppler effect**. Consider a police car with a siren moving from left to right as shown in the following diagram. Sound waves are emitted in all directions from the siren. The frequency at which the sound is produced remains constant. However, as sound waves move forward from the car, the car also moves forward. Thus, the sound waves tend to bunch up so the frequency with which they reach an observer increases. The frequency of the waves heard by the observer in front of the car is therefore greater than the source frequency, so the pitch of the siren sounds higher.

Behind the car, the sound waves reach an observer with a lower frequency as the car moves away from the observer. The frequency of the waves heard by the observer in back of the car is, therefore, lower than the source frequency, so the pitch of the siren sounds lower.

TEST-TAKING HINT

Review the general properties of all waves. Then make sure you know the properties specific to certain kinds of waves. For example, know how the particle motion related to sound waves can be explained. Later on, you will need to know how to explain the properties of electromagnetic waves.

Note that the Doppler effect is also observed if the observer moves relative to a stationary source of sound. The Doppler effect is commonly observed with sound waves, but it occurs for all kinds of waves.

REVIEW QUESTIONS

Questions 1–3 relate to the diagram below, which shows the representation of a wave. Choose the correct term for each blank from the following choices.

(A) Amplitude in meters
(B) Wavelength in meters
(C) Frequency in Hertz
(D) Period in seconds
(E) Number of nodes

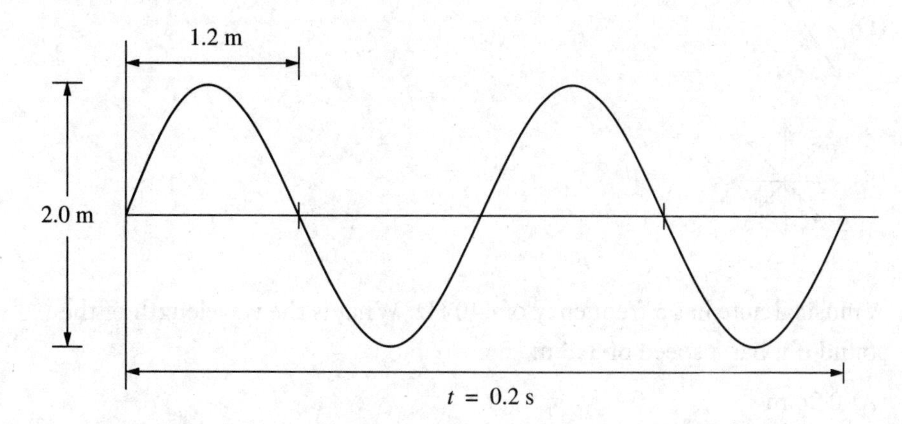

1. The wave shown has a(n) ________________ of 1.0.

2. The wave shown has a(n) ________________ of 2.4.

3. The wave shown has a(n) ________________ of 10.

4. How does a mechanical wave differ from an electromagnetic wave?

(A) A mechanical wave requires a medium through which to travel.
(B) A mechanical wave can travel through a vacuum.
(C) A mechanical wave produces noise as opposed to music.
(D) A mechanical wave is produced as a result of a forced vibration.
(E) A mechanical wave can experience the Doppler effect.

5. A transverse wave moves to the right in the plane of this paper. In which direction will the particles of the medium vibrate?

(A) $\longleftrightarrow$

(B) 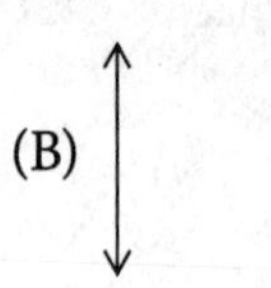

(C) $\rightrightarrows$

(D)

(E)

6. A musical note has a frequency of 440 Hz. What is the wavelength of the sound if it has a speed of 345 m/s?

(A) 0.26 m
(B) 0.78 m
(C) 1.0 m
(D) 1.2 m
(E) 1.5 m

7. A student strikes a tuning fork. How long does it take for the sound to reach the observer? [Use 340 m/s for the speed of sound in air.]

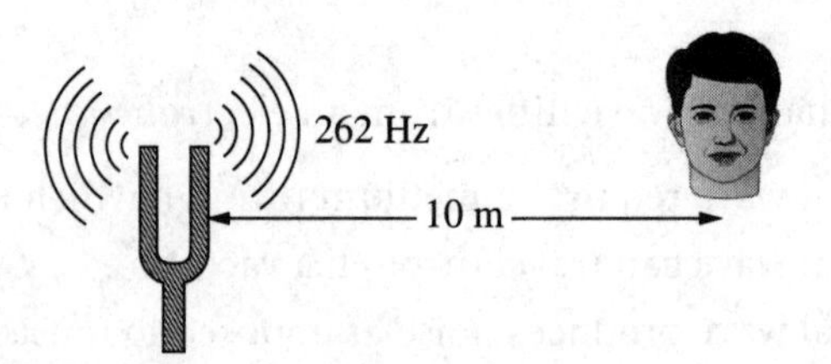

(A) 0.03 s
(B) 0.04 s
(C) 1.3 s
(D) 2.6 s
(E) 3.4 s

8. At which letter do the waves interfere destructively?

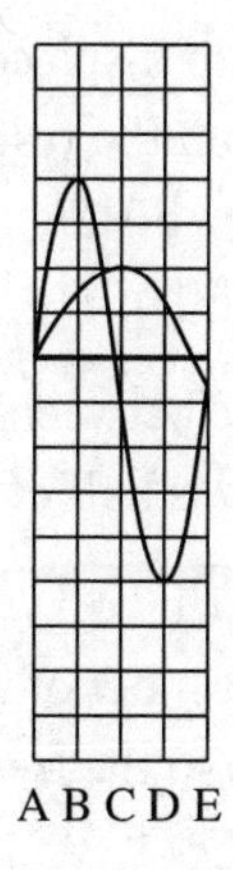

(A) point A
(B) point B
(C) point C
(D) point D
(E) point E

9. How many nodes are present for the wave depicted below?

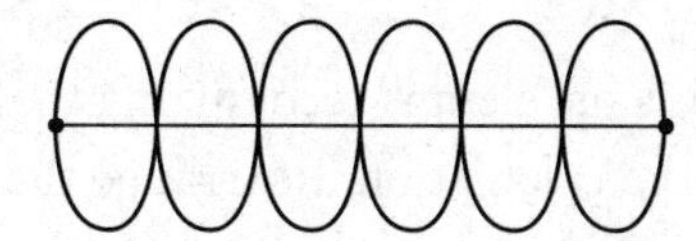

(A) 3
(B) 6
(C) 7
(D) 14
(E) 20

10. A train approaches you as its horn blows. How is the perceived sound of the horn affected by the relative motion between the train and you?

(A) The speed decreases.
(B) The amplitude increases.
(C) The frequency decreases.
(D) The pitch increases.
(E) The direction of the sound shifts.

ANSWERS AND EXPLANATIONS

1. **(A)** The amplitude is the maximum displacement of rest of the particles of the medium. It is the distance from the resting position, or horizontal line, to either a crest or a trough. The total distance from crest to trough is shown as 2.0 m, so the amplitude must be 1.0 m.

2. **(B)** The wavelength is the distance between two consecutive similar points. The distance represented as 1.2 m is half of the wavelength of this wave, so the total wavelength must be 2.4 m.

3. **(C)** The frequency measures the number of waves that pass a point per second. Two complete waves pass in 0.2 s, so the frequency is 2 cycles/0.2 s = 10 Hz.

4. **(A)** A mechanical wave is one that travels through a physical medium, such as water or air. An electromagnetic wave is one that does not require a medium and can therefore travel through a vacuum.

5. **(B)** The particles of the medium vibrate perpendicular to the direction in which a transverse wave travels. The particles, therefore, vibrate at an angle of 90° to the motion of the wave.

6. **(B)** Rearrange the equation for speed equation $v = f\lambda$ as $\lambda = \frac{v}{f} = \frac{345 \text{ m/s}}{440 \text{ Hz}} =$ 0.78 m.

7. **(A)** The key to solving this problem is recognizing that the frequency of the wave is not required in the calculation. Rearrange the speed equation $v = \frac{d}{t}$ as $t = \frac{d}{v} = \frac{10 \text{ m}}{340 \text{ m/s}} = 0.03$ s.

8. **(D)** The waves interfere destructively where the waves are on opposite sides of the axis, which occurs at point D.

9. **(C)** Nodes occur where destructive interference occurs and the net displacement is zero. Therefore, the nodes occur along the horizontal line. The high and low points are antinodes.

10. **(D)** The Doppler effect results in a change in the apparent pitch of the sound because the sound waves pile up in front of the train, causing the frequency to increase.

CHAPTER 16

Light

Light, microwaves, X-rays, and gamma rays all have something in common—they are kinds of electromagnetic waves. The properties of electromagnetic waves can help you understand not only their nature, but their behavior and interactions as well.

Electromagnetic Waves

An electromagnetic wave is produced by a vibration, just as mechanical waves are. However, rather than a hand moving the end of a rope or a pebble dropped in a lake of water, an electromagnetic wave is produced by harmonic motion of charged particles at the atomic level. The result is a wave consisting of vibrating electric and magnetic fields at right angles to one another and to the direction of motion of the wave. An electromagnetic wave, therefore, is a transverse wave that can travel through a vacuum.

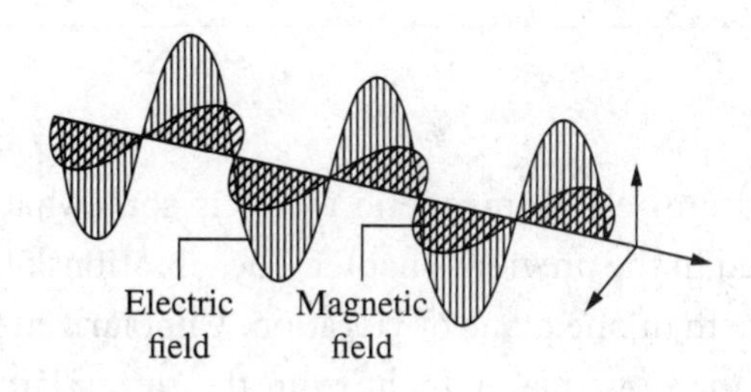

Electromagnetic Spectrum

Like mechanical waves, electromagnetic waves can be described by properties of wavelength, frequency, and speed. In a vacuum, all electromagnetic waves travel at the same speed of 3.0×10^8 m/s. You will often see this speed described as the *speed of light, c*. Note that this speed is used for all electromagnetic waves in air for problems on the SAT Physics test unless otherwise indicated.

Example:

A sample of blue light has a wavelength of 550 nm. What is the frequency of the light?

$$f = \frac{c}{\lambda} = \frac{3.0 \times 10^8 \text{ m/s}}{5.5 \times 10^{-7} \text{ m}} = 5.5 \times 10^{14} \text{ Hz}$$

Electromagnetic waves are commonly arranged in order according to wavelength. This arrangement is known as the **electromagnetic spectrum**. The waves with the longest wavelengths and shortest frequencies are radio waves. Gamma rays have the shortest wavelengths and greatest frequencies. Visible light makes only a very small portion of the spectrum.

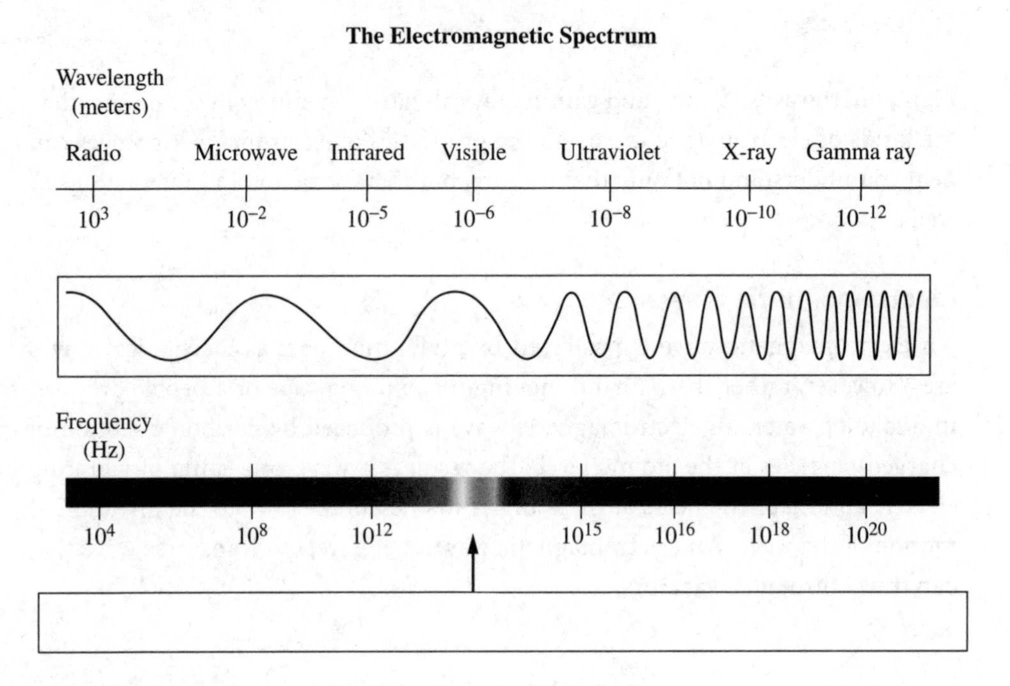

Polarization

The vibrating nature of electromagnetic waves is somewhat different from the waves you examined in the previous chapter. The vibrations of an electromagnetic wave occur in more than one plane of vibration. Unpolarized light describes light vibrating in more than one plane. Light from the sun, a lamp, or a flashlight is unpolarized. It is common to represent unpolarized light as vibrating in all directions. Because roughly half the vibrations can be considered to be in a horizontal plane and the other half in a vertical plane, a more useful representation is simplified as shown.

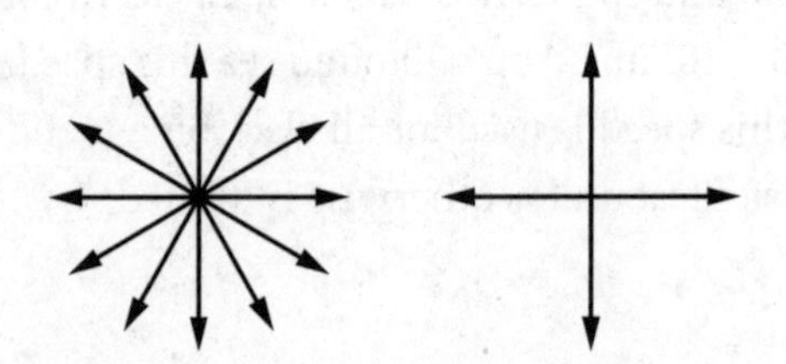

Polarizing filters can be used to block out all but one direction of vibration so that the remaining vibrations are in only one plane. Two filters at 90° angles can block out all of the light.

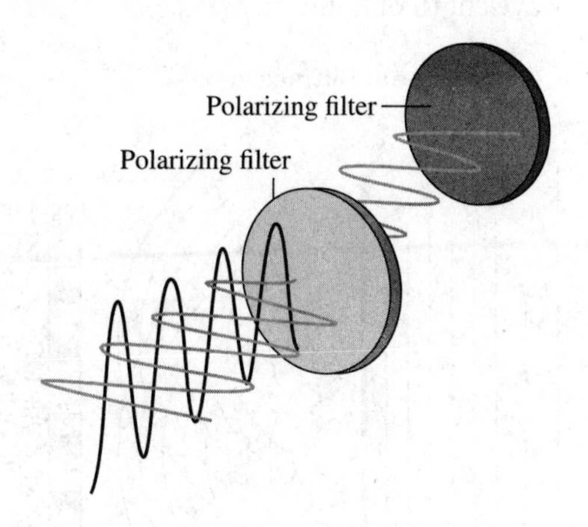

Diffraction

Light waves can bend around an obstacle or through an opening in a process known as **diffraction**. The amount of bending depends on the relationship between the wavelength of light and the size of the obstacle or opening. If the obstacle or opening is much larger than the wavelength, the bending will be unnoticeable. If the sizes are closer to being equal or the wavelength is greater, the bending will be considerable.

The following diagram represents diffraction through a single slit, or aperture. As the light passes through the slit, they become semi-circular. A screen placed a distance from the slit shows a large, bright region in the center directly opposite the slit. In addition, a series of alternating light and dark spots are formed on either side. These regions, known as interference fringes, become smaller and fainter as the distance from the center increases.

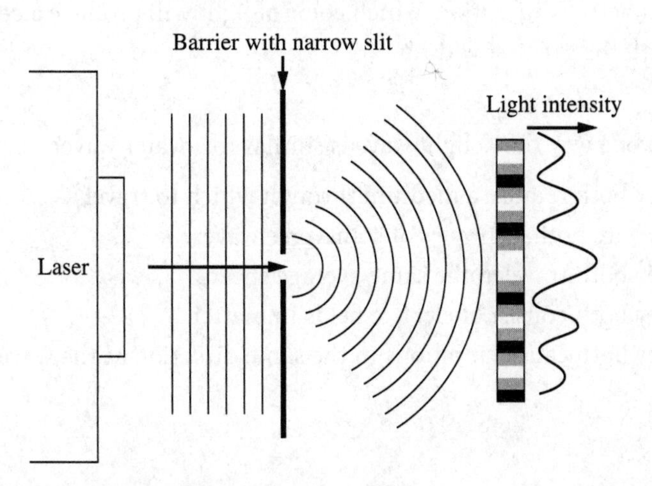

TEST-TAKING HINT

Be familiar with various versions of the same units of measurement. For example, 1 Hz can also be written as 1/sec. This will help you cancel units during calculations and arrive at the correct unit in your answer.

The pattern on the screen changes somewhat if the light is passed through a double slit instead of a single slit. A series of light and dark lines again appears on the screen, but the two patterns overlap. The width of the bright central region is proportional to the wavelength of light.

Double-Slit Diffraction

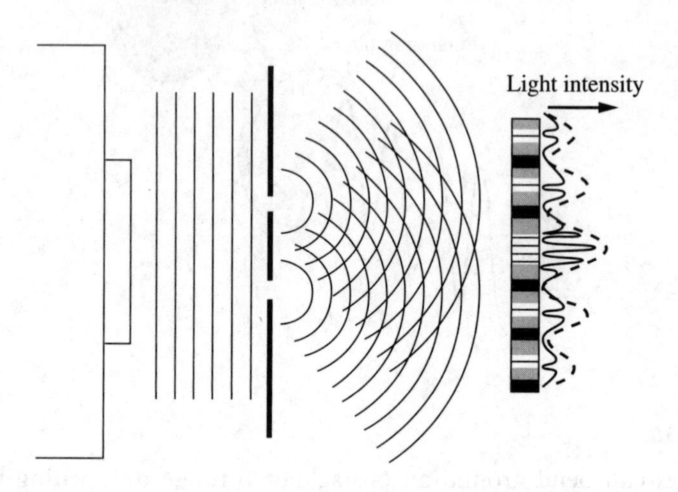

REVIEW QUESTIONS

Questions 1 and 2 refer to the following colors of light.

(A) Red
(B) Orange
(C) Yellow
(D) Green
(E) Blue

1. Which color of light has the shortest wavelength?

2. In a double-slit diffraction, which color of light will produce a central band of light with the greatest width?

3. What is one way that a light wave is similar to a water wave?

(A) They both require a medium through which to travel.
(B) They are both examples of transverse waves.
(C) They both travel at the same average speed.
(D) They both consist of electric fields in matter.
(E) They both cause vibrations in the same direction as the wave travels.

4. Which of the following arranges electromagnetic waves in order of increasing wavelength?

(A) radio wave, infrared, ultraviolet, gamma ray
(B) X-ray, gamma ray, radio wave, visible
(C) ultraviolet, visible, infrared, microwave
(D) microwave, radio wave, visible, infrared
(E) radio wave, visible, X-ray, gamma ray

5. What property is the same for all electromagnetic waves in a vacuum?

(A) frequency
(B) wavelength
(C) amplitude
(D) speed
(E) color

6. A purple light has a frequency of 7.42×10^{14} Hz. What is the wavelength of the light?

(A) 157 nm
(B) 223 nm
(C) 247 nm
(D) 300 nm
(E) 404 nm

7. A weather station broadcasts at 162 MHz. What is the wavelength of the radio waves in meters?

(A) 1.62 m
(B) 1.85 m
(C) 2.59 m
(D) 4.86 m
(E) 5.40 m

8. A color of light in the violet region has a wavelength of 413 nm. What is its frequency?

(A) 4.16×10^{13} Hz
(B) 1.38×10^{14} Hz
(C) 4.10×10^{14} Hz
(D) 7.26×10^{14} Hz
(E) 1.24×10^{15} Hz

9. How does polarization affect light?

(A) It blocks vibrations in all but one plane.
(B) It magnifies the amplitude.
(C) It changes the color.
(D) It converts parallel waves to semi-circular fronts.
(E) It bends the light toward the normal of a surface.

Question 10 relates to the diagram below, which shows a light source placed on one side of two slits in a grating.

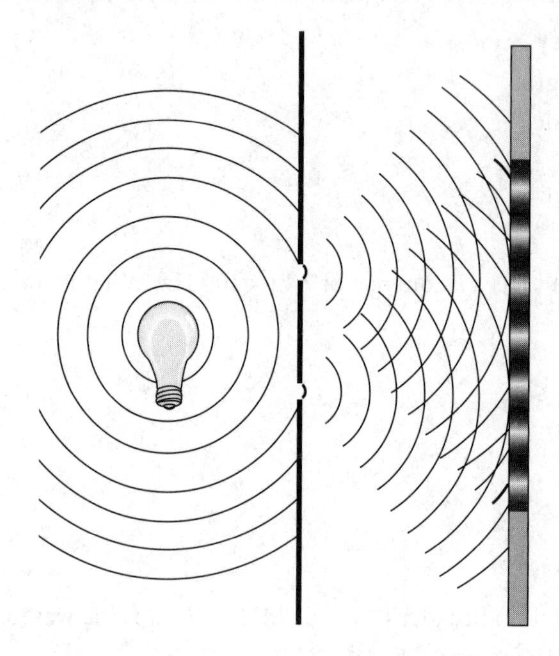

10. Which process results in the bright regions of the diffraction pattern?

(A) virtual images
(B) internal reflection
(C) polarization
(D) constructive interference
(E) destructive interference

ANSWERS AND EXPLANATIONS

1. **(E)** The visible spectrum is arranged from red light, with the longest wavelengths and lowest frequencies, to blue/violet, with the shortest wavelengths and greatest frequencies.

2. **(B)** The width of the central region is directly proportional to the wavelength of light. Therefore, the color with the greatest wavelength will result in the greatest width. Orange has the longest wavelength of the colors listed.

3. **(B)** Both light waves and water waves are classified as transverse waves because the vibration involved is perpendicular to the direction of motion of the wave. However, the light wave consists of vibrating electric and magnetic fields, whereas the water wave consists of vibrating water molecules.

4. **(C)** Consult the electromagnetic spectrum on page 196.

5. **(D)** All electromagnetic waves travel at the same speed in a vacuum. All other properties listed vary with the type of wave.

6. **(E)** $\lambda = \frac{c}{f} = \frac{3.00 \times 10^8 \text{ m/s}}{7.42 \times 10^{14} \text{ Hz}} = 404 \text{ nm}$

7. **(B)** $\lambda = \frac{c}{f} = \frac{3.00 \times 10^8 \text{ m/s}}{162 \times 10^6 \text{ Hz}} = 1.85 \text{ m}$

8. **(D)** $f = \frac{c}{\lambda} = \frac{3.00 \times 10^8 \text{ m/s}}{413 \times 10^{-9} \text{ m}} = 7.26 \times 10^{14} \text{ Hz}$

9. **(A)** Polarization is the process by which some of the vibrations of light are blocked such that the result light vibrates in a single plane.

10. **(D)** The light regions are caused by constructive interference, whereas the dark regions are caused by destructive interference.

CHAPTER 17

Optics

Even though light consists of electromagnetic waves, the properties of light can be investigated by treating the wave as a straight line called a ray. This approach is known as ray approximation and will be the basis of your review of optics in this chapter.

Reflection

When light strikes a surface, it can pass through it, be absorbed by it, or bounce back from it. The process through which light bounces off a surface is known as **reflection**. According to the law of reflection, the angle of incidence is equal to the angle of reflection. Each angle is measured relative to the normal extending from the surface.

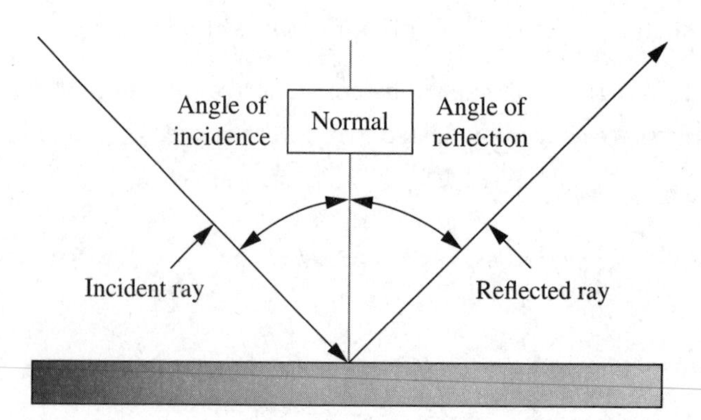

Smooth, shiny surfaces produce specular reflection, in which light is reflected in only one direction. For most surfaces, however, the surface is not perfectly flat. Instead, the surface is rough, even if it is only at the microscopic level. These types of surfaces produce diffuse reflection. Each incident ray still obeys the law of reflection but is reflected in a different direction according to the direction of the surface where the ray struck. Unless otherwise stated, questions on the SAT Physics test will assume specular reflection.

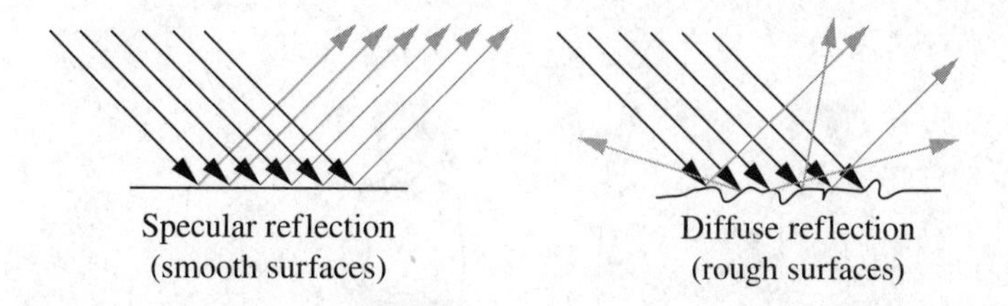

Mirrors

A mirror is an object with a reflective surface. The SAT Physics exam will address both plane mirrors and curved mirrors.

Plane Mirrors

A plane mirror is a flat mirror. When an object is placed in front of a plane mirror, light bounces off the object and is reflected by the mirror. The light rays appear to be coming from behind the mirror for an observer looking at the mirror. This type of image is a **virtual image** because it is formed by rays that appear to come together, but never actually do. A virtual image cannot be projected on a screen.

The image of the object is upright and the same size as the object. The distance between the object and the mirror d_o equals the distance between the mirror and the image d_i. The following diagram is an example of a ray diagram, which is a drawing that uses basic geometry to describe an image formed by a mirror.

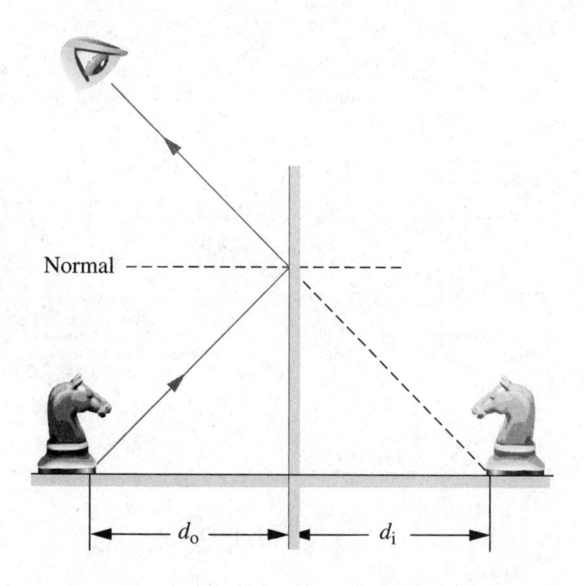

Curved Mirrors

A curved mirror appears to be part of a sphere and therefore is also known as a spherical mirror. A diverging mirror is one in which the reflective surface bulges toward the object. This type of mirror is also known as a **convex mirror**. A converging, or **concave mirror**, curves away from an object.

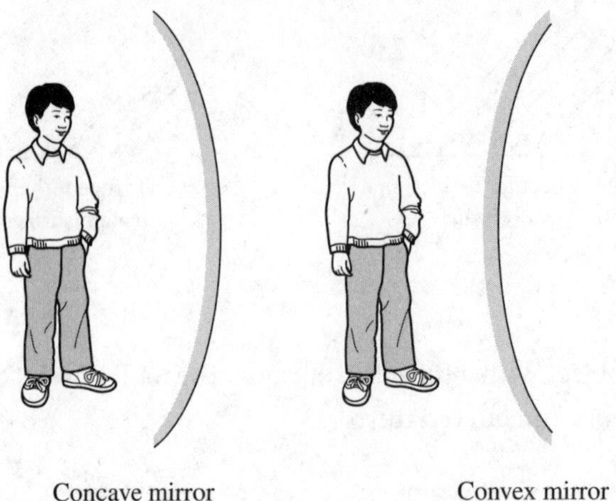

Concave mirror Convex mirror

The point that would be at the center of that sphere is known as the center of curvature C. A line that passes from the surface of the mirror through C is known as the principal axis. The point at which the principal axis meets the mirror is the geographic center of the mirror, known as the vertex. The focal point of the mirror is halfway between the vertex and the center of curvature. Parallel light rays incident on the mirror will meet at the focal point after reflection. The distance between the focal point and the mirror is the focal length. A concave mirror has a positive focal length, whereas a convex mirror has a negative focal length value because the focal point is behind the mirror.

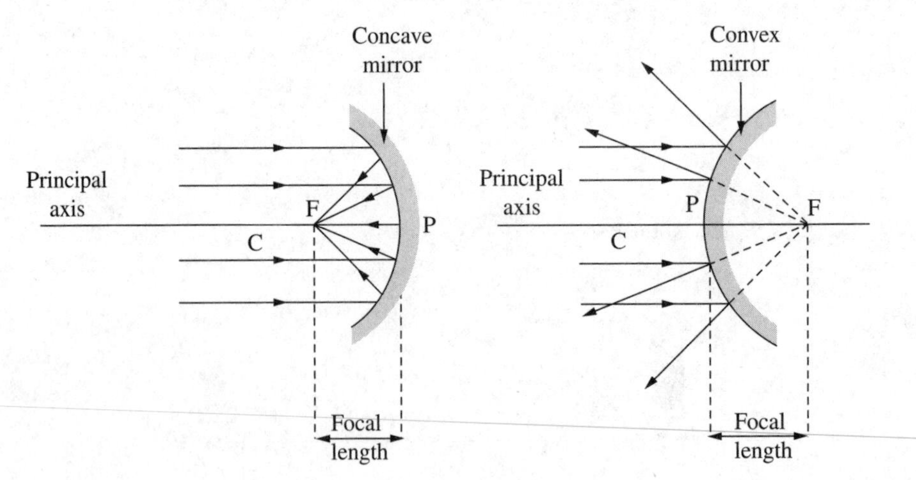

Principal focus or focus

If an object is placed in front of a concave mirror, light rays bounce off the object and are reflected from the mirror. The reflected rays converge at a point in front of the mirror to form an image. Because the light rays do meet where the image is formed, the image is known as a **real image**. If a screen were placed at that point, the image would be projected on the screen.

The location of the image can be determined from the following mirror equation, where d_o is the object distance, d_i is the image distance, and f is the focal length of the mirror.

$$\frac{1}{d_o}+\frac{1}{d_i}=\frac{1}{f}$$

The following ray diagrams show the location of an image when d_o is greater than, equal to, and less than the focal length.

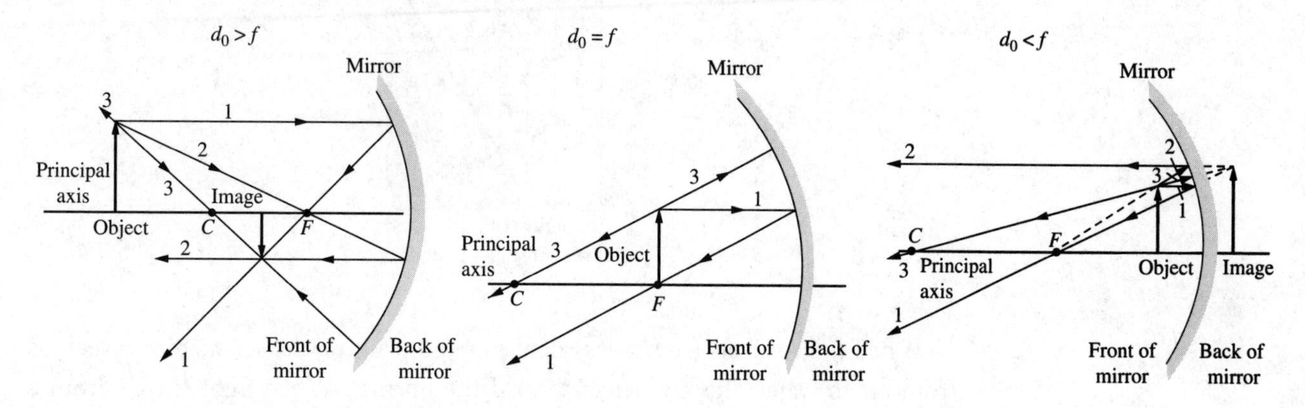

The magnification M of the image compares the size of the image with the size of the object. The magnification can be calculated according to the following equation.

$$M=\frac{h_i}{h_o}=-\frac{d_i}{d_o}$$

Note that the equation includes a negative sign. Distances are considered positive if they are in front of the mirror and negative if they are behind the mirror. If the resulting magnification is positive, the image is upright and virtual. If the resulting magnification is negative, the image is inverted and real. If the magnitude of the magnification is less than 1, the image is smaller than the object. If the magnitude of the magnification is greater than 1, the image is larger than the object.

If an object is placed in front of a convex mirror, light rays again bounce off the object and are reflected from the mirror. The mirror diverges the light rays so the image is formed where the rays appear to be coming from behind the mirror. The resulting image is, therefore, always virtual, and the image distance is always negative.

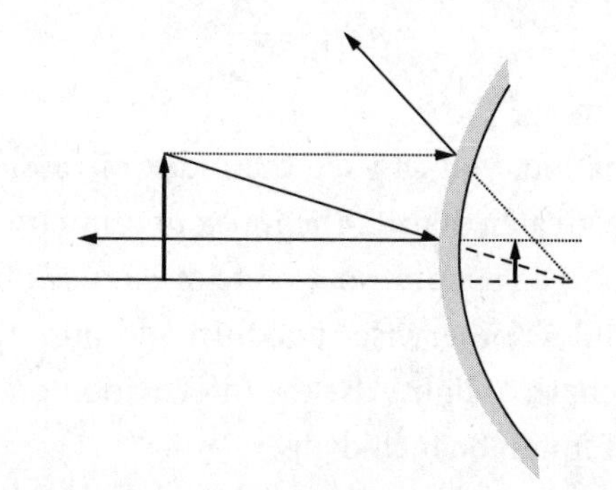

Refraction

If light passes through a medium rather than being reflected from it, the light can be **refracted**, or bent. Refraction can be explained by the fact that the speed of light changes as it travels from one medium to another. If a light ray enters straight into a medium, the entire ray speeds up or slows down together. If, however, the ray enters the medium at an angle, one part of the ray changes speed before the other. This difference causes the ray to bend.

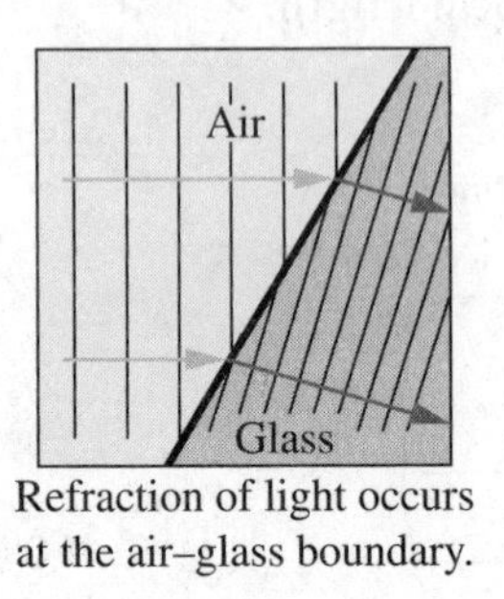

Refraction of light occurs at the air–glass boundary.

When light travels from a less dense medium into a denser medium, such as from air to water, the ray bends toward the normal. When light travels from a denser medium into a less dense medium, such as from water to air, the ray bends away from the normal.

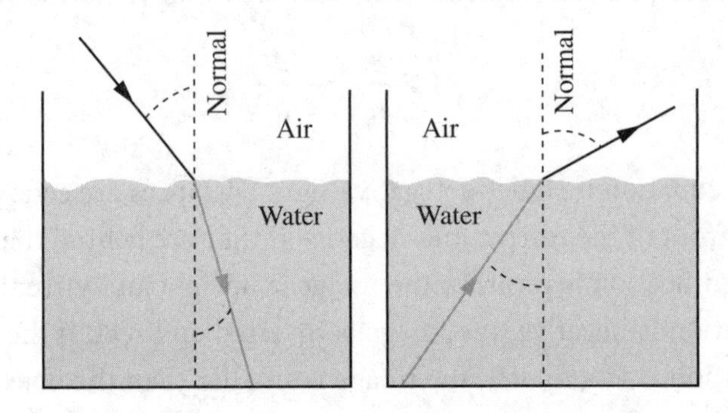

The amount of bending that occurs during refraction depends on the speed of light in the two mediums involved. The **index of refraction** n compares the speed of light in a vacuum with the speed of light in a specific medium.

$$n = \frac{c}{v_{\text{medium}}}$$

According to the equation, you can see the index of refraction is dimensionless and it must always be greater than 1. The index of refraction of light in air is only slightly different than in a vacuum, so a value of 1 is used for problems involving air in SAT Physics, unless otherwise stated. In addition, the index of refraction varies with the wavelength of light. Use the information provided with each question or any tables that might be included.

Example:

The index of refraction for yellow light passing through flint glass is 1.655. At what speed does light pass through flint glass?

$$v_{\text{glass}} = \frac{c}{n} = \frac{3.0 \times 10^8 \text{ m/s}}{1.655} = 1.8 \times 10^8 \text{ m/s}$$

Example:

A glass has a refractive index of 1.5. How does the speed of light through the glass compare to the speed of light in a vacuum?

The index of refraction in a vacuum is 1.00, so $\frac{1.0}{1.5} = 0.67$. Light travels 0.67 times the speed of light in a vacuum when traveling through the glass.

Snell's law of refraction relates the angles of incidence and refraction to the index of refraction. The relationship is summarized by the following equation, in which n_1 and n_2 are the indices of refraction for the two mediums, while θ_1 and θ_2 are the angles of incidence and refraction.

$$n_1 \sin \theta_1 = n_2 \sin \theta_2$$

Example:

A ray of light passes from air to water at an angle of 64° to the normal line. What is the angle of refraction of the light ray? [$n_1 = 1.00$ and $n_2 = 1.333$]

$$n_1 \sin \theta_1 = n_2 \sin \theta_2$$

$$1.00 \sin 64° = 1.333 \sin \theta_2$$

$$(1.00)0.899/1.333 = \sin \theta_2$$

$$\theta_2 = 42.4°$$

Total Internal Reflection

A combination of reflection and refraction can be used to explain a phenomenon known as **total internal reflection**. Consider the light shown in the following diagram with an angle of incidence equal to 0°. At this angle, most of the light passes directly across the boundary between the mediums. As the angle increases, the light experiences both reflection and refraction. You can see that the angle of refraction increases as the angle of incidence increases. The angle of refraction is 90° when the angle of incidence reaches a value known as the **critical angle**. As the angle of incidence increases beyond the critical angle, the light is no longer refracted. Instead, all of the light is reflected back into the medium from which it came.

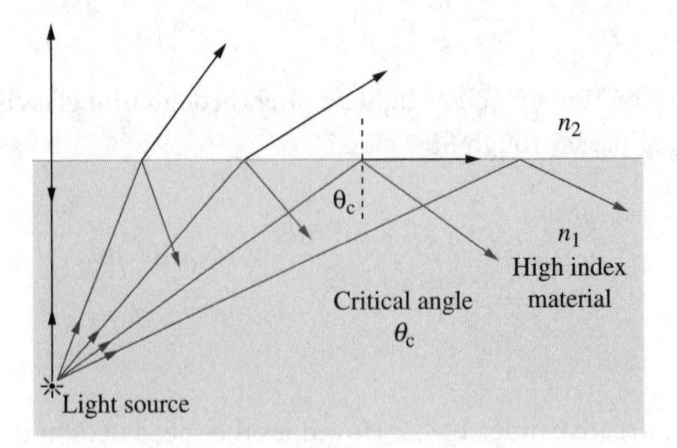

Total internal reflection occurs only when the light source is in the denser medium. This principle makes it possible to transmit light over long distances with little attenuation through optical light fibers.

Lenses

A lens is a clear piece of glass or plastic through which light can be refracted. A diverging or **convex lens** is thicker at the ends than in the middle. A converging or **concave lens** is thicker in the middle than at the ends.

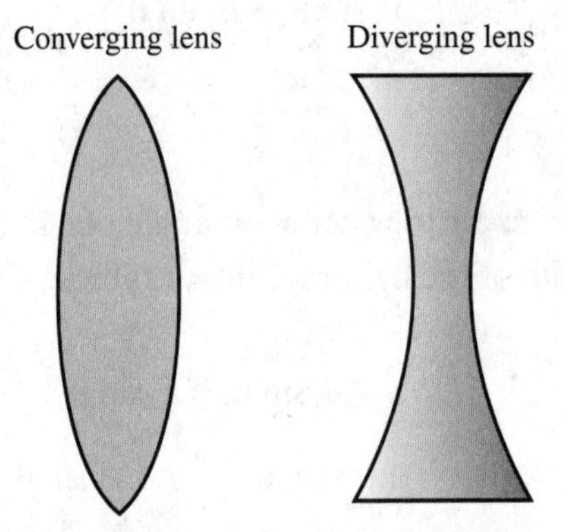

As in the case of curved mirrors, the principal axis connects the center of the lens to the point that would be the center of the sphere related to the lens. The focal point is the point at which the light rays meet after being refracted by the lens. Unlike mirrors, lenses have focal points on either side because light can pass through the lens in either direction.

In a convex lens, the light rays meet at the focal point after passing through the lens. A concave lens diverges the light rays so they do not meet after refraction. However, they can be traced back to the point where they intersect.

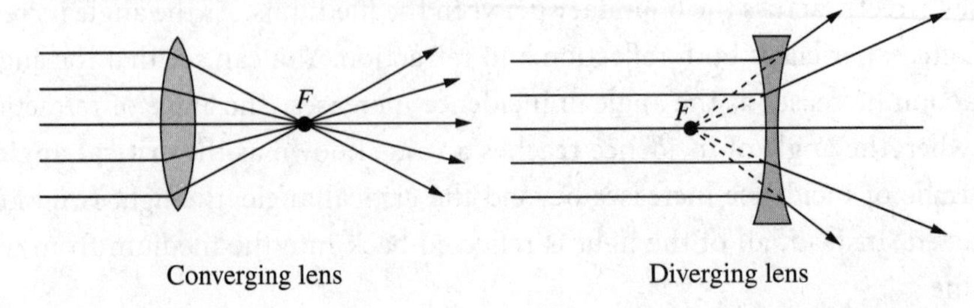

The following illustrations show the ray diagrams for a converging lens with the object at different distances from the lens. You can see that when the object is at infinity, the image is at the focal point. As the object is moved closer to the lens, but still outside $2F$, a real image is formed that is smaller than the object. The image is between F and $2F$. When the object is brought to $2F$, a real image is formed that is the same size as the object. The object distance equals the image distance. If the object is brought between $2F$ and F, a real image is formed that is larger than the object. The image is formed beyond $2F$.

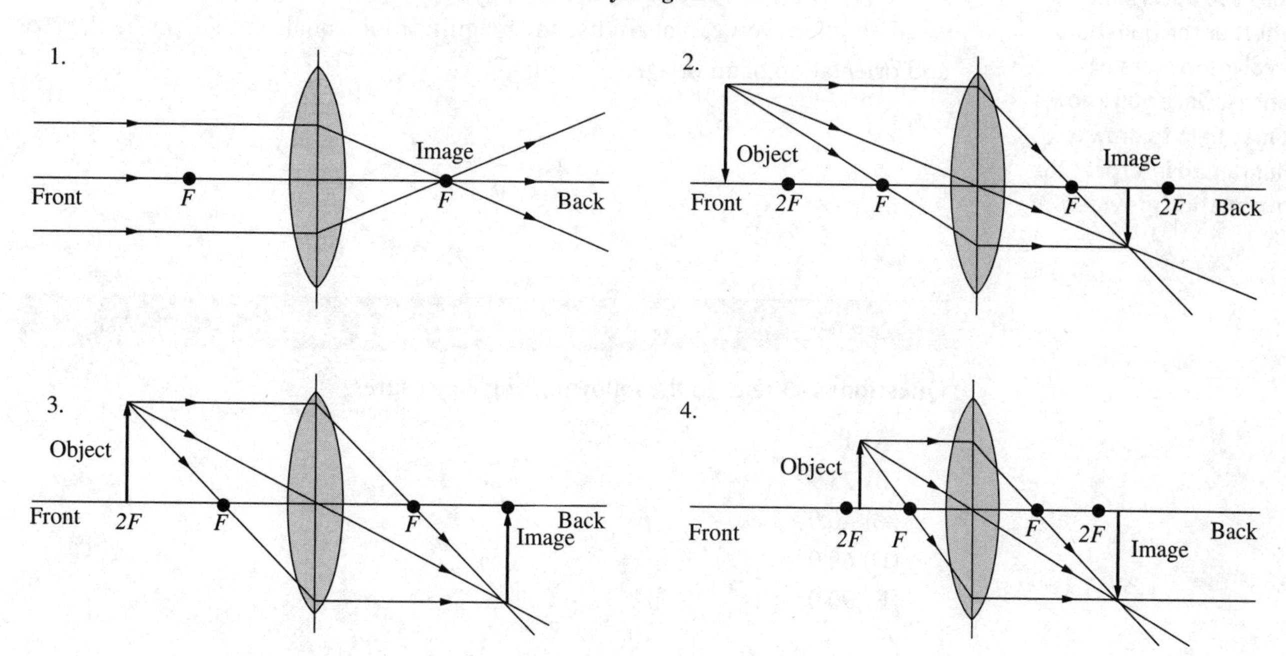

A diverging lens always produces a virtual image. The image is upright and smaller than the object. It is located inside the focal length for any position of the object.

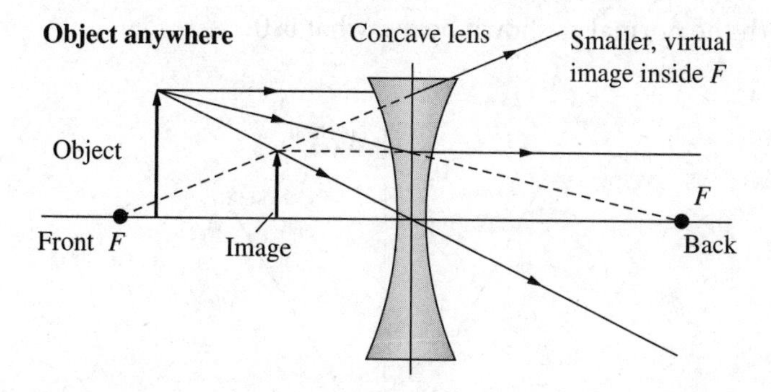

If the lens is assumed to be very thin, such that the thickness of the lens is much smaller than the focal length, the following thin-lens equation can be used to determine the location of the image. As you can see, the equation is identical to the mirror equation you used earlier.

$$\frac{1}{d_o} + \frac{1}{d_i} = \frac{1}{f}$$

As before, using the equation requires that you pay attention to sign conventions. The object and images distances are positive in front of the lens and negative behind the lens. The focal length is positive for a converging lens and negative for a diverging lens.

In addition, you can again use the magnification equation to describe the size and orientation of an image.

$$M = \frac{h_i}{h_o} = -\frac{d_i}{d_o}$$

TEST-TAKING HINT

The rules for interpreting mirrors and lenses are similar, but not identical. If you see the terms *converging*, *diverging*, *convex*, or *concave*, make sure you determine whether the question involves mirrors or lenses. Once you know, it may help to draw a diagram to interpret the information provided.

REVIEW QUESTIONS

Questions 1–3 refer to the following angle measures.

(A) 16.3°
(B) 23.9°
(C) 40.0°
(D) 60.0°
(E) 90.0°

1. A ray of light strikes a plane mirror. The angle formed by the incident ray and the reflected ray equals 80°. What is the angle of incidence?

2. Light strikes a plane mirror as shown. If the incident ray makes an angle of 30° with the normal as shown below, what is θ?

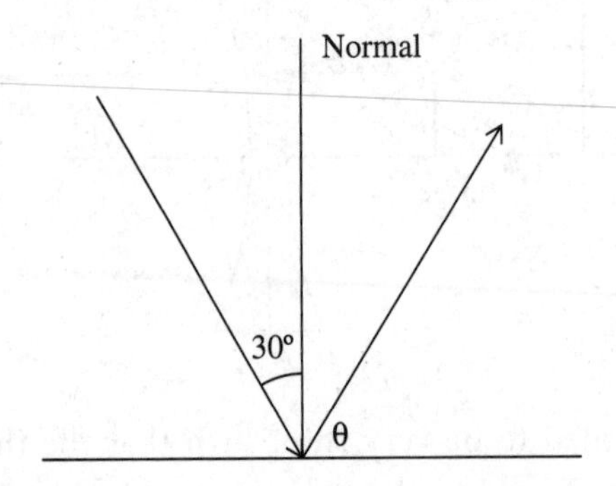

3. A ray of light in air strikes a boundary with crown glass at an angle of 38.0 degrees with the normal. What is the angle of refraction upon entering the glass? [$n_1 = 1.00$ and $n_2 = 1.52$]

4. The diagram below shows two objects *P* and *Q* in front of a plane mirror. Each object is 0.5 m tall, but *P* is twice as far from the mirror as *Q*.

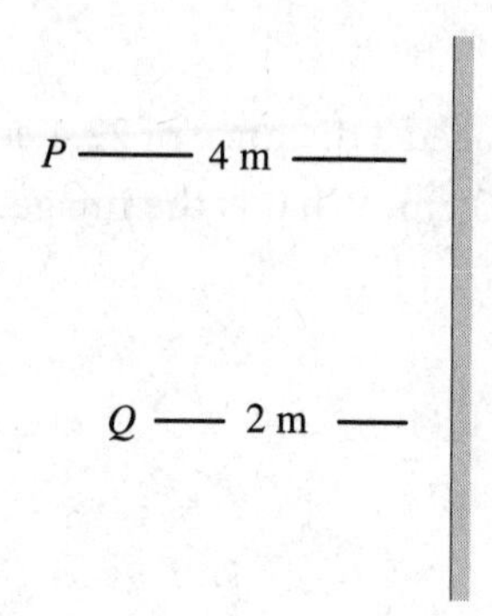

How does the size of *P*'s image compare to the size of *Q*'s image?

(A) *P*'s image is one-fourth the size of *Q*'s image.
(B) *P*'s image is half the size of *Q*'s image.
(C) *P*'s image is the same size as *Q*'s image.
(D) *P*'s image is twice the size of *Q*'s image.
(E) *P*'s image is four times the size of *Q*'s image.

5. Which characteristic do all images produced by convex mirrors have in common?

(A) They are magnified.
(B) They are at infinity.
(C) They are inverted.
(D) They are real.
(E) They are virtual.

6. The speed of light through a particular diamond is 1.24×10^8 m/s. What is the index of refraction of the diamond?

(A) 1.24
(B) 1.76
(C) 2.42
(D) 3.72
(E) 4.13

7. Total internal refraction occurs when light reaches the critical angle of 34.4° in sapphire. What is the index of refraction in this material?

(A) 0.56
(B) 1.77
(C) 2.62
(D) 4.40
(E) 5.64

8. A 6.00-cm-tall object is placed a distance of 22.4 cm from a diverging lens with a focal length of –14.3 cm. What is the image distance?

(A) –8.70 cm
(B) –1.04 cm
(C) –0.70 cm
(D) 6.99 cm
(E) 8.10 cm

Questions 9 and 10 relate to the diagram below, which shows a 5.00-cm-tall object placed 30.0 cm from a concave mirror having a focal length of 15.0 cm.

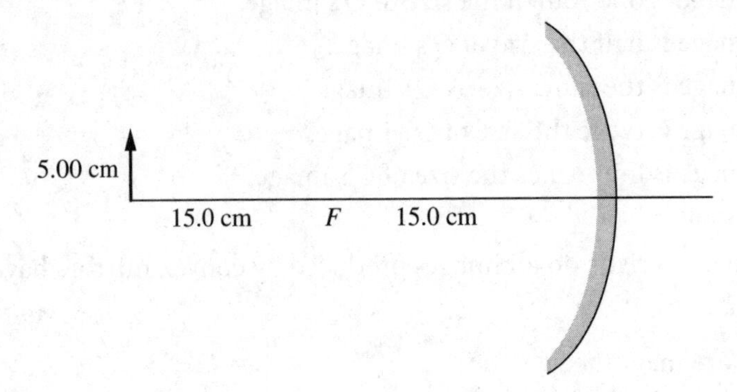

9. What is the image distance?

(A) 7.5 cm
(B) 15.0 cm
(C) 22.5 cm
(D) 30.0 cm
(E) 45.0 cm

10. The image produced by the object in the diagram is

(A) smaller than the object and erect
(B) the same size as the object and erect
(C) the same size as the object and inverted
(D) larger than the object and inverted
(E) larger than the object and erect

ANSWERS AND EXPLANATIONS

1. **(C)** The angle of incidence equals the angle of reflection. Therefore, the angle of incidence = 80° ÷ 2, which is 40°.

2. **(D)** The angle of reflection also equals 30°, so θ equals 90° – 30°, which is 60°.

3. **(B)** $n_1 \sin \theta_1 = n_2 \sin \theta_2$

$$1.00 \sin 38.0° = 1.52 \sin \theta_2$$
$$(1.00)0.616/1.52 = \sin \theta_2$$
$$\theta_2 = 23.9°$$

4. **(C)** The size of the image produced by a plane mirror is the same as the size of the object. Because both objects are the same size, the images will also be the same size.

5. **(E)** The images formed by convex mirrors are located at the point to which the reflected rays can be traced. The image is virtual, erect, and smaller than the object.

6. **(C)** The index of refraction compares the speed of light in a vacuum to the speed in the diamond.

$$n = \frac{3.0 \times 10^8 \text{ m/s}}{1.24 \times 10^8 \text{ m/s}} = 2.42$$

7. **(B)** Use Snell's law to find the solution, recognizing that the angle of refraction is 90° at the critical angle.

$$n_1 \sin \theta_1 = n_2 \sin \theta_2$$
$$n_1 \sin 34.4° = 1.00 \sin 90°$$
$$n_1 \sin 34.4° = 1$$
$$n_1 = 1.77$$

8. **(A)**

$$\frac{1}{d_o} + \frac{1}{d_i} = \frac{1}{f}$$
$$\frac{1}{22.4 \text{ cm}} + \frac{1}{d_i} = \frac{1}{-14.3 \text{ cm}}$$
$$d_i = -8.70 \text{ cm}$$

9. (D)

$$\frac{1}{d_o}+\frac{1}{d_i}=\frac{1}{f}$$

$$\frac{1}{30.0\text{ cm}}+\frac{1}{d_i}=\frac{1}{15.0\text{ cm}}$$

$$\frac{1}{30.0\text{ cm}}-\frac{1}{15.0\text{ cm}}=-\frac{1}{d_i}$$

$$d_i = 30.0\text{ cm}$$

10. (C)

$$M=\frac{h_i}{h_o}=-\frac{d_i}{d_o}$$

$$\frac{h_i}{5.0\text{ cm}}=-\frac{30.0\text{ cm}}{30.0\text{ cm}}$$

$$h_i = -5.0\text{ cm}$$

CHAPTER 18

Atomic Physics

Much of your review of physics has involved understanding phenomena above the atomic level. However, many interactions can be understood only by considering matter at the subatomic level, which involves the particles, forces, and energy within individual atoms. In this chapter, you will review the particle nature of light, atomic models, and nuclear interactions.

The Photoelectric Effect

So far you have treated light as a wave that could be approximated as a ray. In some situations, however, light must be described by a particle nature as well. The particles are individual bundles of energy called **photons**. In the 19th century, Heinrich Hertz discovered that if he shined ultraviolet light on a sample of zinc, the zinc acquired a positive charge. Shortly after, Max Planck defined the photon and related it to the frequency of light according to the following equation.

$$E = hf$$

The letter h represents a value known as Planck's constant, which equals 6.63×10^{-34} J s. Because frequency equals speed divided by wavelength, the equation can be rewritten as follows.

$$E = \frac{hc}{\lambda}$$

Now it becomes obvious that a photon's energy is directly related to its frequency but inversely related to its wavelength.

Putting together Hertz's observation and Planck's idea, Einstein was able to propose an explanation. He suggested that the zinc that Hertz observed became positively charged because it emitted electrons. The electrons were emitted because the metal absorbed photons from the light that struck it. The emitted electrons were, therefore, described as photoelectrons, and the phenomenon became known as the **photoelectric effect**.

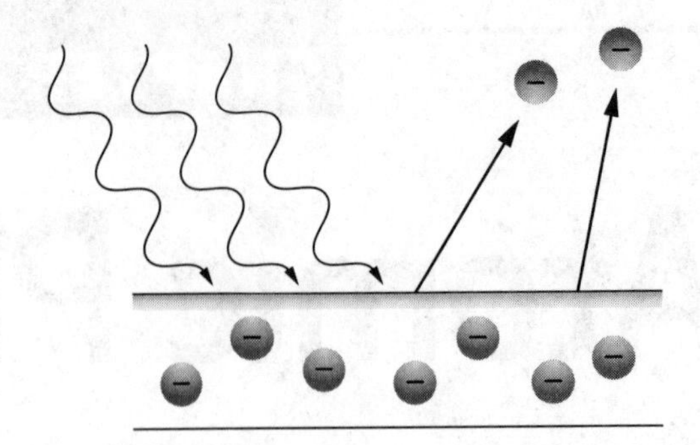

Shining visible light on the same sample of zinc does not have the same effect as shining ultraviolet light. The reason is that each metal has a minimum, or **threshold frequency** (f_0), required for electrons to be emitted. The threshold frequency is the lowest frequency, and therefore the longest wavelength, that enables photoelectrons to be ejected from the surface. The photoelectrons ejected at this frequency have no leftover kinetic energy. At any higher frequency, they have leftover kinetic energy that enables them to travel away from the surface. Einstein rewrote Planck's equation as follows.

$$E = hf = KE_{max} + \phi$$

In this equation, E is the energy supplied by the photon striking the surface of the material, KE_{max} is the maximum kinetic energy of the emitted photoelectron, and ϕ is the energy needed to remove the photoelectron from the surface (also known as the work function). The following diagram shows that red light striking the surface of potassium does not cause photoelectrons to be emitted. Both green and blue light emit photoelectrons. However, blue light provides enough leftover energy to cause the photoelectron to move at a greater speed than green light.

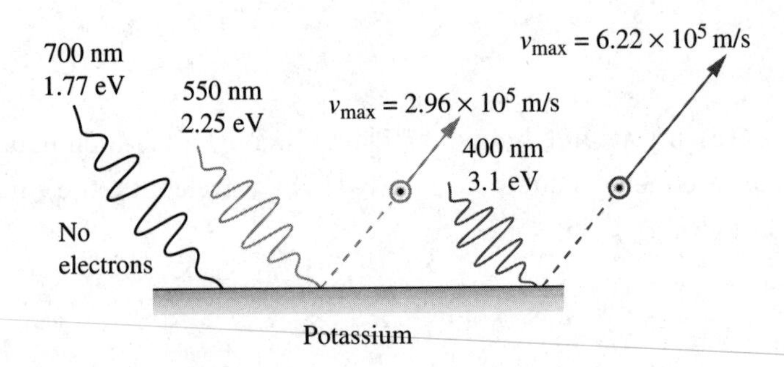

Atomic Models

An understanding of the photoelectric effect involves a useful model of the atomic structure. The earliest description of the nature of matter was provided more than 2000 years ago by the ancient Greeks. In fact, the modern term *atom* comes from the Greek word *atomos*, which means indivisible. The Greek model described matter as consisting of small particles considered to be not only indivisible, but also indestructible.

At the end of the 19th century, J. J. Thompson proved that the atom was divisible by discovering charged particles that would later become known as electrons. As a result, he suggested a model of the atom as a sphere of positive charge with negatively charged electrons embedded in it. It became known as the *plum pudding model* because it resembled a popular dessert at the time.

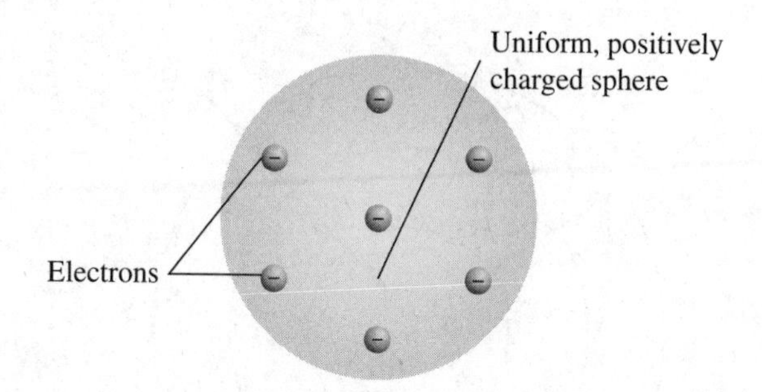

Several years later, Ernest Rutherford conducted his now-famous gold foil experiment. By aiming a stream of positive charges at a thin gold foil and observing the path of the particles upon striking the gold foil, Rutherford was able to develop a new atomic model.

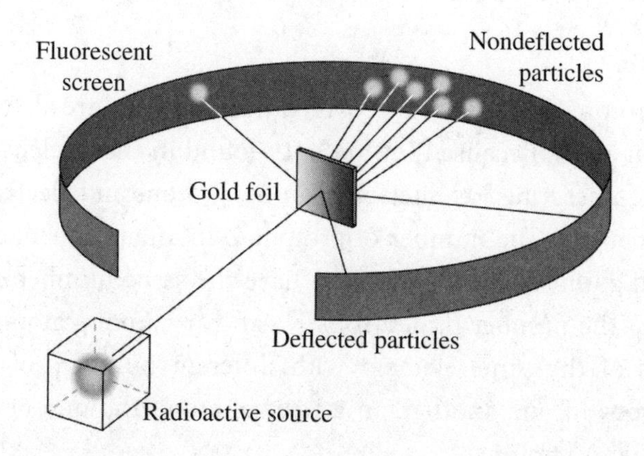

According to Rutherford's model, atoms consist of mostly empty space. Almost all of the atom's mass is concentrated in the center, or nucleus. The nucleus, which repelled the stream of particles, must have a positive charge. Positively charged protons, therefore, are located in the nucleus, whereas negative-charged electrons are located in the region around the nucleus.

The next development in the atomic model came from Niels Bohr. Using the hydrogen atom, Bohr revised the atomic model to show that electrons orbit the nucleus in definite, circular orbits. In the case of hydrogen, there is a single electron orbiting the nucleus, which has a single proton. An electron has a ground state, or orbit, with the least amount of energy with respect to the nucleus. If an atom gains energy, an electron can be raised to a higher energy level. An electron rises to a higher energy level only when it absorbs definite, quantized amounts of energy. When an electron returns to its ground state, it emits photons of energy equal to the amount that had been absorbed. This means that an electron does not emit light while in an energy level, but only when it drops from a higher energy to a lower one.

Energy levels are denoted as $n = 1$, $n = 2$, $n = 3$, and so on. Each level is associated with a different amount of energy.

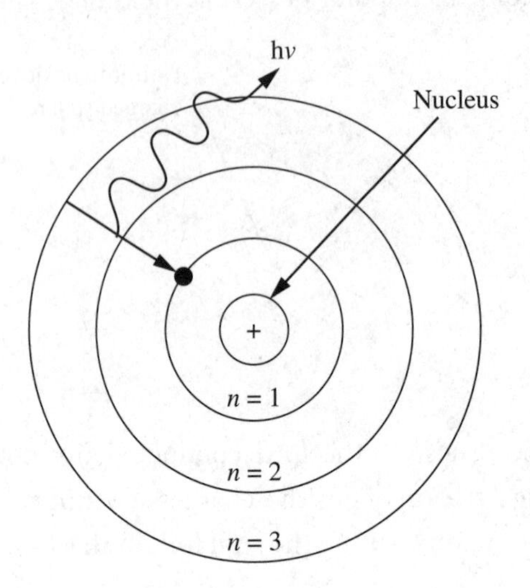

Atomic Nucleus

In addition to protons, it was later discovered that neutrons are also found within the nucleus of an atom. Because they are both found in the nucleus, protons and neutrons are often described as nucleons. Unlike protons and electrons, neutrons are electrically neutral. The number of protons Z in an atom defines the type of element it is. All atoms of a given element have the same number of protons, or **atomic number**. The number of neutrons N can vary among atoms of the same element. Atoms of the same element with different numbers of neutrons are known as **isotopes** of one another. In a neutral atom, the number of electrons equals the number of protons.

The total number of protons and neutrons is the mass number A of an atom. A symbol in the form of the one in the following diagram describes the atomic number and the mass number of an atom.

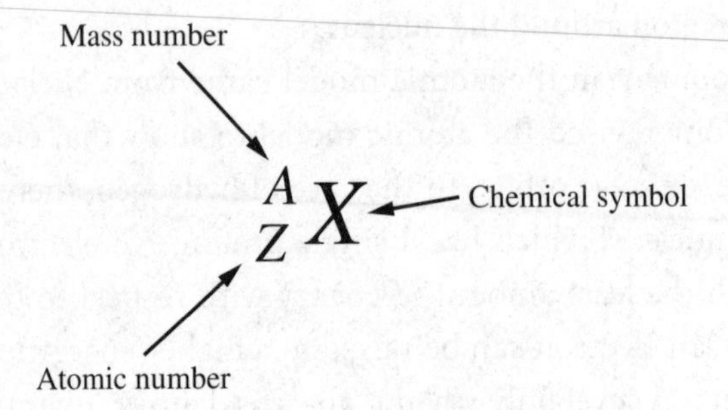

Example:

How many protons and neutrons are in a neutral atom of titanium $^{48}_{22}\text{Ti}$?

The number of protons is the lower number, which is 22. The atomic number of titanium is 22. The mass number is the top number, which is 48. The number of neutrons is $48 - 22 = 26$.

Nuclear Stability

The nucleus consists of tightly packed protons and neutrons. Recall that protons are positively charged and neutrons are electrically neutral. Knowing that positive charges repel one another, you might wonder why the positively charged protons do not repel one another and fly apart. They do indeed repel one another, but another force, known as the **strong force**, overcomes the repulsion. This force is different from other forces you have reviewed. It is extremely short range and acts only across distances of about 10^{-15} m. In addition, it is independent of electric charge. The force has the same magnitude between two protons or two neutrons at the same distance.

The relationship between the number of particles in a nucleus and the strong force determines the stability of a nucleus. In the following graph, the solid line represents nuclei for which the number of neutrons equals the number of protons. Only lighter nuclei fall on the line, whereas heavier nuclei fall above the line. What does this mean? Consider the nucleus in terms of the strong force. When the strong force balances the repulsion between protons, a nucleus is stable. The strong force is short range and acts only between particles and their nearest neighbors. The repulsive electric force, however, is long range and acts between all protons in the nucleus. For the forces to remain balanced, as the number of protons increases, the number of neutrons must also increase to add enough attractive force to maintain stability. Therefore, heavy nuclei are stable only when they have more neutrons than protons.

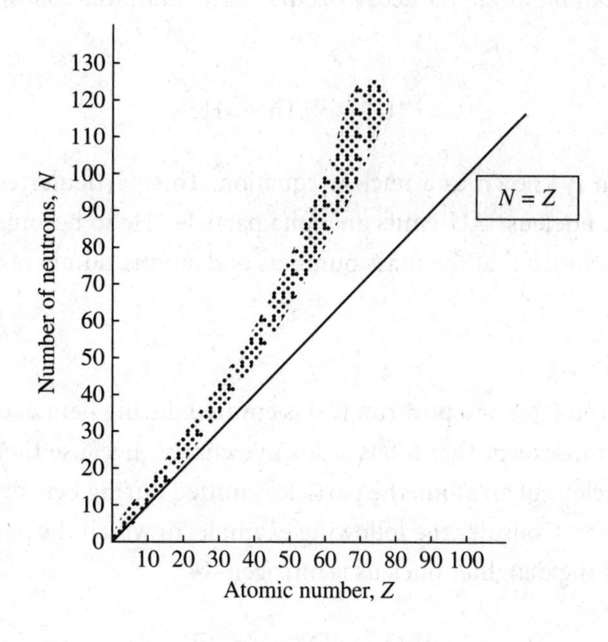

Once the number of protons exceeds 83, additional neutrons cannot compensate for the repulsive forces. As a result, atoms with atomic numbers greater than 83 do not have stable nuclei.

Binding Energy

The particles bound together in a nucleus are at a lower energy state than the particles would have individually. This difference in energy is known as the **binding energy** of the nucleus. The binding energy can be calculated using the following equation.

$$E_{\text{bind}} = \Delta mc^2$$

The quantity Δm is the difference between the mass of the unbound nucleons minus the mass of the bound nucleons. It is known as the **mass defect.**

Nuclear Decay

Nuclei that are not stable tend to break apart into other particles in a process known as **nuclear decay**. This process can occur naturally or be induced artificially. Either way, particles, photons, or both may be emitted. The nucleus that decays is known as the *parent nucleus,* and the nucleus that results is the *daughter nucleus.* The spontaneous emission of radiation is known as **radioactivity**. There are three basic types of radiation: alpha (α), beta (β), and gamma (γ).

Alpha Decay

An alpha particle (${}^4_2\text{He}$) is emitted during **alpha decay**. A nucleus loses two protons and two neutrons, so its mass decreases and its atomic number decreases. A common example of alpha decay occurs when uranium-238 breaks down as shown here.

$${}^{238}_{92}\text{U} \rightarrow {}^{234}_{90}\text{Th} + {}^{4}_{2}\text{He}$$

This expression is known as a nuclear equation. This particular equation shows that the parent nucleus ${}^{238}_{92}\text{U}$ emits an alpha particle ${}^4_2\text{He}$ to become the daughter nucleus ${}^{234}_{90}\text{Th}$. Notice that the mass numbers and atomic numbers balance across the equation.

Beta Decay

Either an electron (${}^{0}_{-1}e$) or a positron (${}^{0}_{1}e$) is emitted during **beta decay**. A positron is like an electron except that it has a positive charge. Because there are no electrons in the nucleus of an atom, the particles emitted during beta decay must have a different source. Consider the following example, in which the parent nucleus is carbon-14, and the daughter nucleus is nitrogen–14.

$${}^{14}_{6}\text{C} \rightarrow {}^{14}_{7}\text{N} + {}^{0}_{-1}e + \overline{\nu}$$

A neutron in the nucleus becomes a proton and an electron. The proton remains in the nucleus, increasing the atomic number by 1, and the electron is emitted. The mass number remains the same because the proton essentially replaced the neutron.

The symbol ${}_{-1}^{0}e$ represents an electron. Unlike the symbols for the nuclei, the symbol for the electron provides information about charge. The −1 indicates that the charge is negative with a magnitude equal to that of the proton. The 0 indicates that the electron does not contribute to the mass number. The type of beta decay in which an electron is emitted is known as beta-minus decay.

The $\bar{\nu}$ represents a particle called an antineutrino. Experimental evidence showed that a new type of particle must be produced in order to conserve energy and momentum. The basic particle, the neutrino ν, has no electric charge and almost no (if any) mass. The antiparticle of the neutrino is the antineutrino.

The following nuclear equation shows another type of beta decay. In this decay, known as beta-plus decay, a positron and a neutrino are emitted.

$${}_{6}^{10}\mathrm{C} \rightarrow {}_{5}^{10}\mathrm{B} + {}_{1}^{0}e + \nu$$

In this process, a proton becomes a positron and a neutron. The positron is emitted along with a neutrino. The atomic number decreases by 1 because a proton is lost, but the mass number remains the same because a neutron is gained. The symbol ${}_{1}^{0}e$ represents a positron, which resembles an electron with a positive charge. The symbol ν represents a neutrino. Beta-plus decay is typically not tested on the SAT Physics exam.

Gamma Decay

When one or more nucleons in an unstable nucleus transition from a higher energy state to a lower one, photons known as gamma rays are emitted. During this type of decay, known as **gamma decay**, energy is released but the atomic number and mass number are not changed.

Gamma decay is often associated with alpha and beta decay. During these processes, the nucleus is left in an excited state, which then returns to normal through gamma decay.

Half-Life

One way to measure the rate of radioactive decay is by identifying half-life. The **half-life** of a substance is the time it takes for half of the radioactive nuclei in a sample to decay. You may often see a decay curve to represent the number of radioactive nuclei remaining in a sample as a function of time. The decay curve below represents the amount of a sample remaining after several half-lives of iodine-131.

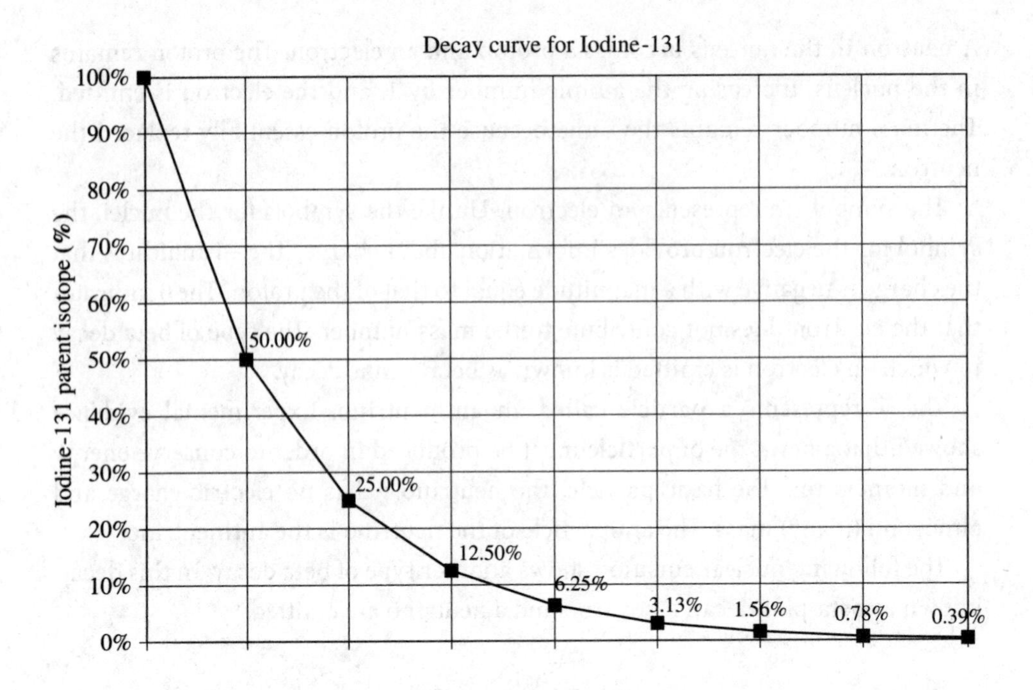

The half-life of iodine-131 is about 8 days. You can see that half of a sample remains after 8 days (1 half-life). Another half of that, or 25%, remains after 16 days (2 half-lives). Half of that, or 12.50%, remains after 24 days (3 half-lives). The amount of radioactive sample remaining continues to decrease exponentially over time.

Half-lives are useful for identifying substances because they are generally unique. In addition, measuring amounts of substances of known half-lives, such as carbon-14, can be used to determine the ages of objects, such as fossils.

Nuclear Reactions

In chemical reactions, only the electrons are involved. As a result, the identity of an element does not change. In nuclear reactions, the composition of the nucleus changes. When the atomic number changes, the identity of the element changes. Two types of nuclear reactions are fusion and fission.

Fusion

Two smaller nuclei combine to form a larger one during **nuclear fusion**. Energy is released during this process. The following example shows fusion between two forms of hydrogen, deuterium and tritium. Helium is formed along with the release of a neutron and energy.

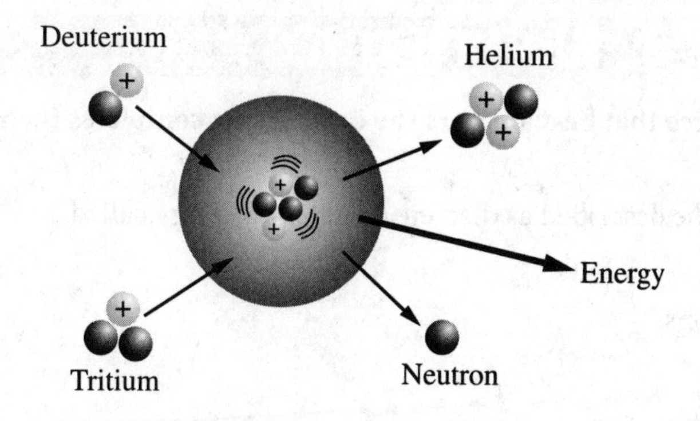

$${}^{3}_{1}H + {}^{2}_{1}H \rightarrow {}^{4}_{2}He + {}^{1}_{0}n$$

Fission

One large nucleus splits to form smaller nuclei during nuclear fission. To initiate fission, a large nucleus is bombarded with a slow neutron. The large nucleus then splits into smaller nuclei as additional neutrons and energy are released. The following example shows the one fission reaction of uranium-235.

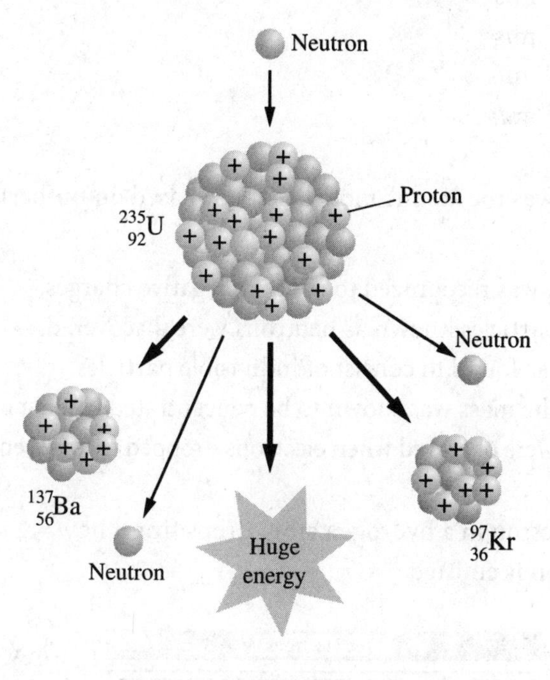

$${}^{235}_{92}U + {}^{1}_{0}n \rightarrow {}^{142}_{56}Ba + {}^{91}_{36}Kr + 3\,{}^{1}_{0}n$$

TEST-TAKING HINT

Practice balancing nuclear equations as you do for chemical equations. This will help you identify missing particles if incomplete nuclear equations are provided to you.

REVIEW QUESTIONS

Select the choice that best answers the question or completes the statement.

1. Light can be described as discrete bundles of energy called

   (A) nuclei
   (B) neutrons
   (C) waves
   (D) protons
   (E) photons

2. Light with a wavelength of 1.20×10^{15} Hz is directed at a sample of pure silver. If the work function is 4.73 eV, what will be the speed of the photoelectrons that are emitted? [$h = 4.14 \times 10^{-15}$ eV · s and the mass of an electron is 9.11×10^{-31} kg]

   (A) 3.81×10^{-20} m/s
   (B) 2.89×10^{5} m/s
   (C) 2.89×10^{5} m/s
   (D) 2.54×10^{14} m/s
   (E) 5.68×10^{15} m/s

3. In what way was the atomic model changed based on Rutherford's gold foil experiment?

   (A) The atom was recognized to contain negative charges.
   (B) Neutral particles known as neutrons were discovered.
   (C) Matter was found to consist of indivisible particles.
   (D) Most of the mass was shown to be concentrated in the nucleus.
   (E) Photons were observed when electrons dropped to lower energy levels.

4. When an electron in a hydrogen atom drops from the $n = 2$ state to the $n = 1$ state, a photon is emitted.

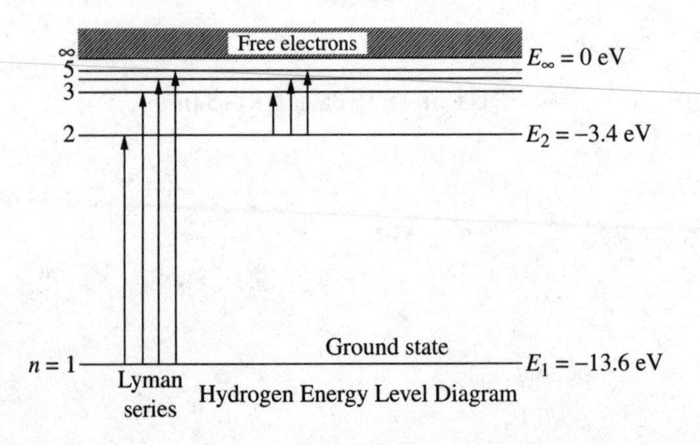

Hydrogen Energy Level Diagram

What is the wavelength of the photon? [6.63×10^{-34} J · s and 1 eV = 1.60×10^{-19} J]

(A) 41.4 nm
(B) 122 nm
(C) 163 nm
(D) 199 nm
(E) 406 nm

5. How many neutrons are in an atom of $^{18}_{8}O$?

(A) 8
(B) 10
(C) 16
(D) 18
(E) 26

6. Which of the following always occurs when a hydrogen atom undergoes nuclear fusion?

(A) It shares an electron with another atom.
(B) It transfers its electron to another atom.
(C) It emits a positron.
(D) It absorbs a neutron.
(E) It combines with another light nucleus.

7. A substance undergoes alpha decay according to the following equation.

$$\_\_\_\_\_ \rightarrow {}^{20}_{11}Na + {}^{4}_{2}He$$

What is the missing substance?

(A) $^{24}_{13}Al$
(B) $^{28}_{14}Si$
(C) $^{21}_{11}Ne$
(D) $^{24}_{12}Mg$
(E) $^{24}_{12}Na$

8. A scientist places 200 g of a radioactive substance in a container at 3:00 P.M. At 9:00 P.M. the next night, only 25 g of the sample remain. What is the half-life of the substance?

(A) 10 hours
(B) 12 hours
(C) 15 hours
(D) 18 hours
(E) 30 hours

Questions 9 and 10 relate to the diagram below, which represents the decay curve for carbon-14.

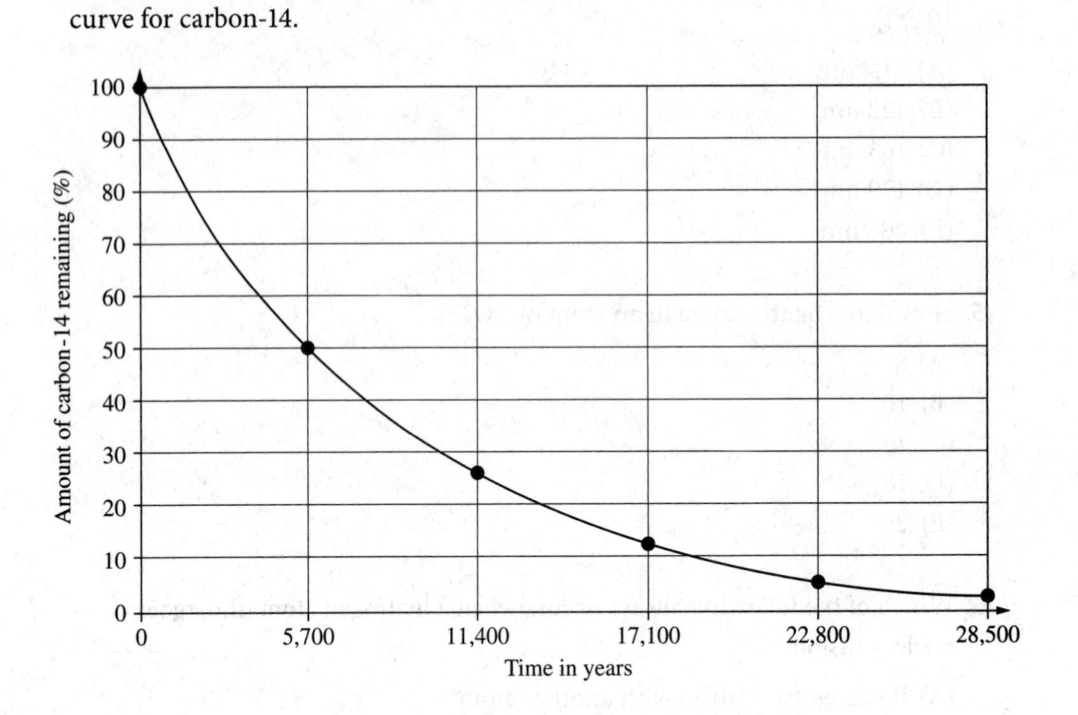

9. Based on the graph, what is the approximate half-life of carbon-14?

(A) 500 years
(B) 2,850 years
(C) 5,700 years
(D) 11,400 years
(E) 28,500 years

10. How much of a 500.0-g sample will remain in 22,800 years?

(A) 250.0 g
(B) 125.0 g
(C) 65.20 g
(D) 31.25 g
(E) 15.60 g

ANSWERS AND EXPLANATIONS

1. **(E)** According to the particle theory of light, light consists of discrete bundles of energy called photons.

2. **(B)** Rearrange Einstein's equation $hf = KE_{max} + \phi$ to solve for the maximum kinetic energy: $KE_{max} = hf - W = (4.14 \times 10^{-15}$ eV.s$)(1.20 \times 10^{15}$ Hz$) -$ 4.73 eV $= 0.238$ eV $= 3.81 \times 10^{-20}$ J. Then use the equation for kinetic energy to solve for speed:

$$KE_{max} = \frac{1}{2}mv^2, \text{ so } v = \sqrt{\frac{2KE}{m}}$$
$$= \sqrt{\frac{2(3.81 \times 10^{-20}\text{ J})}{9.11 \times 10^{-31}\text{ kg}}} = 2.89 \times 10^5 \text{ m/s.}$$

3. **(D)** Rutherford's experiment involved aiming a stream of alpha particles at gold atoms. Most of the particles went directly through the empty space of the atoms, but some were deflected by the positively charged nucleus. The nucleus was defined as containing most of the mass of the atom, while negative charged electrons orbited around it.

4. **(B)** When the electron makes the transition, the electron loses 10.2 eV of energy. Convert this energy to joules: 10.2 eV $\times 1.60 \times 10^{-19}$ J/eV $= 1.632 \times 10^{-18}$ J. Then use the energy to calculate the wavelength:

$$E = hf = \frac{hc}{\lambda} \text{ so } \lambda = \frac{hc}{E} = \frac{(6.63 \times 10^{-34}\text{ J})(3.00 \times 10^8\text{ m/s})}{(1.632 \times 10^{-18}\text{ J})} = 122 \text{ nm.}$$

5. **(B)** In a nuclear symbol, the top number is the mass number, or total number of protons and neutrons. The bottom number is the number of neutrons. Therefore, the number of neutrons is the difference between the top and bottom numbers.

6. **(E)** Nuclear fusion occurs when light nuclei combine to form a heavier nucleus.

7. **(A)** The mass number should be the sum of the mass numbers of the particles produced. The atomic number should be the sum of the atomic numbers of the particles produced. Mass number = 20 + 4 and atomic number = 11 + 2.

8. **(A)** If 200 g have decayed such that only 25 g remain, three half-lives have passed. The time required for this change is 30 hours, so the half-life must be 10 hours.

9. **(C)** The half-life occurs when half, or 50%, of the original sample remains. This occurs at 5700 years on the graph. The actual half-life is 5730 years.

10. **(D)** Four half-lives pass in 22,800 years. At four half-lives, 6.25% of the original sample remains. 6.25% × 500 g = 31.25 g.

CHAPTER 19

Special Relativity

For everyday observations, Newton's laws of motion provide good approximations for describing and predicting motion. If the speed of a particle approaches the speed of light, however, a different approach is required to interpret motion. In this chapter, you will review Einstein's approach to describing motion.

Einstein's Special Theory of Relativity

Objects traveling at speeds that are significant fractions of the speed of light are known as relativistic objects. Einstein proposed his **special theory of relativity** to describe the motion of relativistic objects. His theory results from two basic postulates:

1. The speed of light is constant in all reference frames, despite any relative motion between an observer and the light source.
2. The laws of physics are the same in all inertial reference frames.

First Postulate

The first postulate can be a little difficult to accept at first. After all, if a car drives away from a parked car at 50 km/h, you describe the speed of the car accordingly. If, instead, two cars are moving in opposite directions at 50 km/h, each one appears to be moving away from the other at 100 km/h. The speed of each car relative to the other depends on the motion of both cars.

What the first postulate says is that relative motion does not affect the speed of light. If you turn on a light toward an observer, the observer will measure the speed of light c as 3.0×10^8 m/s. If the observer approaches you at a high rate of speed, the observer would still measure the same speed of light. If you move toward the observer, the observer would again measure the same speed of light.

Second Postulate

The second postulate is a lot easier to accept because it is a basic rule of science. An inertial reference frame is one that is not accelerating. In other words, it is at rest or moving at a constant speed in a straight line. In this type of reference frame, the same laws describe motion. For example, a force will have the same effect on an object regardless of what caused the force and where the object is.

As you review Einstein's special theory of relativity, keep two points in mind. The first is that Einstein's theory did not replace Newton's laws of motion. Newton's laws continue to serve as a particular case of Einstein's theory, which is motion at speeds that do not approach the speed of light. The second is that Einstein's law of general relativity is different from special relativity, and it involves the gravitational force.

Length Contraction

An interesting result of special relativity is that although observers moving relative to one another must agree on the speed of light, they do not have to agree on measurements of length. An observer at rest relative to a moving object traveling at relativistic speeds would observe the length of the object to be shorter than it would be at rest relative to the observer. This phenomenon is known as **length contraction**, and the amount of contraction depends on the object's speed relative to the observer. The faster the speed is, the greater the contraction is. At normal everyday speeds, the effect is very small. Near the speed of light, however, the contraction becomes more pronounced.

Spaceship moving at 10% of the speed of light

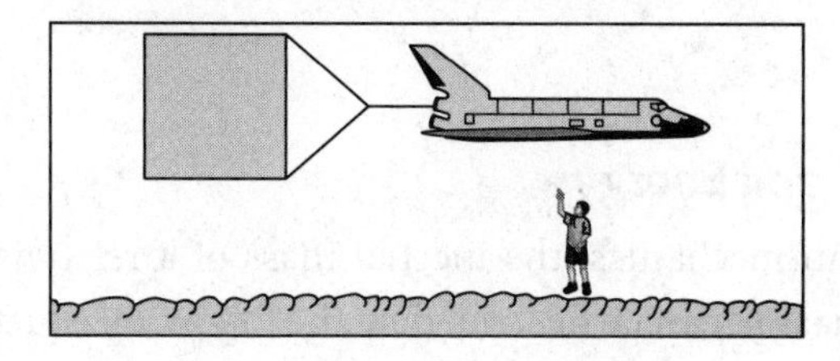

Spaceship moving at 86.5% of the speed of light

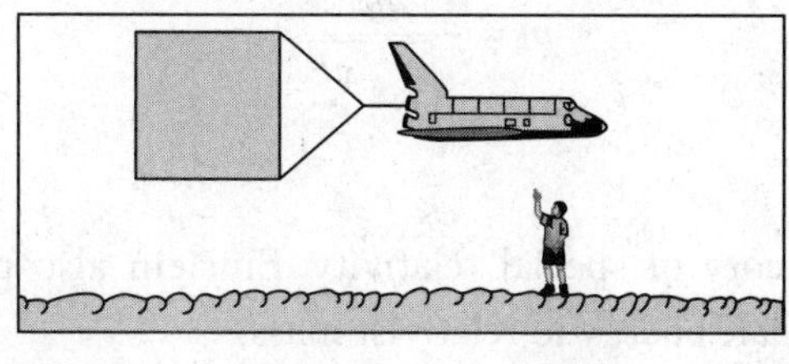

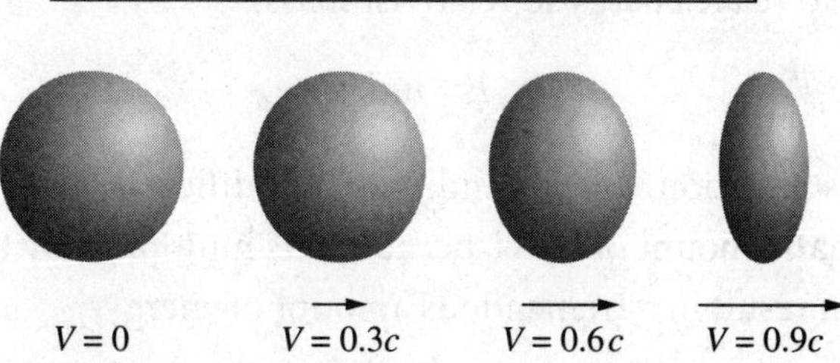

Consider a rod of length L in a stationary frame of reference and a moving frame of reference. The observer will measure the length of the rod in the moving frame of reference L_0 as shorter than in the stationary frame of reference. The difference can be found according to the following equation.

$$L = L_0\sqrt{1 - \frac{v^2}{c^2}}$$

Be aware that the dilation is only in the direction of motion. If, for example, motion is from left to right, length contraction occurs in this direction only. It does not occur in the vertical direction.

Time Dilation

Just as length is contracted, time is stretched out under relativistic conditions. In other words, moving clocks run slow. This phenomenon is known as **time dilation.**

If the length between two ticks of a clock is measured, an observer moving relative to the clock will measure the time as longer than an observer who is stationary when compared with the clock. The relationship between time in a moving frame of reference T to time in a stationary frame of reference T_0 is described by the following equation.

$$T = \frac{T_0}{\sqrt{1 - \frac{v^2}{c^2}}}$$

TEST-TAKING HINT

Plan out your study schedule so that you have enough time to review all the concepts and study any topics with which you have difficulty. If you leave too much until the last minute, you may become frustrated and confused by more complex topics such as relativity.

Relativistic Mass and Energy

Unlike in Newtonian mechanics, the inertial mass of a relativistic object varies with velocity. The mass m can be described by the following equation, where m_0 is the mass in a stationary frame of reference.

$$m = \frac{m_0}{\sqrt{1 - \frac{v^2}{c^2}}}$$

As part of this theory of special relativity, Einstein also presented his now famous equation to relate energy to relativist mass.

$$E = mc^2$$

According to this equation, energy and mass are different manifestations of the same thing. A small amount of mass, because it is multiplied by the square of the speed of light, can result in a tremendous amount of energy.

REVIEW QUESTIONS

Questions 1–5 refer to the following ideas. Which does each scenario best exemplify?

(A) Time dilation
(B) Length contraction
(C) Mass-energy equivalence
(D) Newton's first law
(E) Newton's second law

1. A person applies 4.25 N of force to kick a 0.75-kg ball, which rolls 10 yards. The same force is applied to a 2-kg ball, which rolls only 1 yard.

2. In an argon atom undergoing gamma decay, the total number of nucleons does not change, but gamma radiations are produced.

3. Astronauts on the International Space Station age more slowly than do people on Earth.

4. When the driver of a car going 55 m/h slams on the brakes, she is thrown forward against the steering wheel.

5. To an observer on the ground, a 2-kg meteor moving at 11 km/s appears larger than a 2-kg meteor moving at 72 km/s.

6. Which of the following is a postulate of Einstein's special theory of relativity?

(A) The acceleration of an object is directly proportional to the force exerted on it.
(B) The speed of light is constant in all reference frames.
(C) An object's inertia is a measure of its mass.
(D) Time is constant for all observers regardless of relative motion.
(E) Objects do not have mass at the speed of light.

7. A rod is moving at a velocity of $0.6c$. How does its length in a stationary frame of reference L_0 compare to its length in a moving frame of reference L?

(A) $L = 8L_0$
(B) $L = 6L_0$
(C) $L = 0.2L_0$
(D) $L = 0.6L_0$
(E) $L = 0.8L_0$

8. What is the length of a meter stick if it is moving with a velocity of $0.8c$ relative to Earth?

(A) 0.36 m
(B) 0.48 m
(C) 0.60 m
(D) 0.80 m
(E) 1.2 m

9. A muon at rest has a lifetime of about 2.2 microseconds. How long is the lifetime of a muon traveling at $0.995c$?

(A) 2.3 μs
(B) 10 μs
(C) 22 μs
(D) 115 μs
(E) 220 μs

10. What is the rest energy of a 5.0-kg rock?

(A) 1.5×10^{9} J
(B) 6.0×10^{16} J
(C) 9.0×10^{16} J
(D) 6.5×10^{17} J
(E) 4.5×10^{17} J

ANSWERS AND EXPLANATIONS

1. **(E)** Newton's second law states that the rate of change of momentum of a body is directly proportional to the force applied, so a larger mass moves a shorter distance when the same amount of force is applied.

2. **(C)** Gamma decay is the emission of a particle of electromagnetic radiation from an electron, which releases energy as photons. $E = mc^2$ can be used to calculate the amount of energy.

3. **(A)** This is an example of time dilation because a person on Earth and a person on the International Space Station are moving at different speeds relative to one another.

4. **(D)** This is an example of Newton's first law, the law of inertia. The driver's body is still moving at 55 m/h when the car stops abruptly.

5. **(B)** This is an example of length contraction. A faster object will appear smaller to a stationary observer.

6. **(B)** Only choice B is correct. The other statements do not describe relativistic motion.

7. **(E)** Substitute the given information into the equation for length dilation.

$$L = L_o\sqrt{1 - \frac{(0.6c)^2}{c^2}} = 0.8L_o$$

8. **(C)** Substitute the given velocity and the length of a meter stick into the equation for length dilation:

$$L = (1\text{ m})\sqrt{1 - \frac{(0.8c)^2}{c^2}} = 0.6\text{ m}$$

9. **(C)**

$$T = \frac{T_o}{\sqrt{1 - \frac{v^2}{c^2}}} = \frac{T_o}{\sqrt{1 - \frac{(0.995c)^2}{c^2}}} = 10T_o = 22\ \mu\text{s}$$

10. **(E)**

$$E = mc^2 = (5.0\text{ kg})(3.0 \times 10^8\text{ m/s}^2) = 4.5 \times 10^{17}\text{ J}$$

PART IV

Practice Tests

Practice Test 1

Treat this practice test as the actual test, and complete it in one 60-minute sitting. Use the following answer sheet to fill in your multiple-choice answers. Once you have completed the practice test:

1. Check your answers using the Answer Key.
2. Review the Answers and Explanations.
3. Complete the Score Sheet on page 290 and see how well you did.

Answer Sheet

Tear out this answer sheet and use it to complete the practice test. Determine the BEST answer for each question. Then fill in the appropriate oval.

1. Ⓐ Ⓑ Ⓒ Ⓓ Ⓔ	21. Ⓐ Ⓑ Ⓒ Ⓓ Ⓔ	41. Ⓐ Ⓑ Ⓒ Ⓓ Ⓔ	61. Ⓐ Ⓑ Ⓒ Ⓓ Ⓔ
2. Ⓐ Ⓑ Ⓒ Ⓓ Ⓔ	22. Ⓐ Ⓑ Ⓒ Ⓓ Ⓔ	42. Ⓐ Ⓑ Ⓒ Ⓓ Ⓔ	62. Ⓐ Ⓑ Ⓒ Ⓓ Ⓔ
3. Ⓐ Ⓑ Ⓒ Ⓓ Ⓔ	23. Ⓐ Ⓑ Ⓒ Ⓓ Ⓔ	43. Ⓐ Ⓑ Ⓒ Ⓓ Ⓔ	63. Ⓐ Ⓑ Ⓒ Ⓓ Ⓔ
4. Ⓐ Ⓑ Ⓒ Ⓓ Ⓔ	24. Ⓐ Ⓑ Ⓒ Ⓓ Ⓔ	44. Ⓐ Ⓑ Ⓒ Ⓓ Ⓔ	64. Ⓐ Ⓑ Ⓒ Ⓓ Ⓔ
5. Ⓐ Ⓑ Ⓒ Ⓓ Ⓔ	25. Ⓐ Ⓑ Ⓒ Ⓓ Ⓔ	45. Ⓐ Ⓑ Ⓒ Ⓓ Ⓔ	65. Ⓐ Ⓑ Ⓒ Ⓓ Ⓔ
6. Ⓐ Ⓑ Ⓒ Ⓓ Ⓔ	26. Ⓐ Ⓑ Ⓒ Ⓓ Ⓔ	46. Ⓐ Ⓑ Ⓒ Ⓓ Ⓔ	66. Ⓐ Ⓑ Ⓒ Ⓓ Ⓔ
7. Ⓐ Ⓑ Ⓒ Ⓓ Ⓔ	27. Ⓐ Ⓑ Ⓒ Ⓓ Ⓔ	47. Ⓐ Ⓑ Ⓒ Ⓓ Ⓔ	67. Ⓐ Ⓑ Ⓒ Ⓓ Ⓔ
8. Ⓐ Ⓑ Ⓒ Ⓓ Ⓔ	28. Ⓐ Ⓑ Ⓒ Ⓓ Ⓔ	48. Ⓐ Ⓑ Ⓒ Ⓓ Ⓔ	68. Ⓐ Ⓑ Ⓒ Ⓓ Ⓔ
9. Ⓐ Ⓑ Ⓒ Ⓓ Ⓔ	29. Ⓐ Ⓑ Ⓒ Ⓓ Ⓔ	49. Ⓐ Ⓑ Ⓒ Ⓓ Ⓔ	69. Ⓐ Ⓑ Ⓒ Ⓓ Ⓔ
10. Ⓐ Ⓑ Ⓒ Ⓓ Ⓔ	30. Ⓐ Ⓑ Ⓒ Ⓓ Ⓔ	50. Ⓐ Ⓑ Ⓒ Ⓓ Ⓔ	70. Ⓐ Ⓑ Ⓒ Ⓓ Ⓔ
11. Ⓐ Ⓑ Ⓒ Ⓓ Ⓔ	31. Ⓐ Ⓑ Ⓒ Ⓓ Ⓔ	51. Ⓐ Ⓑ Ⓒ Ⓓ Ⓔ	71. Ⓐ Ⓑ Ⓒ Ⓓ Ⓔ
12. Ⓐ Ⓑ Ⓒ Ⓓ Ⓔ	32. Ⓐ Ⓑ Ⓒ Ⓓ Ⓔ	52. Ⓐ Ⓑ Ⓒ Ⓓ Ⓔ	72. Ⓐ Ⓑ Ⓒ Ⓓ Ⓔ
13. Ⓐ Ⓑ Ⓒ Ⓓ Ⓔ	33. Ⓐ Ⓑ Ⓒ Ⓓ Ⓔ	53. Ⓐ Ⓑ Ⓒ Ⓓ Ⓔ	73. Ⓐ Ⓑ Ⓒ Ⓓ Ⓔ
14. Ⓐ Ⓑ Ⓒ Ⓓ Ⓔ	34. Ⓐ Ⓑ Ⓒ Ⓓ Ⓔ	54. Ⓐ Ⓑ Ⓒ Ⓓ Ⓔ	74. Ⓐ Ⓑ Ⓒ Ⓓ Ⓔ
15. Ⓐ Ⓑ Ⓒ Ⓓ Ⓔ	35. Ⓐ Ⓑ Ⓒ Ⓓ Ⓔ	55. Ⓐ Ⓑ Ⓒ Ⓓ Ⓔ	75. Ⓐ Ⓑ Ⓒ Ⓓ Ⓔ
16. Ⓐ Ⓑ Ⓒ Ⓓ Ⓔ	36. Ⓐ Ⓑ Ⓒ Ⓓ Ⓔ	56. Ⓐ Ⓑ Ⓒ Ⓓ Ⓔ	
17. Ⓐ Ⓑ Ⓒ Ⓓ Ⓔ	37. Ⓐ Ⓑ Ⓒ Ⓓ Ⓔ	57. Ⓐ Ⓑ Ⓒ Ⓓ Ⓔ	
18. Ⓐ Ⓑ Ⓒ Ⓓ Ⓔ	38. Ⓐ Ⓑ Ⓒ Ⓓ Ⓔ	58. Ⓐ Ⓑ Ⓒ Ⓓ Ⓔ	
19. Ⓐ Ⓑ Ⓒ Ⓓ Ⓔ	39. Ⓐ Ⓑ Ⓒ Ⓓ Ⓔ	59. Ⓐ Ⓑ Ⓒ Ⓓ Ⓔ	
20. Ⓐ Ⓑ Ⓒ Ⓓ Ⓔ	40. Ⓐ Ⓑ Ⓒ Ⓓ Ⓔ	60. Ⓐ Ⓑ Ⓒ Ⓓ Ⓔ	

PART A

Directions: Each set of lettered choices refers to the numbered questions of statements immediately following it. Select the one lettered choice that best answers each question or best fits each statement, and then fill in the corresponding oval on the answer sheet. A choice may be used once, more than once, or not at all in each set.

Questions 1–4 refer to the following nuclear equations.

(A) ${}^{1}_{0}n + {}^{235}_{92}U \rightarrow {}^{140}_{56}Ba + {}^{93}_{36}Kr + 3{}^{1}_{0}n$

(B) ${}^{12}_{6}C \rightarrow {}^{12}_{6}C + \gamma$

(C) ${}^{239}_{94}Pu \rightarrow {}^{235}_{92}U + {}^{4}_{2}He$

(D) ${}^{2}_{1}H + {}^{3}_{1}H \rightarrow {}^{4}_{2}He + {}^{1}_{0}n$

(E) ${}^{234}_{90}Th \rightarrow {}^{234}_{91}Pa + {}^{0}_{-1}e$

1. Which of the equations is an example of alpha decay?

2. Which of the equations is an example of beta decay?

3. Which of the equations is an example of nuclear fusion?

4. Which of the equations is an example of nuclear fission?

Questions 5–8 refer to the following physical principles. Identify the principle that best explains each observation.

(A) Newton's first law of motion
(B) Newton's second law of motion
(C) Newton's third law of motion
(D) Law of universal gravitation
(E) Kepler's Laws of Motion

5. An acorn speeds up as it falls from a tree.

6. An astronaut moves backward upon throwing a pack of tools in space.

7. In a tug-of-war, the rope does not move when the two teams are pulling equally hard in opposite directions.

8. Jupiter holds many moons in orbit around it.

Questions 9–12 refer to the images described below. F represents focal distance.

(A) real, upright image
(B) real, inverted image
(C) virtual, upright image
(D) virtual, inverted image
(E) no image

9. Which can result from an object placed between the center of curvature and the focal point of a converging mirror?

10. Which can result from an object placed inside the focal length of a diverging lens?

11. Which can result from an object placed a distance of 2F in front of a converging lens?

12. Which can result from an object placed between the focal point and the surface of a converging mirror?

Practice Test

GO ON TO THE NEXT PAGE

PART B

Directions: Each of the questions or incomplete statements below is followed by five suggested answers or completions. Select the one that is best in each case, and then fill in the corresponding oval on the answer sheet. Do not use a calculator.

13. What is the magnitude of 2**A** – **B** in the figure below?

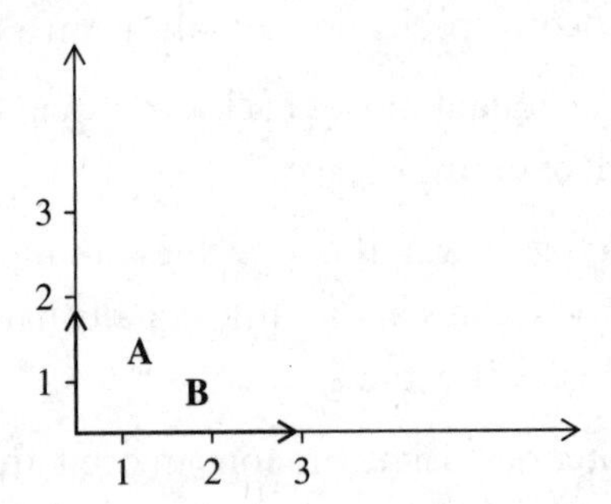

(A) 2
(B) 5
(C) 6
(D) 7
(E) 9

14. The three lowest energy levels of an atom are shown. In a single transition, an atom in the $n = 3$ state can spontaneously emit a photon having an energy of

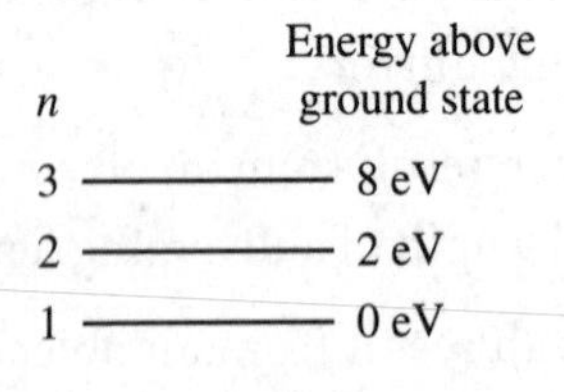

(A) 8 eV only
(B) 2 eV only
(C) 6 eV only
(D) 2 eV and 6 eV only
(E) 6 eV and 8 eV only

15. The angle of incidence in the diagram below is θ. Which number represents the angle of reflection?

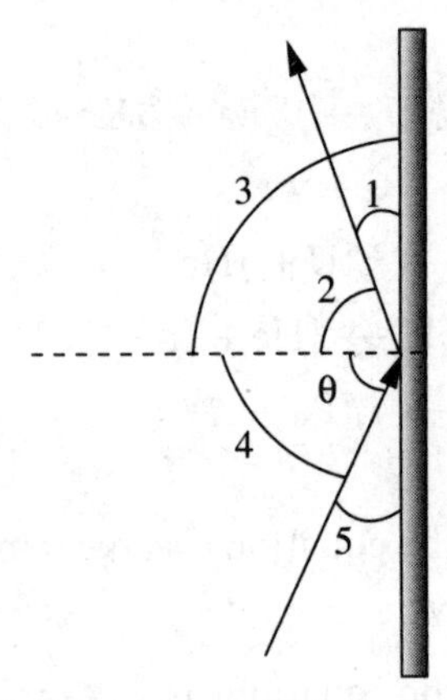

(A) 1
(B) 2
(C) 3
(D) 4
(E) 5

Questions 16 and 17 refer to the diagram of a wave shown below.

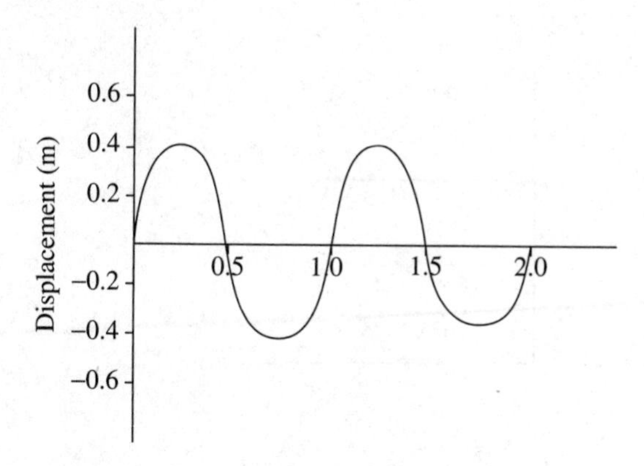

16. What is the wavelength of the wave?

(A) 0.4 m
(B) 0.5 m
(C) 0.8 m
(D) 1.0 m
(E) 2.0 m

17. What is the period of the wave if the speed of the wave is 10 m/s?

(A) 0.1 s
(B) 0.2 s
(C) 0.01 s
(D) 0.02 s
(E) 0.5 s

18. A sound wave from a guitar string travels at a speed of 200 m/s. If the frequency of the wave is 440 Hz, what is its wavelength?

(A) 0.20 m
(B) 0.45 m
(C) 2.2 m
(D) 4.0 m
(E) 8.8 m

19. An astronomer discovers a planet with two moons. One moon is 2 times as far from the center of the planet as the other and 3 times the mass of the other moon. What is the ratio of the gravitational force on the first moon to the gravitational force on the second moon?

(A) 0.33
(B) 0.40
(C) 0.66
(D) 0.75
(E) 0.90

Questions 20 and 21 refer to a 2.0-kg block of ice to which heat is added. The specific heat of water is 4.2×10^3 J/kg°C and the heat of fusion of ice is 3.3×10^5 kJ/kg.

20. How much heat is required to melt the block of ice?

(A) 1.2×10^4 J
(B) 6.5×10^4 J
(C) 1.7×10^5 J
(D) 6.6×10^5 J
(E) 8.4×10^5 J

21. Once the ice melts, heat is added until the temperature of the water reaches 30°C. How much heat is required to raise the temperature from 0°C to 30°C?

(A) 8.4×10^4 J
(B) 2.5×10^5 J
(C) 3.0×10^5 J
(D) 6.0×10^5 J
(E) 1.9×10^6 J

22. Which of the graphs below can be used to represent the magnetic field established at the center of a coil by a steady current in the coil?

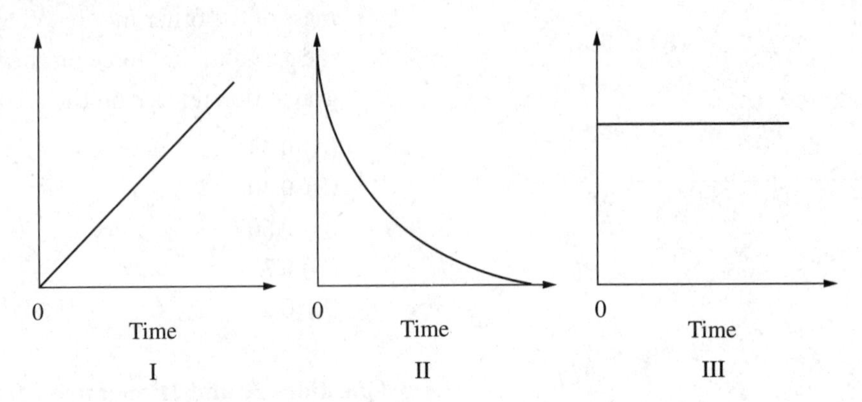

(A) I only
(B) II only
(C) III only
(D) I and III only
(E) none of these

23. A 1-newton force and a 4-newton force act in opposite directions. What is the magnitude of the resultant force, in newtons?

(A) 0
(B) 1
(C) 3
(D) 4
(E) 5

24. All of the following are kinds of electromagnetic waves EXCEPT

(A) blue light
(B) gamma rays
(C) ultraviolet radiation
(D) alpha rays
(E) X-rays

Questions 25 and 26 refer to the diagram below of an airplane traveling at 400 kph at an angle of 25° from north.

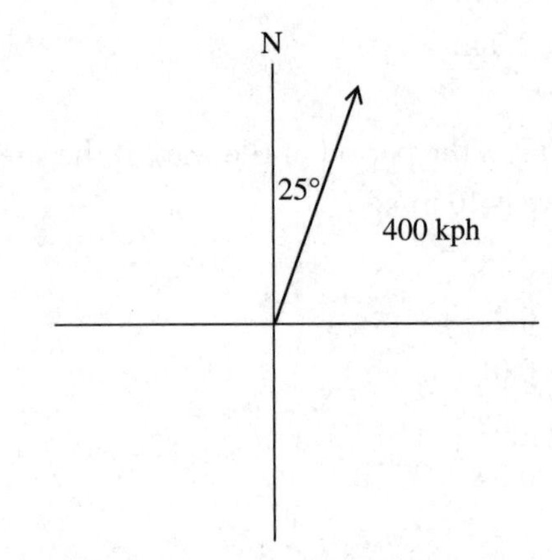

25. What is the plane's northward speed in kilometers per hour?

(A) 400 tan 25°
(B) 400 cos 65°
(C) 400 sin 25°
(D) 400 sin 65°
(E) 400

GO ON TO THE NEXT PAGE

26. What is the plane's eastward speed in kilometers per hour?

(A) $400 \cos 65°$
(B) $400 \sin 65°$
(C) $400 \tan 25°$
(D) $400 \sin 25°$
(E) 400

27. The cross-sectional area of an engine cylinder is 0.020 m^2. If a gas in the cylinder exerts a constant pressure of 6.5×10^5 Pa on the piston to move it a distance of 0.050 m, the amount of work done by the gas is

(A) 43 J
(B) 52 J
(C) 5.5×10^2 J
(D) 6.5×10^2 J
(E) 7.5×10^2 J

28. Which statements(s) about the circuit shown is true?

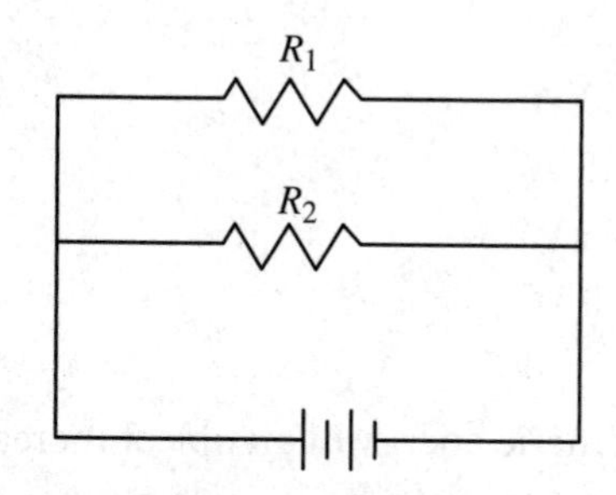

I. The sum of the currents in R_1 and R_2 equals the total current.
II. R_1 and R_2 have the same potential different across them.
III. The sum of R_1 and R_2 gives the equivalent resistance of the circuit.

(A) I only
(B) II only
(C) III only
(D) I and II only
(E) II and III only

29. Which particle is formed when radium–226 undergoes alpha decay according to the equation below?

$$^{226}_{88}\text{Ra} \rightarrow X + ^{4}_{2}\text{He}$$

(A) $^{222}_{86}\text{Rn}$
(B) $^{226}_{88}\text{Ra}$
(C) $^{230}_{90}\text{Th}$
(D) $^{222}_{89}\text{Ac}$
(E) $^{220}_{85}\text{At}$

30. The mass of an object is doubled. How must the magnitude of the force change if the magnitude of the acceleration is also doubled?

(A) It is quartered.
(B) It is halved.
(C) It is unchanged.
(D) It is doubled.
(E) It is quadrupled.

31. What is the missing particle in the nuclear equation below?

$$^{27}_{13}\text{Al} + X \rightarrow ^{24}_{11}\text{Na} + ^{4}_{2}\text{He}$$

(A) $^{0}_{-1}\text{e}$
(B) $^{1}_{0}\text{Al}$
(C) $^{1}_{0}\text{n}$
(D) $^{1}_{0}\text{H}$
(E) $^{4}_{2}\text{He}$

32. A hiker walks due north for 3 km and then due east for 4 km. How much shorter would the walk be if the hiker walked in a straight line from the starting point to the displacement?

(A) 0 km
(B) 1 km
(C) 2 km
(D) 4 km
(E) 5 km

Practice Test

GO ON TO THE NEXT PAGE ➡

33. Two small conducting spheres are identical except that sphere *X* has a charge of −6 microcoulombs and sphere *Y* has a charge of +8 microcoulombs. After the spheres are brought in contact and then separated, what is the charge on each sphere, in microcoulombs?

(A) sphere *X* (+1), sphere *Y* (+1)
(B) sphere *X* (0), sphere *Y* (+2)
(C) sphere *X* (−1), sphere *Y* (+1)
(D) sphere *X* (−2), sphere *Y* (−1)
(E) sphere *X* (+2), sphere *Y* (0)

34. Which of the following will increase the strength of an electromagnet?

I. Increase the density of coils of wire.
II. Remove the iron core from the solenoid.
III. Change the direction of the current through the wire.

(A) I only
(B) II only
(C) III only
(D) I and II only
(E) I and III only

35. Which of the following is the largest quantity?

(A) 0.35 m
(B) 35×10^{-3} m
(C) 3.5×10^{-6} m
(D) 0.000035×10^{2} m
(E) 0.0000035×10^{3} m

36. A process adds 10 joules of heat to an ideal gas, which causes the gas to do 6 joules of work. Which of the following is true about the internal energy of the gas during this process?

(A) It decreases by 16 joules.
(B) It decreases by 4 joules.
(C) It remains the same.
(D) It increases by 4 joules.
(E) It increases by 16 joules.

Questions 37 and 38 refer to the circuit diagram below, which includes a 12-V battery, three resistors, and a current of 1.5 A.

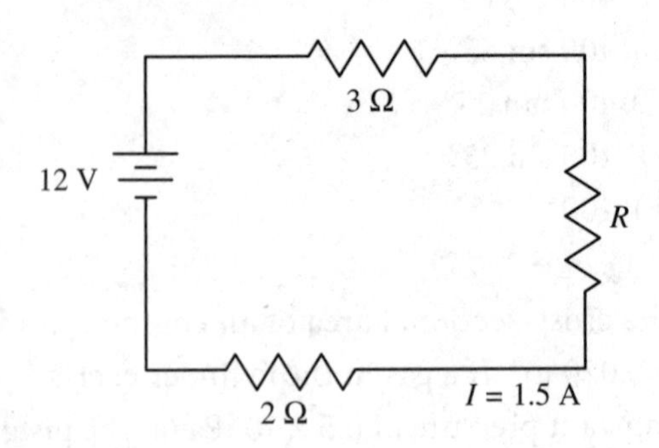

37. What is the value of the unknown resistor?

(A) 3.0 Ω
(B) 4.0 Ω
(C) 5.0 Ω
(D) 6.0 Ω
(E) 8.0 Ω

38. What amount of power is dissipated through the 2-Ω resistor?

(A) 1.3 W
(B) 3.0 W
(C) 4.5 W
(D) 6.5 W
(E) 9.0 W

39. The kinetic energy of an object increases by a factor of 16. What happens to its speed?

(A) decreases by 4 times
(B) decreases by 2 times
(C) remains the same
(D) increases by 4 times
(E) increases by 16 times

GO ON TO THE NEXT PAGE ➡

40. In a system, Q represents the energy transferred to or from a system by heat and W represents the energy transferred to or from a system by work.

I. $Q > 0$ and $W = 0$
II. $Q < 0$ and $W = 0$
III. $W > 0$ and $Q = 0$
IV. $W < 0$ and $Q = 0$

Which condition(s) will lead to an increase in the internal energy of the system?

(A) I only
(B) I and IV only
(C) II only
(D) II and III only
(E) II and IV only

41. Which best describes the shape of the magnetic field lines established by a straight wire carrying a strong current?

(A) straight lines parallel to the wire pointing in the same direction as the current
(B) straight lines parallel to the wire pointing opposite to the current
(C) straight lines perpendicular to the wire
(D) concentric circles surrounding the wire
(E) circles on either side of the wire

Questions 42 and 43 relate to the diagram below, which shows a point charge +Q is fixed in the position shown. Each letter represents a charge in the plane of the page.

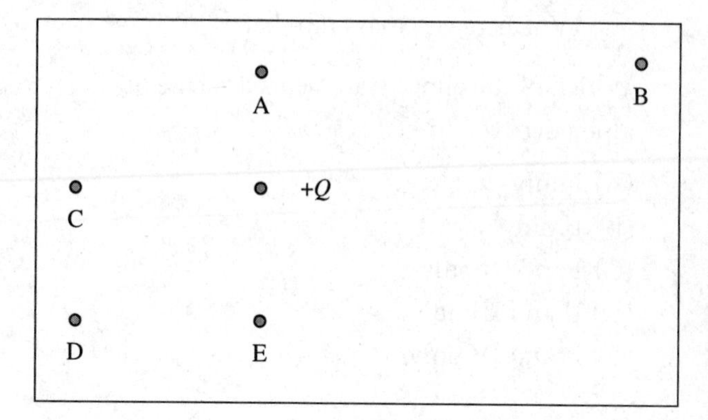

42. Which point will experience an electric field with the least magnitude?

(A) A
(B) B
(C) C
(D) D
(E) E

43. At which point would an electron experience a force directed toward the bottom of the page?

(A) A
(B) B
(C) C
(D) D
(E) E

GO ON TO THE NEXT PAGE ➡

44. The statements below relate to nuclear reactions.

I. Increases atomic number.
II. Decreases atomic number.
III. Increases mass number.
IV. Decreases mass number.

Which of the above statements are true of alpha decay?

(A) I only
(B) II only
(C) I and III only
(D) II and III only
(E) II and IV only

45. Force A and Force B act on an object. Let theta equal the angle between the directions of the two forces. At what value of theta is the resultant force the greatest?

(A) 0°
(B) 45°
(C) 60°
(D) 90°
(E) 180°

46. The diagram shows an electric circuit with a resistor *R* connected to a battery.

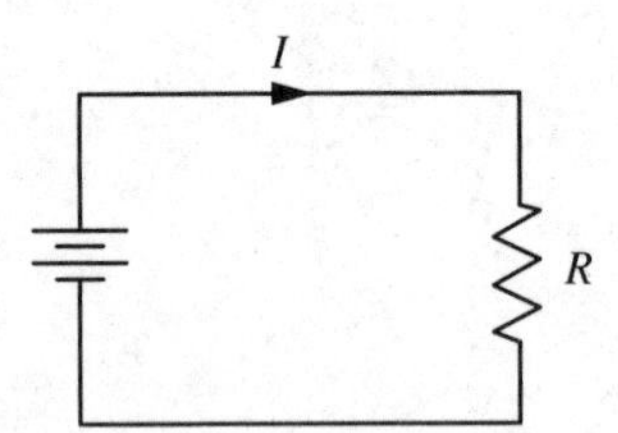

If the emf is tripled while *R* is held constant, the current will be

(A) 1/9 *I*
(B) 1/3 *I*
(C) *I*
(D) 3*I*
(E) 9*I*

47. What particles make up an atom of potassium $^{42}_{19}K$?

(A) 42 protons, 19 electrons, 42 neutrons
(B) 23 protons, 23 electrons, 19 neutrons
(C) 23 protons, 19 electrons, 42 neutrons
(D) 19 protons, 23 electrons, 19 neutrons
(E) 19 protons, 19 electrons, 23 neutrons

48. The diagram shows a 6-kilogam block suspended from the ceiling. The force exerted on the block by the cord is approximately

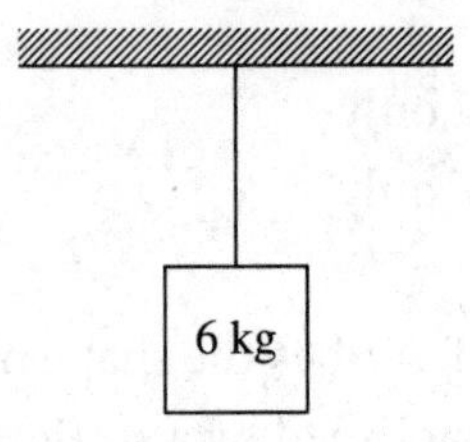

(A) 0 N
(B) 10 N
(C) 26 N
(D) 60 N
(E) 90 N

49. The work done in holding a weight of 20 newtons at a height of 1 meter above the ground for 5 seconds is, in joules,

(A) 0
(B) 4
(C) 25
(D) 50
(E) 100

50. An object starts from rest and accelerates at 6.0 m/s^2 for 4.0 seconds. How far does it travel?

(A) 8.0 m
(B) 12 m
(C) 48 m
(D) 96 m
(E) 288 m

GO ON TO THE NEXT PAGE ➡

51. The Celsius temperature of an object is increased by 40°C. Its Kelvin temperature increases by

(A) 8 K
(B) 10 K
(C) 40 K
(D) 72 K
(E) 233 K

52. A concave spherical mirror has a focal length of 12.0 cm. A candle is placed upright at a distance of 36.0 cm from the mirror. Where is the image of the candle located?

(A) 6.0 cm
(B) 8.0 cm
(C) 9.0 cm
(D) 18.0 cm
(E) 24.0 cm

53. A 2000-newton object falls freely from rest. If it drops 15 meters, its kinetic energy just before it strikes the ground is

(A) 1,300 J
(B) 15,250 J
(C) 25,000 J
(D) 30,000 J
(E) 45,000 J

54. An optical fiber is made up of silica with an index of refraction of about 1.50. How fast does light travel in this optical fiber?

(A) 2×10^6 m/s
(B) 4×10^6 m/s
(C) 2×10^8 m/s
(D) 4×10^8 m/s
(E) 5×10^8 m/s

55. If the addition of 4,000 joules of heat to 20 kilograms of a substance raises its temperature 2°C, the specific heat of the substance is

(A) 0.20 J/kg·°C
(B) 0.40 J/kg·°C
(C) 40 J/kg·°C
(D) 100 J/kg·°C
(E) 400 J/kg·°C

56. During one cycle, a gasoline engine receives 200 J of energy from combustion and loses 150 J by heat to exhaust. The efficiency of the engine is about

(A) 15%
(B) 25%
(C) 33%
(D) 75%
(E) 125%

57. If the speed of sound in air on a hot day is 350 m/s, what is the third harmonic of a 1.75-m-long pipe that is open at both ends?

(A) 100 Hz
(B) 150 Hz
(C) 200 Hz
(D) 250 Hz
(E) 300 Hz

Questions 58 and 59 refer to an inclined plane with a length of 8.0 m that is lifted 2.0 meters above the ground on one end. A person needs to push a 700-newton object up the plane. The force of friction is 50 newtons.

58. What is the potential energy gained by the object when it is raised to the top of the plane?

(A) 700 J
(B) 900 J
(C) 1000 J
(D) 1400 J
(E) 1500 J

GO ON TO THE NEXT PAGE ➡

59. Approximately what is the minimum force required to push the object up the plane?

(A) 175 N
(B) 225 N
(C) 500 N
(D) 650 N
(E) 750 N

60. A 90-kg object moves around a flat, circular track. The track has a radius of 20 meters and the object completes one revolution every 8.0 seconds. The speed of the object is

(A) 2.5 m/s
(B) 7.9 m/s
(C) 11.3 m/s
(D) 15.7 m/s
(E) 23.1 m/s

61. A 6-kg object is at rest on a frictionless horizontal plane.

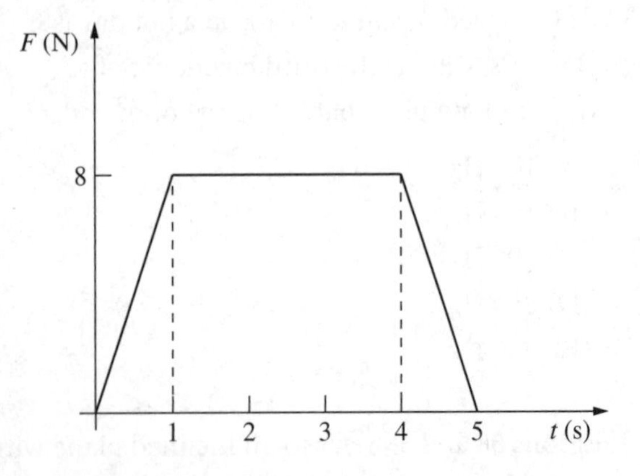

If the object is moved from rest by the force shown, what is the velocity of the object at time $t = 4$ s?

(A) 2 m/s
(B) 6 m/s
(C) 12 m/s
(D) 16 m/s
(E) 32 m/s

62. The half-life of an isotope is 6 years. What fraction of a sample of that isotope will remain after 18 years?

(A) 1/18
(B) 1/12
(C) 1/8
(D) 1/6
(E) 1/3

63. How many meters will a 4.00-kilogram ball travel if it starts from rest and falls freely for 3.00 seconds?

(A) 14.7 m
(B) 15.0 m
(C) 29.4 m
(D) 44.1 m
(E) 88.2 m

64. Which quantity is equivalent to 1 $kg \cdot m^2/s^2$?

(A) 1 Calorie
(B) 1 calorie
(C) 1 British thermal unit
(D) 1 joule
(E) 1 therm

65. An incoming ray of light in a vacuum with a wavelength of 589 nm strikes a diamond at an angle of 30.0° to the normal. If the index of refraction for the diamond is 2.00, the angle of refraction will be

(A) $\sin^{-1}(0.5)$
(B) $\cos^{-1}(0.5)$
(C) $\cos^{-1}(0.43)$
(D) $\sin^{-1}(0.25)$
(E) $\cos^{-1}(0.25)$

66. What voltage is required to establish a current of 0.5 A through a resistance of 0.8 Ω?

(A) 0.3 V
(B) 0.4 V
(C) 0.6 V
(D) 1.3 V
(E) 1.6 V

67. A current-carrying wire is placed in the magnetic field shown. Toward which point does the wire experience a magnetic force?

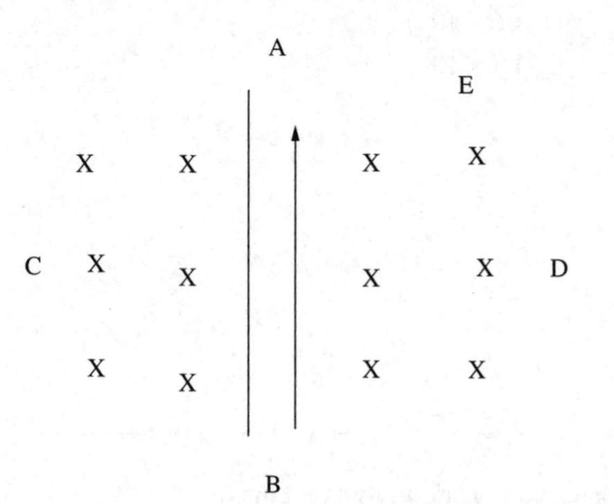

(A) A
(B) B
(C) C
(D) D
(E) E

68. The units used to measure potential difference are

(A) joules
(B) volts
(C) newtons
(D) coulombs
(E) amperes

69. An object traveling around a curve at a speed of 5 meters per second has a centripetal force of 45 newtons. What will happen to the centripetal force if the speed of the object increases to 10 meters per second?

(A) It will be divided by 4.
(B) It will be halved.
(C) It will be unchanged.
(D) It will be doubled.
(E) It will be quadrupled.

70. A process in which volume remains constant is described as

(A) isothermic
(B) isochoric
(C) isometric
(D) isobaric
(E) isosceles

Questions 71 and 72 can be solved by applying the Bohr radius, which is 5.3×10^{-11} m. It is equal to the most probable distance between the proton and electron in a hydrogen atom in its ground state. The charge on an electron is equal and opposite to the charge on a proton, which is 1.60×10^{-19} C. The Coulomb constant has an approximate value of $8.99 \times 10^{9}\ \text{N} \cdot \text{m}^2/\text{C}^2$.

71. The electric force between an electron and a proton in a hydrogen atom has a value closest to

(A) 10^{-47} N
(B) 10^{-28} N
(C) 10^{-18} N
(D) 10^{-7} N
(E) 10^{-3} N

72. To what value is the electrical potential energy of a hydrogen atom closest?

(A) 10^{-18} N
(B) 10^{-7} N
(C) 10^{2} N
(D) 10^{7} N
(E) 10^{28} N

GO ON TO THE NEXT PAGE ➡

73. Which range of the electromagnetic spectrum consists of waves with the shortest wavelengths?

(A) radio waves
(B) infrared waves
(C) gamma rays
(D) ultraviolet light
(E) microwaves

74. At $t = 0$, a ball is at rest at the top of a 20-m inclined plane. Then the ball rolls down the plane with constant acceleration. If the ball rolls 2 meters in the first second, how far will it have rolled at $t = 3$ seconds?

(A) 6 m
(B) 9 m
(C) 12 m
(D) 18 m
(E) 20 m

75. A 2.0-kg sample of water is cooled from steam at 120°C to a liquid at 90°C. If $c_{p,\text{steam}} = 2.0 \times 10^3 \dfrac{\text{J}}{\text{kg} \cdot {}^\circ\text{C}}$, $c_{p,\text{water}} = 4.2 \times 10^3 \dfrac{\text{J}}{\text{kg} \cdot {}^\circ\text{C}}$, and $L_v = 2.3 \times 10^6 \dfrac{\text{J}}{\text{kg}}$, and $L_v = 2.3 \times 10^6 \dfrac{\text{J}}{\text{kg}}$, the amount of energy removed during the process is

(A) 3.4×10^3 J
(B) 6.4×10^2 J
(C) 8.2×10^2 J
(D) 1.6×10^4 J
(E) 1.7×10^5 J

Practice Test

STOP

If you finish before time is called, you may check your work on this test only.

Do not turn to any other test in this book.

Answer Key

1. C
2. E
3. D
4. A
5. B
6. C
7. A
8. D
9. B
10. C
11. B
12. C
13. B
14. E
15. B
16. D
17. A
18. B
19. D
20. D
21. B
22. C
23. C
24. D
25. D
26. A
27. D
28. D
29. A
30. E
31. C
32. C
33. A
34. A
35. A
36. D
37. A
38. C
39. D
40. B
41. D
42. B
43. A
44. E
45. A
46. D
47. E
48. D
49. A
50. C
51. C
52. D
53. D
54. C
55. D
56. B
57. E
58. D
59. B
60. D
61. B
62. C
63. D
64. D
65. D
66. B
67. C
68. B
69. E
70. B
71. D
72. A
73. C
74. D
75. E

Answers and Explanations

1. **(C)** Plutonium–239 decays by alpha particle emission.

2. **(E)** Thorium–234 decays by beta particle emission.

3. **(D)** Nuclear fusion occurs when lighter nuclei combine to form a heavier nucleus.

4. **(A)** Nuclear fission occurs when a heavy nucleus, such as U–235, breaks into lighter nuclei.

5. **(B)** According to Newton's second law of motion, an object accelerates when an unbalanced force is applied to it. The gravitational force is the unbalanced force acting on the acorn.

6. **(C)** According to Newton's third law of motion, when a force is exerted on an object, the object exerts an equal and opposite force. When the astronaut exerts a force on the pack, the pack exerts an equal and opposite force on the astronaut.

7. **(A)** According to Newton's first law of motion, an unbalanced force is required to change the motion of an object. Because the teams are tied, the forces are balanced and there is no change in motion.

8. **(D)** The law of universal gravitation states that there is a force of attraction between every pair of objects in the universe. Jupiter exerts a force of attraction on its moons. The moons, in turn, attract a force of attraction on Jupiter.

9. **(B)** An object placed between the center of curvature and the focal point of a converging mirror will form a real, inverted image.

10. **(C)** An object placed within the focal length of a diverging mirror will form a virtual, upright image.

11. **(B)** An object placed at 2F of a converging lens will form a real, inverted image.

12. **(C)** An object placed between the focal point and the surface of a converging mirror will form a virtual, upright image.

13. **(B)** The magnitude of the vector 2**A** is 4. The magnitude of the vector **B** is 3. The angle between the vectors is 90°. Use the Pythagorean theorem to find the magnitude of 2**A** – **B** as 5, as shown.

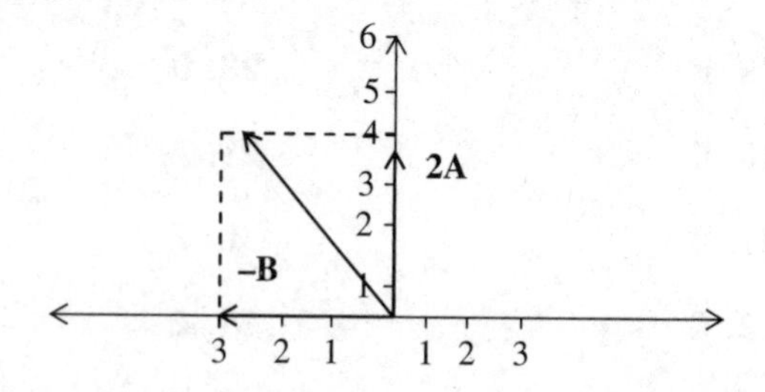

14. **(E)** When the atom in the $n = 3$ state makes just a single transition, it can end up in either the $n = 2$ state or the $n = 1$ state. The energy of the photon released is just the difference in the energy of the initial and final atomic states. In this case, the photon energy is either 8 eV – 2 eV = 6eV or 8 ev – 0 eV = 8 eV.

15. **(B)** Both the angle of incidence and the angle of reflection are measured relative to the normal. They are equal in magnitude, and on opposite sides of the normal.

16. **(D)** The wavelength is one full cycle, or 1.0 m.

17. **(A)** The period is the inverse of the frequency.

$f = \frac{v}{\lambda} = \frac{10\text{ m/s}}{1.0\text{ m}} = 10\text{ Hz}$. Then $T = \frac{1}{f} = \frac{1}{10\text{ Hz}} =$ 0.1 s.

18. **(B)** Wavelength = speed/frequency, so

$$\lambda = \frac{200\text{ m/s}}{440\text{ s}^{-1}} = 0.45\text{ m}.$$

Practice Test

19. **(D)** The force is directly proportional to the masses and indirectly proportional to the square of the distance. This results in a ratio of ¾, or 0.75.

20. **(D)** The amount of energy required to melt the ice is found by $Q = mH_f = (2 \text{ kg})(3.3 \times 10^5 \text{ kJ/kg}) = 6.6 \times 10^5$ J.

21. **(B)** The heat required can be found using $Q = mc\Delta T = (2 \text{ kg})(4.2 \times 10^3 \text{ J/kg°C})(30°\text{C}) = 2.5 \times 10^5$ J.

22. **(C)** The magnetic field at the center of a coil is directly proportional to the current in the coil. If the current remains constant over time, so does the magnetic field.

23. **(C)** Because the forces act in opposite directions, the magnitude of the resultant force is the difference between them. The direction is in the direction of the greater force.

24. **(D)** Alpha rays are made up of streams of alpha particles, which are helium nuclei.

25. **(D)** Draw vectors to form a right triangle. Then use trigonometric relationships to find the northward-pointing vector. The answer makes sense because the plane is traveling more toward the north than toward the east.

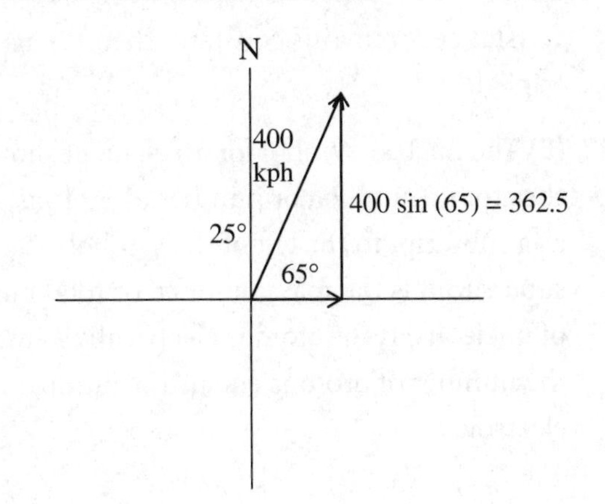

26. **(A)** Draw vectors to form a right triangle. Then use trigonometric relationships to find the eastward-pointing vector.

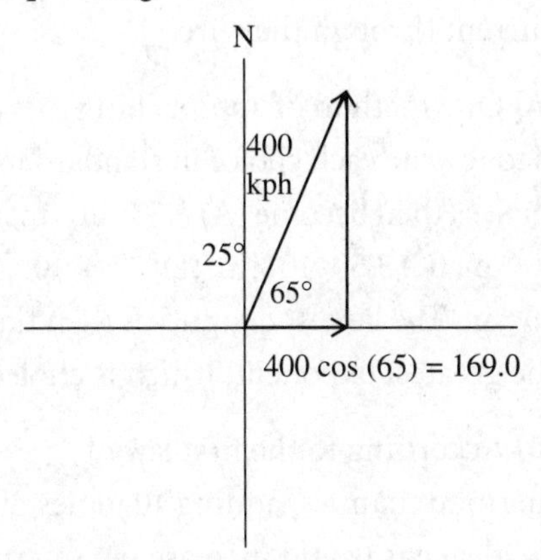

27. **(D)** Find the volume change using the formula $\Delta V = Ad$. $\Delta V = (0.020 \text{ m}^2)(0.050 \text{ m}) = 1.0 \times 10^{-3} \text{ m}^3$. Then use the formula $W = P\Delta V$ to find the work. $W = (6.5 \times 10^5 \text{ N/m}^2)(1.0 \times 10^{-3} \text{ m}^3) = 6.5 \times 10^2$ J.

28. **(D)** The equivalent resistance is calculated using a reciprocal relationship rather than a direct sum. Therefore, III is incorrect.

29. **(A)** The mass numbers must balance on both sides of the equation: 226 – 4 = 222. The atomic numbers must also balance: 88 –2 = 86.

30. **(E)** Acceleration is directly proportional to the net force and inversely proportional to the mass: $a = \frac{F}{m} \cdot \frac{?F}{2m} = 2a$ and ? = 4.

31. **(C)** The mass numbers must balance on both sides of the equation: 28 – 27 = 1. The atomic numbers must also balance: 13 – 13 = 0.

32. **(C)** The displacement is the resultant of the two vectors. $d = \sqrt{3^2 + 4^2} = 5$ km. Since the hiker originally walked a distance of 3 + 4 = 7 km, a straight path would have been 2 km shorter.

33. **(A)** The net excess charge on these two spheres is +2 microcoulombs (+8 + (–6)). Because charge is conserved, each sphere will carry +1 microcoulomb when the excess charge is shared equally.

34. (A) The strength of an electromagnet is increased by increasing the number of coils of wire, adding an iron core to a solenoid, and increasing the current through the wire.

35. (A) One method of approaching this question is to rewrite each choice in standard form. The choices then become (A) 3.5×10^{-1} m, (B) 3.5×10^{-2} m, (C) 3.5×10^{-6} m, (D) 3.5×10^{-3} m, (E) 3.5×10^{-3} m. The largest quantity is then the one with the greatest exponent, which is choice A.

36. (D) According to the first law of thermodynamics, adding 10 joules of heat to an ideal gas would increase the internal energy of the gas by 10 joules. However, the internal energy of the gas decreases by the 6 joules converted to work done by the gas. The net change in the internal energy of the gas is 10 joules – 6 joules, which is a net increase of 4 joules.

37. (A) The total resistance is found by Ohm's Law, $R = V/I$. $R = 12\text{ V}/1.5\text{ A} = 8.0\ \Omega$. The total resistance is the sum of the three individual resistances. $8.0\ \Omega = 3.0\ \Omega + 2.0\ \Omega + R$, and $R = 3.0\ \Omega$.

38. (C) The power dissipated equals the product of the square of the current and the resistance, $P = I^2R$. Substitute the given values to find that $P = (1.5\text{ A})^2(2\ \Omega) = 4.5\text{ W}$.

39. (D)

$$KE = \frac{1}{2}mv^2,\ \frac{KE_f}{KE_i} = \left(\frac{v_f}{v_i}\right)^2,\ 16 = \left(\frac{v_f}{v_i}\right)^2,\ \frac{v_f}{v_i} = 4$$

40. (B) According to the first law of thermodynamics, $\Delta U = Q - W$. Only I and IV will lead to a positive ΔU.

41. (D) The magnetic field established by a current-carrying wire takes the shape of circles surrounding the wire.

42. (B) The magnitude of the electric field at any location due to a point charge is inversely proportional to the square of the distance r from the location to the charge. As the distance increases, the magnitude of the electric field decreases, so the farthest point has the least magnitude.

43. (A) The electron is negatively charged, so it will experience a force directed straight toward the charge $+Q$, which is fixed in position. The only labeled point where a force toward $+Q$ is directed toward the bottom of the page is point A.

44. (E) An alpha particle ${}^{4}_{2}He$ is made up of 2 neutrons and 2 protons. When they are emitted, the atomic number decreases by 2 and the mass number decreases by 4.

45. (A) The resultant force is greatest when the magnitudes of the two forces combine. This occurs when they act in the same direction and, therefore, have an angle of 0° between them.

46. (D) In this series circuit, the emf across the battery is equal to the voltage V across the resistor. If the emf is tripled, V is also tripled. The current I in the resistor is given by Ohm's Law, $I = V/R$, so if V is tripled while the resistance R remains constant, then I is also tripled.

47. (E) The nuclear symbol for an element shows the atomic number, or number of protons, as a subscript to the left of the symbol. The superscript is the mass number, or total number of nucleons. If the atom is electrically neutral, the number of protons equals the number of electrons.

48. (D) The block is in equilibrium, so the force acting on it must be balanced. The force exerted by the cord on the block must be equal in magnitude and opposite in direction to the weight of the block. The weight of the block is equal to m times g. If g is approximated as 10 m/s², the force is 60 newtons.

49. (A) Work is determined by the product of the force exerted on an object and the distance the object moves as a result of that force. Although a force is exerted, the object does not move as a result of that force, so no work is done.

50. (C)

$$\Delta x = v_i \Delta t + \frac{1}{2} a \Delta t^2 = 0 + \frac{1}{2}(6 \text{ m/s}^2)(4 \text{ s})^2 = 48 \text{ m}$$

51. (C) Each unit on the Celsius scale is the same as a unit on the Kelvin scale. Therefore, an increase of 40°C on the Celsius scale is equal to an increase of 40 kelvins on the Kelvin scale.

52. (D) Use the mirror equation to relate the object distance, image distance, and focal length: $\frac{1}{f} = \frac{1}{p} + \frac{1}{q}$. Rearrange to solve for q, the image distance $\frac{1}{q} = \frac{1}{f} - \frac{1}{p} = \frac{1}{12.0 \text{ cm}} - \frac{1}{36.0 \text{ cm}} =$ 0.06 cm. $q = 18.0$ cm.

53. (D) The kinetic energy just before it strikes the ground is equal to the potential energy it had just before falling.

$$PE = (2{,}000 \text{ N})(15 \text{ m}) = 30{,}000 \text{ J}.$$

54. (C) The angle of refraction equals the ratio of the speed of light in a vacuum to the speed of light in the silica.

$$v = \frac{c}{n} = \frac{3.00 \times 10^8 \text{ m/s}}{1.5} = 2.00 \times 10^8 \text{ m/s}$$

55. (D) The change in temperature ΔT of a mass m of a substance is related to the amount of heat added to it by the equation $Q = mc\Delta T$, where c is the specific heat of the substance. Solving for the specific heat and substituting the given quantities,

$$c = \frac{4000 \text{ J}}{(20 \text{ kg})(2°\text{C})} = 100 \text{ J/kg} \cdot °\text{C}$$

56. (B) The efficiency of a heat engine is determined by finding the ratio of the work done by the engine to the energy transferred by heat, $eff = \frac{W_{net}}{Q_h} = 1 - \frac{Q_c}{Q_h}$. Substituting in the given values gives $1 - \frac{150}{200} = 0.25$, which is 25%.

57. (E) For a pipe that is open at both ends, all harmonics are present. Use the equation for the harmonic series to find the fundamental frequency: $f_1 = n\frac{v}{2L} = (1)\frac{350 \text{ m/s}}{2(1.75 \text{ m})} = 100$ Hz. The third harmonic is three times the fundamental frequency: $f_1 = 3(100 \text{ Hz}) = 300$ Hz.

58. (D) The increase in potential energy is equal to the product of the weight and the height. PE = (700 N)(2 m) = 1400 J.

59. (B) The ratio of the height of the plane to the length of the plane is $\frac{2}{8}$. The parallel component is the product of this ratio and the weight of the object. The parallel component = $\frac{2}{8}(700 \text{ N}) = 175$ N. The person needs to overcome the parallel component, as well as the force of friction. So, 175 N + 50 N = 225 N.

60. (D) The speed depends on the circumference and the period. $v = \frac{2\pi r}{T} = \frac{2(3.14)(20 \text{ m})}{8.0 \text{ s}} = 15.7$ m/s.

Practice Test

61. **(B)** Find the impulse by determining the area under the graph, and then use it to find the velocity.

$$p = \frac{8}{2} + (4 \times 8) = 36,\ v = \frac{p}{m} = \frac{36}{6} = 6 \text{ m/s}$$

62. **(C)** The half-life is 6 years, so 18 years is 3 half-lives. After 1 half-life, ½ of the sample remains. After 2 half-lives, ¼ of the sample remains. After 3 half-lives, ⅛ of the sample remains.

63. **(D)** The distance $d = 1/2gt^2 = ½(10 \text{ m/s}^2)(3.00 \text{ s})^2 =$ 45 m. Choice D is closest.

64. **(D)** The joule is the SI unit of energy equivalent to the quantity described.

65. **(D)** By rearranging Snell's Law, you find that $\theta_r = \sin^{-1}\left[\frac{n_i}{n_r}(\sin\theta_i)\right]$. Substitute in the known values and solve: $\theta_r = \sin^{-1}\left[\frac{1.00}{2.00}(\sin 30°)\right] = \sin^{-1}(0.25)$.

66. **(B)** According to Ohm's Law, $V = IR$. Therefore, $V = (0.5 \text{ A})(0.8\ \Omega) = 0.4$ V.

67. **(C)** The right-hand rule can be used to determine the direction of the force. Place your thumb in the direction of the current. Your straightened fingers point in the direction of the magnetic force.

68. **(B)** Potential difference is measured in volts and is, therefore, often known as voltage.

69. **(E)** The centripetal force is directly related to the square of the velocity. If the velocity is doubled, the force will be multiplied by 4.

70. **(B)** The term *isochoric* describes a process at constant volume.

71. **(D)** Use Coulomb's Law to find the magnitude of the electric force. The magnitude of the charge on each particle is the same. The signs of the charges determine the direction of the force and need not be included in the calculation to find the magnitude of the force.

$$F_{\text{electric}} = k_c\frac{q_1q_2}{r^2} \approx 9\times10^9\,\frac{\text{Nm}^2}{\text{C}^2} \times \frac{(2\times10^{-19}\text{ C})^2}{(5\times10^{-11}\text{ m})^2} \approx \frac{9\times4}{25}\times10^{9-19-19+11+11} \approx 1\times10^{-7}\text{ N.}$$

72. **(A)** The electrical potential energy for a pair of charges is found by $F_{\text{electric}} = k_c\frac{q_1q_2}{r}$. Substitute the approximate values into the equation to solve:

$$F_{\text{electric}} = k_c\frac{q_1q_2}{r} \approx 9\times10^9\,\frac{\text{Nm}^2}{\text{C}^2} \times \frac{(2\times10^{-19}\text{ C})^2}{(5\times10^{-11}\text{ m})} \approx \frac{9\times4}{5}\times10^{9-19-19+11} \approx 7\times10^{-18}$$

73. **(C)** Gamma rays have the shortest wavelengths and the greatest frequencies of the waves listed.

74. **(D)** Because the initial speed of the ball is zero, distance x down the plane as a function of time t is related to acceleration a by the equation $x = \frac{1}{2}at^2$. If the ball rolls 2 meters in 1 second, you can find that the acceleration is 4 m/s². Using the equation again with this acceleration and $t = 3$ seconds gives a distance of 18 meters.

75. **(E)** Find the energy change of the cooling steam:

$$Q_{\text{cooling steam}} = mc_{p,\text{steam}}\Delta T = (2\text{ kg})\left(2.0\times10^3\,\frac{\text{J}}{\text{kg}\cdot°\text{C}}\right)(20°\text{C}) = 8.0\times10^4\text{ J}$$

Find the energy change of the condensing steam:

$$Q_{\text{condensing steam}} = mL_v = (2\text{ kg})\left(2.3\times10^3\,\frac{\text{J}}{\text{kg}}\right) = 4.6\times10^3\text{ J}$$

Find the energy change of the cooling water:

$$Q_{\text{cooling water}} = mc_{p,\text{water}}\Delta T = (2\text{ kg})\left(4.2\times10^3\,\frac{\text{J}}{\text{kg}\cdot°\text{C}}\right)(10°\text{C}) = 8.4\times10^4\text{ J}$$

$$Q_{\text{total}} = 8.0\times10^4\text{ J} + 0.46\times10^4\text{ J} + 8.4\times10^4\text{ J} = 16.86\times10^4 = 1.7\times10^5$$

Practice Test 2

Treat this practice test as the actual test, and complete it in one 60-minute sitting. Use the following answer sheet to fill in your multiple-choice answers. Once you have completed the practice test:

1. Check your answers using the Answer Key.
2. Review the Answers and Explanations.
3. Complete the Score Sheet on page 290 and see how well you did.

Answer Sheet

Tear out this answer sheet and use it to complete the practice test. Determine the BEST answer for each question. Then fill in the appropriate oval.

1. A B C D E	21. A B C D E	41. A B C D E	61. A B C D E
2. A B C D E	22. A B C D E	42. A B C D E	62. A B C D E
3. A B C D E	23. A B C D E	43. A B C D E	63. A B C D E
4. A B C D E	24. A B C D E	44. A B C D E	64. A B C D E
5. A B C D E	25. A B C D E	45. A B C D E	65. A B C D E
6. A B C D E	26. A B C D E	46. A B C D E	66. A B C D E
7. A B C D E	27. A B C D E	47. A B C D E	67. A B C D E
8. A B C D E	28. A B C D E	48. A B C D E	68. A B C D E
9. A B C D E	29. A B C D E	49. A B C D E	69. A B C D E
10. A B C D E	30. A B C D E	50. A B C D E	70. A B C D E
11. A B C D E	31. A B C D E	51. A B C D E	71. A B C D E
12. A B C D E	32. A B C D E	52. A B C D E	72. A B C D E
13. A B C D E	33. A B C D E	53. A B C D E	73. A B C D E
14. A B C D E	34. A B C D E	54. A B C D E	74. A B C D E
15. A B C D E	35. A B C D E	55. A B C D E	75. A B C D E
16. A B C D E	36. A B C D E	56. A B C D E	
17. A B C D E	37. A B C D E	57. A B C D E	
18. A B C D E	38. A B C D E	58. A B C D E	
19. A B C D E	39. A B C D E	59. A B C D E	
20. A B C D E	40. A B C D E	60. A B C D E	

PART A

Directions: Each set of lettered choices refers to the numbered questions of statements immediately following it. Select the one lettered choice that best answers each question or best fits each statement, and then fill in the corresponding oval on the answer sheet. A choice may be used once, more than once, or not at all in each set.

Questions 1–4 refer to the following graphs.

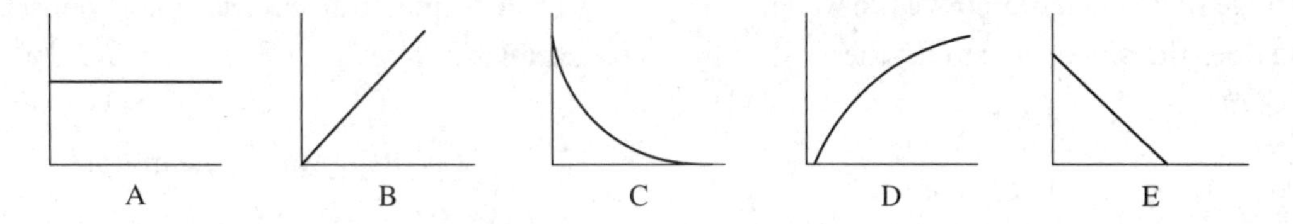

1. Which graph best represents the relationship between pressure and volume at constant temperature?
2. Which graph best represents the relationship between the volume of a gas and the temperature at constant pressure?
3. Which graph best represents the relationship between the average kinetic energy of the molecules of a gas and its temperature?
4. Which graph best represents the relationship between the product PV and P?

Questions 5–9 Which example most closely relates to each example of the interactions between waves and matter?

(A) polarization
(B) absorption
(C) refraction
(D) diffraction
(E) reflection

5. Light striking a surface bounces off the surface at an angle equal to the angle at which it approached.
6. Light passing from one medium to another at an angle is bent because of which phenomenon?
7. Light is passed through a filter that blocks waves vibrating in all directions except one. Which phenomenon explains this observation?
8. A light ray strikes the surface of a medium and is neither returned from the surface nor passed through it.
9. Light passing through small openings forms patterns of light and dark regions as a result of interference.

GO ON TO THE NEXT PAGE

PART B

Directions: Each of the questions or incomplete statements below is followed by five suggested answers or completions. Select the one that is best in each case, and then fill in the corresponding oval on the answer sheet. Do not use a calculator.

10. An object is thrown horizontally at a velocity v from a bridge that is height y above the water. How long does the object take to hit the water below?

(A) $\sqrt{2gy}$

(B) $\sqrt{\frac{2y}{g}}$

(C) $\frac{y}{v}$

(D) $\frac{2y}{v}$

(E) $\frac{2y}{v^2}$

11. An object is moving toward the right in a straight line at constant speed. Which conclusion(s) about the object can you reach?

I. There are no forces acting on the object.
II. Any forces acting on the object are balanced.
III. A net force toward the right is acting on the object.

(A) I only
(B) II only
(C) III only
(D) I and II only
(E) II and III only

Questions 12 and 13 refer to a 12-V battery that delivers a 1.5-A current to an electric motor connected across its terminals.

12. What is the power of the motor?

(A) 6 W
(B) 8 W
(C) 10 W
(D) 14 W
(E) 18 W

13. The amount of electric energy delivered in 5.0 minutes is

(A) 450 J
(B) 600 J
(C) 1200 J
(D) 2100 J
(E) 5400 J

14. A person exerts an upward force of 110 newtons to accelerate a 10.0-kg stone upward. The acceleration of the stone is

(A) 1.2 m/s^2
(B) 1.3 m/s^2
(C) 2.2 m/s^2
(D) 6.8 m/s^2
(E) 9.8 m/s^2

15. An airplane starting from rest undergoes a uniform acceleration of 5 m/s^2 for time of 20 s from one end of a runway until takeoff. What is the final velocity of the airplane just before takeoff?

(A) 50 m/s
(B) 100 m/s
(C) 120 m/s
(D) 150 m/s
(E) 200 m/s

16. What particles make up an atom of cobalt-60 ($^{60}_{27}$Co)?

(A) 27 protons, 33 electrons, 60 neutrons
(B) 27 protons, 27 electrons, 33 neutrons
(C) 33 protons, 27 electrons, 27 neutrons
(D) 60 protons, 27 electrons, 27 neutrons
(E) 33 protons, 33 electrons, 27 neutrons

Questions 17–19 relate to the graph below, which represents the motion of Trains A and B in parallel tracks. Train A passes Train B as Train B starts from rest at $t = 0$.

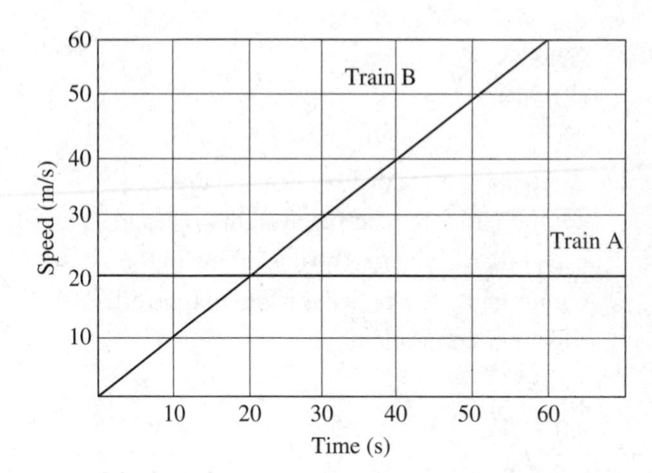

17. Which of the following is represented by the intersection of the two lines?

(A) the distance traveled before the two trains meet
(B) the difference in the distances traveled by the two trains
(C) the time at which the two trains have traveled the same distance
(D) the time at which the two trains begin to move in opposite directions
(E) the time at which the two trains travel at the same speed

18. How far does Train A travel between $t = 0$ and $t = 50$?

(A) 50 m
(B) 100 m
(C) 625 m
(D) 1250 m
(E) 2500 m

Practice Test

GO ON TO THE NEXT PAGE ➡

19. How long does it take Train B to catch up with Train A?

(A) 20 s
(B) 40 s
(C) 45 s
(D) 60 s
(E) 70 s

20. A resultant force of 10 N is made up of two component forces acting at right angles to each other. If the magnitude of one of the components is 8 N, what is the magnitude of the other component?

(A) 2 N
(B) 6 N
(C) 9 N
(D) 12 N
(E) 18 N

21. A gas in a container with a constant volume has a pressure of 250,000 Pa at 25°C. When the temperature of a gas is increased to 50°C, what is the pressure?

(A) 130,000 Pa
(B) 160,000 Pa
(C) 230,000 Pa
(D) 270,000 Pa
(E) 320,000 Pa

22. Two atoms have the same number of protons and electrons, but different numbers of neutrons. The atoms are different

(A) ions
(B) isotopes
(C) positrons
(D) elements
(E) states

Questions 23 and 24 refer to a ball that is thrown straight up with an initial velocity of 20 m/s. The ball is caught at the same height above the ground from which it started.

23. The ball rises to a height of about

(A) 10 m
(B) 20 m
(C) 40 m
(D) 60 m
(E) 80 m

24. What length of time does the ball remain in the air?

(A) 1 s
(B) 2 s
(C) 4 s
(D) 6 s
(E) 8 s

25. The diagram shows an electric circuit with a resistor *R* connected to a battery. The current is represented by *I*, and the power dissipated as heat is *P*. If the emf of the battery is tripled while *R* is held constant, the power dissipated in *R* is

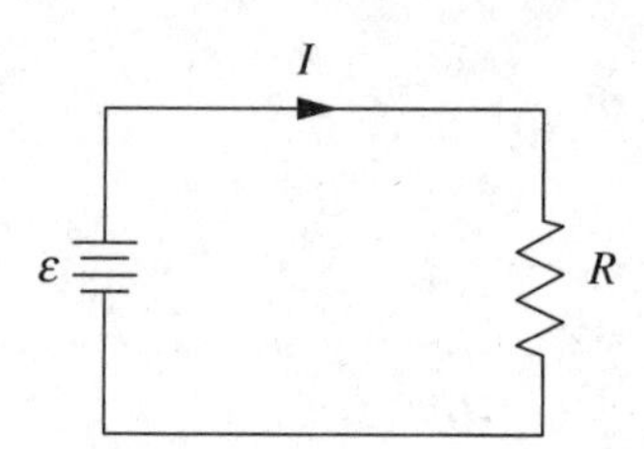

(A) 1/9*P*
(B) 1/3*P*
(C) 3*P*
(D) 6*P*
(E) 9*P*

GO ON TO THE NEXT PAGE

26. A 30.0-N wooden block is at rest on a smooth wooden table. A force of 12.0 N is required to keep the block moving at constant velocity. What is the coefficient of friction for the block and the table?

(A) 0.180
(B) 0.200
(C) 0.400
(D) 0.600
(E) 0.825

27. A satellite moving in a circular orbit with respect to the Earth's orbit is put into a new circular orbit of larger radius. Which of the following changes, if any, will the satellite experience?

(A) The gravitational force will increase and the speed will decrease.
(B) The gravitational force will decrease and the speed will increase.
(C) Both the gravitational force and the speed will decrease.
(D) Both the gravitational force and the speed will increase.
(E) Neither the gravitational force nor the speed will change.

28. When a vector of magnitude 4 units is added to a vector of magnitude 9 units, the magnitude of the resultant vector could be

(A) 5 units only
(B) 13 units only
(C) 9 units only
(D) 0 units, 9 units, or some value in between
(E) 5 units, 13 units, or some value in between

29. The frequency of a sound wave is heard as its

(A) pitch
(B) loudness
(C) resonance
(D) echo
(E) amplitude

30. What is the missing particle in the nuclear equation below?

$$^{99}_{43}\text{Tc} \rightarrow X + ^{0}_{-1}\text{e}$$

(A) $^{99}_{43}\text{Tc}$
(B) $^{99}_{42}\text{Mo}$
(C) $^{99}_{45}\text{Rh}$
(D) $^{99}_{44}\text{Ru}$
(E) $^{98}_{41}\text{Nb}$

31. An object is placed 20.0 cm in front of a converging lens with a focal length of 10.0 cm. How far from the lens is the image formed?

(A) 5 cm
(B) 15 cm
(C) 20 cm
(D) 25 cm
(E) 30 cm

Questions 32 and 33 refer to the energy level diagram for hydrogen.

$n = \infty$	$E = 0$
$n = 4$	$E = -0.85$ eV
$n = 3$	$E = -1.51$ eV
$n = 2$	$E = -3.40$ eV
$n = 1$	$E = -13.6$ eV

32. Which of the following transitions will produce the photon with the highest frequency?

(A) $n = 2$ to $n = 1$
(B) $n = 4$ to $n = 2$
(C) $n = 3$ to $n = 1$
(D) $n = 3$ to $n = 2$
(E) $n = 4$ to $n = 1$

GO ON TO THE NEXT PAGE ➡

33. Which of the following photon energies could not be emitted from the atom after an electron has been excited to the fourth energy level?

(A) 0.66 eV
(B) 1.89 eV
(C) 4.91 eV
(D) 10.2 eV
(E) 13.6 eV

34. A force of 3 newtons and a force of 4 newtons are exerted on a point at right angles to one another. What is the magnitude of the resultant force, in newtons?

(A) 1
(B) 4
(C) 5
(D) 7
(E) 12

35. A change in temperature of 42°C is equivalent to a temperature change of

(A) −231 K
(B) 42 K
(C) 84 K
(D) 142 K
(E) 315 K

36. A negative charge P is located between two unknown charges Q_1 and Q_2. The net electric force acting on P is zero. Which conclusion may correctly be reached about the unknown charges?

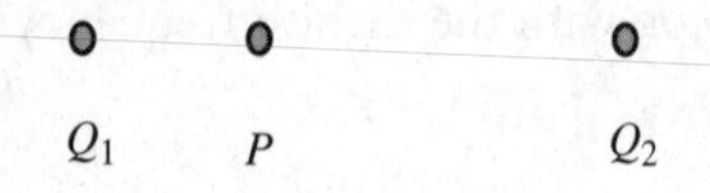

(A) Both Q_1 and Q_2 are negative.
(B) Both Q_1 and Q_2 are positive.
(C) Q_1 and Q_2 have opposite charges.
(D) Q_1 and Q_2 have the same sign, but Q_1 is greater in magnitude than Q_2.
(E) Q_1 and Q_2 have the same sign, but Q_2 is greater in magnitude than Q_1.

37. If a simple pendulum oscillates with a small amplitude and its length is doubled, what happens to the frequency of its motion?

(A) It doubles.
(B) It becomes $1/\sqrt{2}$ as large.
(C) It is halved.
(D) It becomes $\sqrt{2}$ times as large.
(E) It remains the same.

38. Which examples describe an object moving with nonzero acceleration?

I. A raindrop speeds up as it falls to the ground.
II. A bus slows down as it moves toward a stop.
III. A satellite orbits Earth at a constant speed.

(A) I only
(B) II only
(C) III only
(D) I and II only
(E) I, II, and III

39. What is the current through the circuit shown?

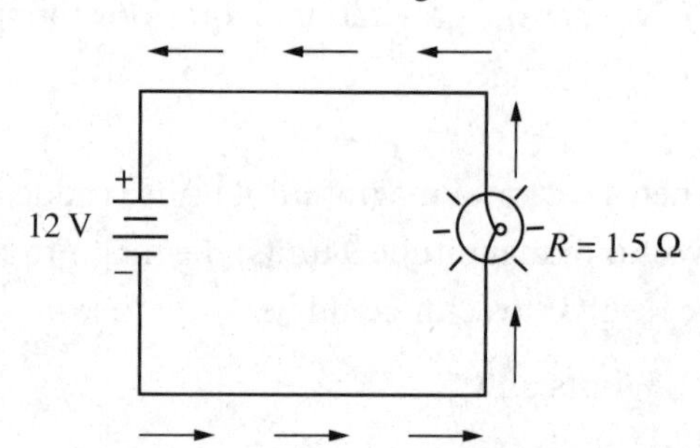

(A) 2.0 A
(B) 4.5 A
(C) 8.0 A
(D) 10.5
(E) 12 A

40. A vibrating sting produces a standing wave. What is the speed of the wave if the length of the string is 2.1 m and the frequency is 60 Hz?

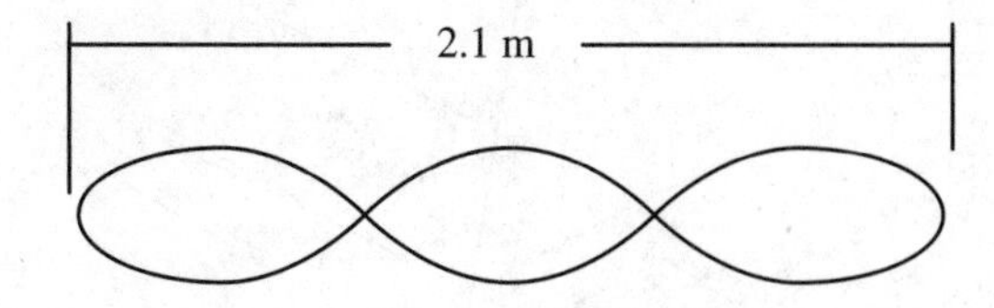

(A) 23 m/s
(B) 29 m/s
(C) 43 m/s
(D) 61 m/s
(E) 84 m/s

41. Which of the following actions will induce a current in a wire?

I. Move the circuit back and forth through a magnetic field.
II. Rotate the circuit in a magnetic field.
III. Vary the intensity of a magnetic field while the circuit is in the field.

(A) I only
(B) II only
(C) III only
(D) I and II only
(E) I, II, and III only

Questions 42 and 43 refer to a 20-Ω resistor which is connected across a 60-V battery.

42. The current in the circuit is

(A) 2.0 A
(B) 3.0 A
(C) 4.0 A
(D) 10 A
(E) 20 A

43. The energy used by the resistor in 10.0 s is

(A) 0.8 kJ
(B) 1.2 kJ
(C) 1.8 kJ
(D) 2.0 kJ
(E) 6.0 kJ

Questions 44 and 45 refer to the velocity-time graph below.

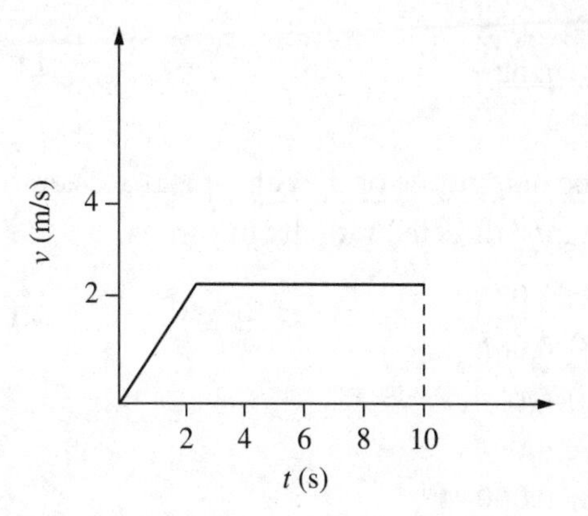

44. The displacement of the object between $t = 0$ and $t = 10$ is

(A) ½ m
(B) 2 m
(C) 16 m
(D) 18 m
(E) 20 m

45. The acceleration of the object between $t = 0$ and $t = 2$ s is

(A) ½ m/s^2
(B) 1 m/s^2
(C) 2 m/s^2
(D) 4 m/s^2
(E) 10 m/s^2

GO ON TO THE NEXT PAGE ➡

Questions 46–48 refer to a 0.02-kg rubber stopper that is attached to a 1-m length of string and swung in a horizontal circle. The stopper completes one revolution in 1 s.

46. The speed of the stopper in m/s is

(A) $\pi/2$
(B) π
(C) 2π
(D) 0.02π
(E) 0.04π

47. The magnitude of the centripetal acceleration in m/s^2 directed radially inward is

(A) 0.08π
(B) $0.08\pi^2$
(C) 4π
(D) $4\pi^2$
(E) $0.0004\pi^2$

48. The force exerted radially inward by the string is

(A) 0.48 N
(B) 0.52 N
(C) 0.65 N
(D) 0.96 N
(E) 1.10 N

49. An amount of energy is added to ice, raising its temperature from –20°C to –10°C. A larger amount of energy is added to the same mass of water, raising its temperature from 10°C to 20°C. These results demonstrate that

(A) overcoming the latent heat of fusion of ice requires an input of energy
(B) the latent heat of fusion of ice delivers some energy to the system
(C) the specific heat of ice is less than that of water
(D) the specific heat of ice is greater than that of water
(E) more information is needed to demonstrate anything

50. A current-carrying wire is placed between the poles of two magnets.

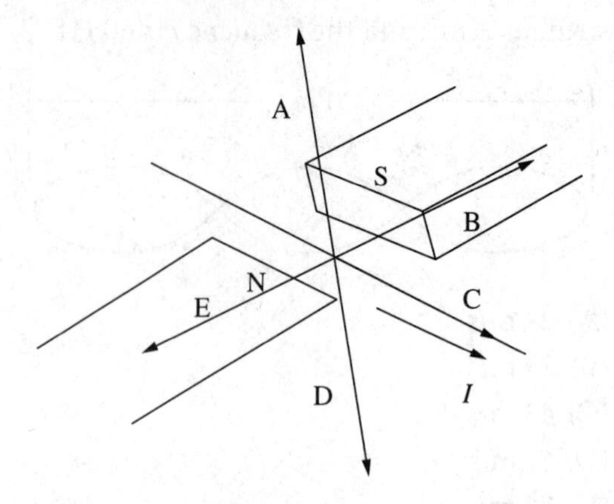

Which arrow shows the direction of the force exerted on the wire?

(A) A
(B) B
(C) C
(D) D
(E) E

Questions 51 and 52 refer to a parallel circuit with four resistors shown below.

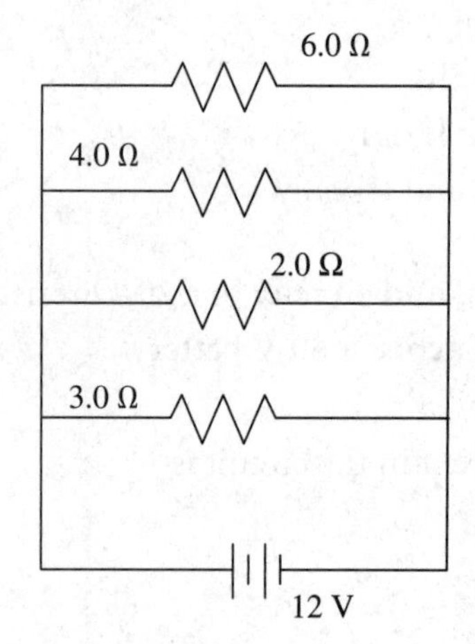

51. What is the equivalent resistance for the circuit?

(A) 0.10 Ω
(B) 0.20 Ω
(C) 0.80 Ω
(D) 0.90 Ω
(E) 1.0 Ω

52. What is the total current in the circuit?

(A) 13 A
(B) 15 A
(C) 20 A
(D) 30 A
(E) 45 A

53. How long does it take for a car to slow from 18 m/s to 2 m/s with a constant acceleration of -1.8 m/s^2?

(A) 2.8 s
(B) 3.3 s
(C) 8.9 s
(D) 11 s
(E) 16 s

Questions 54 and 55 refer to the wave shown, which has a frequency of 30 Hz.

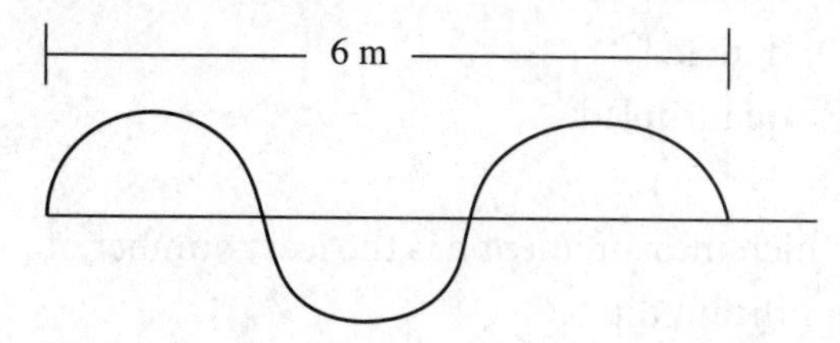

54. What is the wavelength of the wave?

(A) 2 m
(B) 3 m
(C) 4 m
(D) 6 m
(E) 12 m

55. What is the speed of the wave?

(A) 7.5 m/s
(B) 34 m/s
(C) 60 m/s
(D) 120 m/s
(E) 160 m/s

56. A basketball is thrown directly upward. After reaching some peak height, the ball descends and then is caught at the same height from which it was thrown. The upward direction is considered positive for position, velocity, and acceleration. Which signs describe the position, velocity, and acceleration as the basketball rises toward its peak?

(A) position +, velocity +, acceleration +
(B) position +, velocity +, acceleration –
(C) position +, velocity –, acceleration –
(D) position –, velocity –, acceleration +
(E) position –, velocity –, acceleration –

57. Which statement about electric circuits is true?

I. An ammeter must be connected in series.
II. An ammeter cannot be connected in the same circuit as a voltmeter.
III. A voltmeter must be connected in series.

(A) I only
(B) II only
(C) III only
(D) I and II only
(E) I and III only

Practice Test

GO ON TO THE NEXT PAGE ➡

58. The graph represents the motion of four runners. Which statement about the runners is supported by the graph?

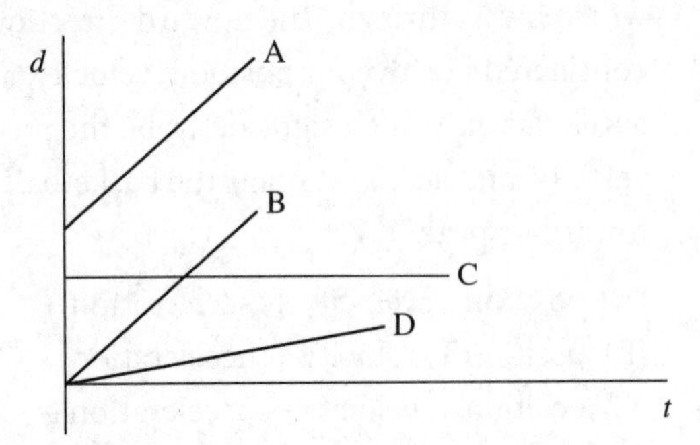

(A) Runner D moves away from the origin with constant velocity.
(B) Runner C starts at a positive coordinate and travels at constant velocity.
(C) Runners B and D have the same velocity.
(D) Runners C and D start from the same point.
(E) Runner A travels with the greatest acceleration.

59. The following processes can be explained by the kinetic molecular theory.

I. evaporation
II. condensation
III. sublimation

Which process(es) involve an increase in energy?

(A) I only
(B) II only
(C) III only
(D) I and II only
(E) I and III only

60. A space station moves in a circular orbit at a constant speed around the Earth. Which of the following statements is true?

(A) No force acts on the space station.
(B) The force of gravity works on the space station.
(C) The space station moves at a constant speed and does not accelerate.
(D) The space station has an acceleration directed away from the Earth.
(E) The space station has an acceleration directed toward the Earth.

61. The mass of an object and the net force acting on it is halved. The acceleration of the object is

(A) quartered
(B) halved
(C) unchanged
(D) doubled
(E) quadrupled

62. Which measurement has the least number of significant digits?

(A) 3206 m
(B) 3.26 m
(C) 0.003206 m
(D) 0.0033 m
(E) 3.03×10^{-2} m

GO ON TO THE NEXT PAGE ➡

63. Which statement about a simple pendulum is true?

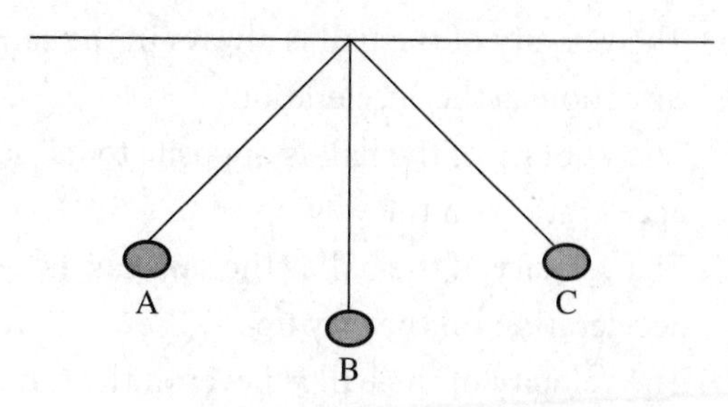

I. The period of swing is dependent on the mass of the bob.
II. The kinetic energy of the bob is greatest at point C.
III. The speed of the bob is greatest at point B.

(A) I only
(B) II only
(C) III only
(D) I or II only
(E) II or III only

64. The diagram shows a simple series circuit. Which of the following would change the current in the circuit to 0.2 A?

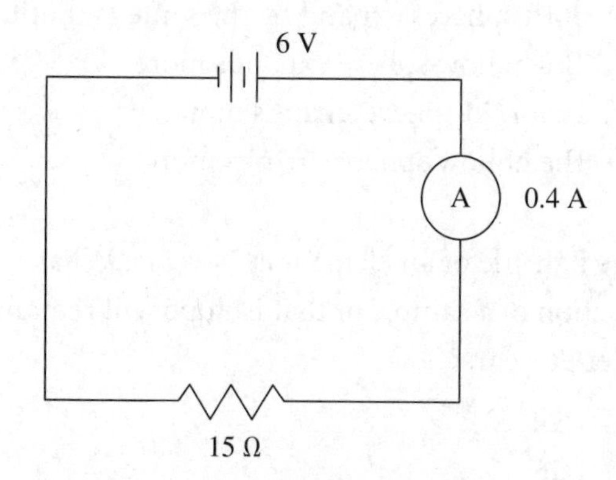

I. Decrease the voltage to 3 V.
II. Increase the resistance to 30 Ω.
III. Double the voltage to 12 V.

(A) I only
(B) II only
(C) III only
(D) I and II only
(E) I and III only

65. A sound wave traveling through the air has a frequency f and wavelength λ. If a second sound wave traveling through the air has a wavelength $\lambda/3$, what is the frequency of the second sound wave in terms of f?

(A) $f/9$
(B) $f/3$
(C) f
(D) $3f$
(E) $9f$

66. A car traveling at a velocity of 2.0 m/s at $t = 0$ accelerates at a rate of 3.0 m/s^2 for 4.0 s. The velocity of the car at $t = 4.0$ s is

(A) 10 m/s
(B) 12 m/s
(C) 14 m/s
(D) 16 m/s
(E) 20 m/s

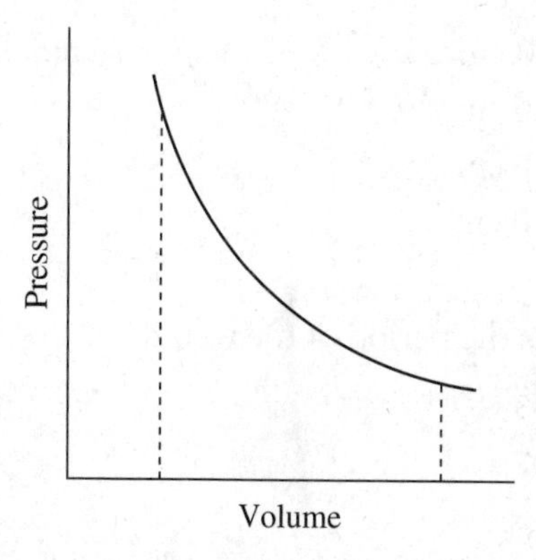

67. During an isothermal process, if the pressure is ___________, then the volume is ___________.

(A) doubled; halved
(B) halved; halved
(C) doubled; doubled
(D) halved; the same
(E) doubled; the same

GO ON TO THE NEXT PAGE

Practice Test

68. How does the emission of a beta particle affect the nucleus of an atom?

(A) Increases the atomic number by 1
(B) Decreases the atomic number by 1
(C) Increases the mass number by 1
(D) Decreases the mass number by 1
(E) Decreases the charge by 1

Questions 69–71 refer to the wave represented below, which is traveling in the positive *x* direction with a frequency of 32.0 Hz.

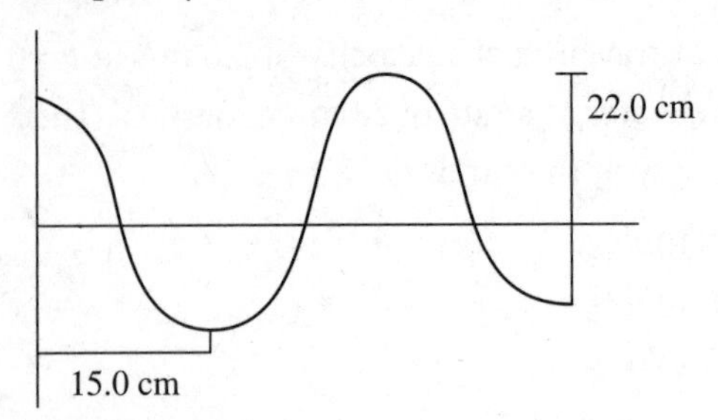

69. The wavelength of the wave is

(A) 11.0 cm
(B) 15.0 cm
(C) 22.0 cm
(D) 30.0 cm
(E) 44.0 cm

70. What is the period of the wave?

(A) 0.03 s
(B) 0.07 s
(C) 6.1 s
(D) 22.0 s
(E) 30.0 s

71. What is the speed of the wave?

(A) 0.09 m/s
(B) 2.9 m/s
(C) 3.0 m/s
(D) 9.6 m/s
(E) 10.7 m/s

72. When a ball is thrown straight up in the air, which statement is true?

(A) The velocity of the ball is always in the same direction as the acceleration.
(B) The velocity of the ball is opposite to its acceleration on the way up.
(C) The velocity of the ball is the same as its acceleration on the way up.
(D) The velocity of the ball is never in the same direction as its acceleration.
(E) The acceleration of the ball is zero.

73. Which of the following is the smallest quantity?

(A) 6.2×10^{-1} g
(B) 62×10^{-2} g
(C) 0.0062×10^{2} g
(D) 0.0062 g
(E) 0.062×10^{-2} g

74. Two spheres are made of steel and have the same radius, but one is hollow and the other is a solid. Both spheres are heated from 20°C to 30°C. Which of the following results?

(A) The solid sphere expands more.
(B) Both spheres expand by the same amount.
(C) The hollow sphere expands more.
(D) The solid sphere shrinks more.
(E) The hollow sphere shrinks more.

75. The half-life of an element is 24 years. What fraction of a sample of that isotope will remain after 96 years?

(A) 1/24
(B) 1/16
(C) 1/12
(D) 1/8
(E) 1/4

STOP

If you finish before time is called, you may check your work on this test only.

Do not turn to any other test in this book.

Answer Key

1. C	20. B	39. C	58. A
2. B	21. D	40. E	59. E
3. B	22. B	41. E	60. E
4. A	23. B	42. B	61. C
5. E	24. C	43. C	62. D
6. C	25. E	44. D	63. C
7. A	26. C	45. B	64. D
8. B	27. C	46. C	65. D
9. D	28. E	47. D	66. C
10. B	29. A	48. A	67. A
11. B	30. D	49. C	68. A
12. E	31. C	50. A	69. D
13. E	32. E	51. C	70. A
14. A	33. C	52. B	71. D
15. B	34. C	53. C	72. B
16. B	35. B	54. C	73. E
17. E	36. E	55. D	74. B
18. D	37. B	56. B	75. B
19. B	38. E	57. A	

Answers and Explanations

1. **(C)** Boyle's Law states that the product of pressure and volume remains constant. Therefore, the two quantities have an inverse relationship.

2. **(B)** Charles' Law states that volume is proportional to Kelvin temperature when pressure is held constant. This direct relationship can be represented by a line with a constant positive slope.

3. **(B)** The temperature is proportional to the average kinetic energy. This direct relationship can be represented by a line with a constant positive slope.

4. **(A)** The product of pressure and volume plotted against either pressure or volume forms a straight, horizontal line.

5. **(E)** Reflection occurs when light strikes a surface, changes direction, and then travels away from the surface. This process is known as reflection. According to the law of reflection, the angle the incident light ray makes with the normal is equal to the angle the reflected ray makes with the normal.

6. **(C)** When light strikes a new medium at an angle, one part of the ray changes speed before the other. This causes the ray to bend in a process known as refraction.

7. **(A)** In polarized light, the waves vibrate in a single plane.

8. **(B)** The energy of light is taken into a medium when the light is absorbed.

9. **(D)** Diffraction is the apparent bending of light waves around obstacles or through small openings. A diffraction pattern forms on a screen according to the constructive and destructive interference of the waves.

10. **(B)** The horizontal velocity does not affect motion in the *y*-direction. Rearrange the equation $y = v_y t + \frac{1}{2} gt^2$ and set the initial position and v_y to zero to get $t^2 = \frac{2y}{g}$. So $t = \sqrt{\frac{2y}{g}}$.

11. **(B)** The motion of the object is constant and, therefore, unchanging. An unbalanced, or net, force would change the motion of the object. Only balanced forces do not change motion.

12. **(E)** The power is found by multiplying the potential difference by the current. $P = VI = (12\text{ V})(1.5\text{ A}) = 18\text{ W}$.

13. **(E)** The energy is determined by multiplying the power by the time. $E = Pt = (18\text{ W})(300\text{ s}) = 5400\text{ J}$.

14. **(A)** The weight of the stone is 98.0 N. The person exerts a force of 110 N. The net force acting on the stone is 110 N – 98.0 N = 12 N. Acceleration equals force divided by mass $a = \frac{12\text{ N}}{10\text{ kg}} = 1.2\text{ m/s}^2$.

15. **(B)** Acceleration equals the change in velocity divided by the change in time: $a = \frac{v_f - v_i}{\Delta t}$. Rearrange the equation to solve for $v_f : v_f = v_i + a\Delta t = 0 + (5\text{ m/s}^2)(20\text{ s}) = 100\text{ m/s}$.

16. **(B)** The nuclear symbol for an element shows the atomic number, or number of protons, as a subscript to the left of the symbol. The superscript is the mass number, or total number of nucleons. If the atom is electrically neutral, the number of protons equals the number of electrons.

17. **(E)** Train A travels at constant speed, whereas Train B increases in speed over time. The intersection represents the time at which both trains are traveling at the same speed.

18. **(D)** Train B starts from rest and travels with constant acceleration. The distance traveled is, therefore, found by $d = 1/2at^2 = ½(1\ m/s^2)(50\ s)^2 =$ 1250 m.

19. **(B)** Train A travels with constant speed of 20 m/s. The distance it travels is found according to the equation $d = vt$. Set this equal to the distance Train B travels using the equation $d = 1/2at^2$.

$1/2at^2 = vt$
$½(1\ m/s^2)\ t^2 = 20t$
$1/2t = 20$
$t = 40\ s$

20. **(B)** The component vectors make a right triangle with the resultant. Therefore, you can use the Pythagorean Theorem to find the missing value. $10^2 = 8^2 + x^2$, so $x = 6$ N.

21. **(D)** Change the temperatures to the Kelvin scale by adding 273 to each value. Therefore, $T_1 = 298$ K and $T_2 = 323$ K. Then use the relationship between pressure and temperature, $\frac{P_1}{T_1} = \frac{P_2}{T_2}$. $P = \frac{P_1 T_2}{T_1} = \frac{(250{,}000\ Pa)(323\ K)}{298\ K} = 270{,}000$ Pa.

22. **(B)** The number of protons defines the element so that atoms are the same element. Atoms of the same element with different numbers of neutrons are known as isotopes.

23. **(B)** At the highest point, the velocity of the ball is zero. Therefore, you know the initial velocity, the final velocity, and the acceleration due to gravity, so you can use the equation $v_f^2 = v_i^2 + 2gd$. The equation becomes $d = -\frac{v_i^2}{2g} = -\frac{(20\ m/s)^2}{2(10\ m/s^2)} = 20$ m.

24. **(C)** Rearrange the equation $v_f = v_i + gt$ and solve for t. The time t in the equation is the time needed to get to the highest point, so the total time is double this for going up and then down. Therefore, $t = \frac{-v_i}{-g} = \frac{-20\ m/s}{-(10\ m/s^2)} = 2s$. Thus, the total time is 4 s.

25. **(E)** The power P dissipated by a resistor is given by $P = IV$, where I is the current in the resistor and V is the voltage across it. Power can also be written as $P = \frac{V^2}{R}$, where R is the resistance. If the emf is tripled while R remains the same, P is multiplied by 9.

26. **(C)** The coefficient is the ratio of the force to the weight of the block.

$$\mu = \frac{F_A}{W} = \frac{12.0\ N}{30.0\ N} = 0.400$$

27. **(C)** The gravitational force is inversely proportional to the square of the radius of the orbit. As the radius increases, the gravitational force decreases. The centripetal force on the satellite is provided by the gravitational force and is proportional to the centripetal acceleration $\frac{v^2}{r}$. Therefore, v also decreases as the radius increases.

28. **(E)** The magnitude of the resultant vector depends on the angle between the two vectors. The resultant vector will be greatest when both vectors point in the same direction and the angle between them is zero. The resulting vector will have a magnitude equal to the sum of the vectors, or 13 units. The resultant vector will be least when the vectors point in opposite directions and the angle between them is 180 degrees. The resulting vector will have a magnitude equal to the difference between the vectors, or 5 units. For any other angle, the magnitude of the resultant vector will be between 5 units and 13 units.

29. **(A)** The frequency determines the pitch of a sound. A high pitch is associated with a greater frequency than a low pitch.

30. **(D)** The mass numbers must balance on both sides of the equation: 99 – 0 = 99. The atomic numbers must also balance: 43 – (–1) = 44.

31. **(C)** Use the thin-lens equation $\frac{1}{p}+\frac{1}{q}=\frac{1}{f}$.

Rearrange to solve for q, $\frac{1}{q}=\frac{1}{f}-\frac{1}{p}=\frac{1}{10.0\text{ cm}}+$

$\frac{1}{20.0\text{ cm}}=0.05$ cm, so q = 20 cm.

32. **(E)** The frequency is proportional to the energy, so the photon with the greatest amount of energy will have the highest frequency.

33. **(C)** The energy of the photon can equal the difference between any two levels. None of the energy differences equals choice C.

34. **(C)** The resultant force can be found by using the Pythagorean Theorem. $R^2 = 3^2 + 4^2$, so R = 5 N.

35. **(B)** Each unit on the Celsius scale is equivalent to one unit on the Kelvin scale. Therefore, a change of 42°C is equivalent to 42 K.

36. **(E)** If the net force on charge P is zero, the forces due to the other two charges must be equal in magnitude but opposite in direction. Q_1 and Q_2 can both attract or both repel charge P, so they are either both negative or both positive. The force between two point charges is proportional to $\frac{Q_1Q_2}{r^2}$. Because Q_1 is farther from charge P, its magnitude must be greater than that of Q_2.

37. **(B)** The period for a simple pendulum is $T=2\pi\sqrt{\frac{L}{g}}$, so the frequency is $f=\frac{1}{T}=\frac{1}{2\pi}\sqrt{\frac{g}{L}}$.

If you double L, then f is $1/\sqrt{2}$ as large.

38. **(E)** An object accelerates when it speeds up, slows down, or changes direction. Therefore, all three examples involve acceleration.

39. **(C)** According to Ohm's Law, $I = V/R$.
$I=\frac{12\text{ V}}{1.5\ \Omega}=8.0$ A.

40. **(E)** The wavelength is 1.4 m. Therefore, the speed of the wave is $v = \lambda f$ = (1.4 m)(60 Hz) = 84 m/s.

41. **(E)** Any relative motion between the field and the wire will result in an induced current.

42. **(B)** According to Ohm's Law, $I = V/R$. Substitute the given values and solve. I = 60 V/20 Ω = 3.0 A.

43. **(C)** Find the energy according to the equation $E = I^2Rt$. Substitute the given values and solve. $E = (3.0\text{ A})^2(20\ \Omega)(10.0\text{ s}) = 1.8$ kJ.

44. **(D)** The displacement is equal to the area under the graph. The area from t = 0 to t = 2 s is 1/2(2)(2), or 2 m. The area from t = 2 s to t = 10 s is 16 m. The total area, and therefore the displacement, is 18 m.

45. **(B)** The acceleration is the slope of the graph between t = 0 and t = 2 s, which is 1 m/s^2.

46. **(C)** Use the period of revolution to find the speed.

$$v=\frac{2\pi r}{T}=\frac{2\pi(1\text{ m})}{(1\text{ s})}=2\pi\text{ m/s}.$$

47. **(D)** Centripetal acceleration can be found using the following equation:

$$a_c=\frac{v^2}{r}=\frac{(2\pi\cdot\text{m/s})^2}{1\text{ m}}=4\pi^2\text{ m/s}^2.$$

48. **(A)** The force can be determined from the acceleration and the mass of the stopper: $F_c = ma_c$ = (0.015 kg)(32 m/s^2) = 0.48 N.

49. **(C)** The amount of heat depends on mass, specific heat, and temperature change according to the equation $Q = mC\Delta T$. Since heating water requires more energy Q, the specific heat of water C must be greater.

Practice Test

50. **(A)** Using the right-hand rule, point the thumb in the direction of the current and the fingers in the direction of the magnetic field, which is from the north pole to the south pole. The force is exerted upward from the palm.

51. **(C)** Total resistance for a parallel circuit is found according to the following equation: $\frac{1}{R_{eq}} = \frac{1}{R_1} + \frac{1}{R_2} + \frac{1}{R_3} + \frac{1}{R_4}$. Using the resistance indicated in the diagram, $\frac{1}{R_{eq}} = \frac{1}{6\ \Omega} + \frac{1}{4\ \Omega} + \frac{1}{2\ \Omega} + \frac{1}{3\ \Omega} = 1.25\ \Omega$. Then $R_{eq} = 0.80\ \Omega$.

52. **(B)** The total current is the voltage divided by the equivalent resistance. $I = \frac{12.0\ \text{V}}{0.80\ \Omega} = 15\ \text{A}$

53. **(C)** Rearrange the equation $v_f = v_i + at$ to solve for t. Substitute the given values and solve: $t = \frac{v_f - v_i}{a} = \frac{18 - 2\ \text{m/s}}{1.8\ \text{m/s}^2} = 8.9\ \text{s}$

54. **(C)** The diagram shows that the length of 1.5 waves is 6 m. Of that, 2/3 represents 1 wave, which is 4 m.

55. **(D)** $v = f\lambda = (30\ \text{Hz})(4\ \text{m}) = 120\ \text{m/s}$

56. **(B)** As the ball rises, its position is above the point from which it was thrown, so it is positive. The ball is moving upward, so its velocity is positive. The gravitational force is pulling it downward, so the acceleration of the ball is negative.

57. **(A)** An ammeter must be connected in series, so there is only one path of current. A voltmeter must be connected in parallel, so the potential difference across an element in the same as the potential difference across the voltmeter. Both instruments can be connected in the same circuit.

58. **(A)** For a distance versus time graph, the slope represents the velocity. Runners A and B start at the same positive coordinate and travel with the same velocity. Runner C is at rest. Runner D starts at the origin, but runs at a slower velocity than runner C.

59. **(E)** Processes that increase the kinetic energy of a sample include vaporization, of which evaporation is an example, and sublimation. Condensation involves a loss in energy.

60. **(E)** The force of gravity acts on the space station. No work is done by gravity because the motion of the force is not parallel to the motion of the space station. Centripetal force acts in the direction of the force of gravity, toward the Earth, so the centripetal acceleration is toward the Earth.

61. **(C)** Acceleration is directly proportional to the net force and inversely proportional to the mass: $a = \frac{F}{m} \cdot \frac{\frac{1}{2}F}{\frac{1}{2}m} = a$. The acceleration remains the same.

62. **(D)** Choice A has 4, choice B has 3, choice C has 4, choice D has 2, and choice E has 3. Therefore, choice D has the least.

63. **(C)** The period of swing is independent of the mass of the bob and dependent on the length of the string. The potential energy is greatest as the points at which the bob changes direction. The kinetic energy and speed is greatest as the bob passes through its lowest point.

64. **(D)** According to Ohm's Law, the quantities are related by $V = IR$. Increasing the resistance or decreasing the voltage will both decrease the current in the circuit.

65. **(D)** Frequency depends on wavelength and speed as follows: $f = v/\lambda$, so if λ is reduced by one-third, then frequency is tripled.

66. **(C)** Choose the equation $v_f = v_i + at$, and then solve.

$v_f = 2.0\text{ m/s} + (3.0\text{ m/s}^2)(4.0\text{ s}) = 14\text{ m/s}$.

67. **(A)** In an isothermal process, the internal energy remains constant, and there is no change in temperature. Since $PV = nRT$ and nRT is constant, when pressure increases, volume decreases.

68. **(A)** A beta particle is an electron produced when a neutron becomes a proton and an electron. The addition of a proton increases the atomic number, and therefore the positive charge, by 1, whereas the mass number remains the same.

69. **(D)** The wavelength is one full cycle, or 30.0 cm.

70. **(A)** The period is the inverse of the frequency.

$$T = \frac{1}{f} = \frac{1}{32\text{ s}^{-1}} = 0.03\text{ s}$$

71. **(D)** The speed is the product of the frequency and the wavelength, $v = f\lambda$. $v = (32.0\text{ s}^{-1})(0.30\text{ m}) = 9.6\text{ m/s}$.

72. **(B)** The acceleration of the ball is always toward the Earth. So when the ball is moving upward, the acceleration and velocity are in opposite directions. As the ball moves down, the acceleration and velocity are in the same direction.

73. **(E)** One method of approaching this question is to rewrite each choice in standard form. The choices then become (A) 6.2×10^{-1} g; (B) 6.2×10^{-3} g; (C) 6.2×10^{-1} g; (D) 6.2×10^{-3} g; (E) 6.2×10^{-4} g. The smallest quantity is then the one with the lowest exponent, which is choice E.

74. **(B)** The solid sphere and the hollow sphere expand the same amount because they have the same volume and are made of the same material. Thermal expansion does not depend on mass.

75. **(B)** The half-life is 24 years, so 96 years is 4 half-lives. After 1 half-life, ½ of the sample remains. After 2 half-lives, ¼ of the sample remains. After 3 half-lives, ⅛ of the sample remains. After 4 half-lives, 1/16 of the sample remains.

Appendixes

APPENDIX A

Mathematics Review

It is often said that mathematics is the language of physics. Therefore, it is in your best interest to be comfortable with the basic math knowledge you will need to interpret and solve physics problems.

Algebra

Scientific notation: Any number can be written as a power of 10: 4.8×10^3
The first number must be $1 < n < 10$.

Exponents: Any number can be written in the form of a base number raised to a power.

$$x^0 = 1$$
$$x^1 = x$$
$$x^{-a} = \frac{1}{x^a}$$
$$x^a \bullet x^b = x^{a+b}$$
$$(x^a)^b = x^{ab}$$
$$\frac{x^a}{x^b} = x^{a-b}$$

Undefined values: A fraction is undefined when the denominator equals 0.
$\frac{5}{x-8}$ is undefined when $x = 8$.

Factorial: The symbol $n!$ represents the product of all positive integers from 1 to n.

$$4! = 4 \bullet 3 \bullet 2 \bullet 1$$

Multiplication involving binomials (FOIL method): Distribute each value:

$$(x+2)(x-3) = x \bullet x + x \bullet (-3) + 2 \bullet x + 2 \bullet (-3)$$
$$= x^2 - 3x + 2x - 6 = x^2 - x - 6$$

Logarithms: If $n = b^a$, then $a = \log_b n$.

$$\log(xy) = \log x + \log y$$
$$\log(x/y) = \log x - \log y$$
$$\log(x^n) = n \log x$$

Geometry

Equations of common shapes:

Line with slope m and y-intercept b: $y = mx + b$ (slope-intercept)

$y - y_1 = m(x - x_1)$ (point-slope)

Circle with radius r centered at the origin: $x^2 + y^2 = r^2$

Parabola whose vertex is at $y = b$: $y = ax^2 + b$

Ellipse with semimajor axis a and semiminor axis b: $\frac{x^2}{a^2} + \frac{y^2}{b^2} = 1$

Pythagorean Theorem:
For a right triangle: $c^2 = a^2 + b^2$

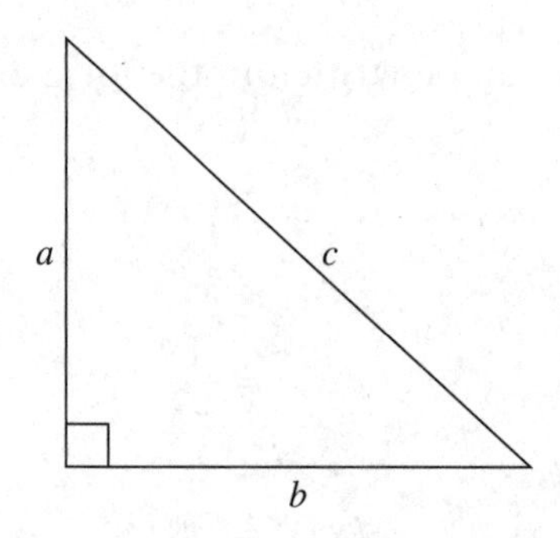

Area:

Triangle: $A = \frac{1}{2}bh$

Rectangle: $A = bh$

Circle: $A = \pi r^2$

Sector of a circle: $A = \frac{n}{360}\pi r^2$

Volume:

Prism: $V = Bh$

Pyramid: $V = \frac{1}{3}Bh$

Cylinder: $V = \pi r^2 h$

Cone: $V = \frac{1}{3}\pi r^2 h$

Sphere: $V = \frac{4}{3}\pi r^3$

Trigonometry

Trigonometric functions:

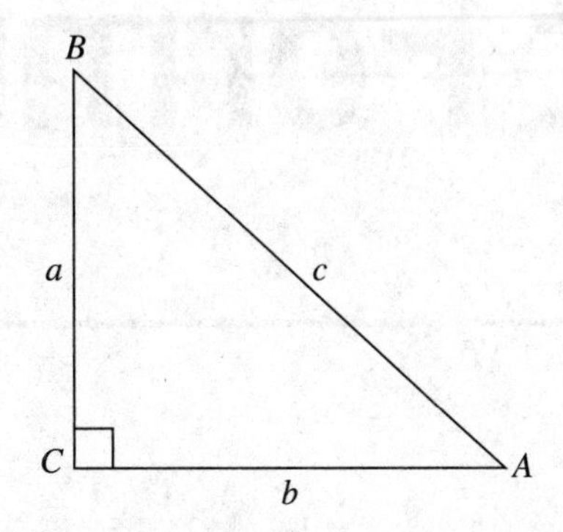

$$\sin A = \frac{a}{c} \qquad \sec A = \frac{c}{a}$$

$$\cos A = \frac{b}{c} \qquad \csc A = \frac{c}{b}$$

$$\tan A = \frac{a}{b} \qquad \cot A = \frac{b}{a}$$

Laws:

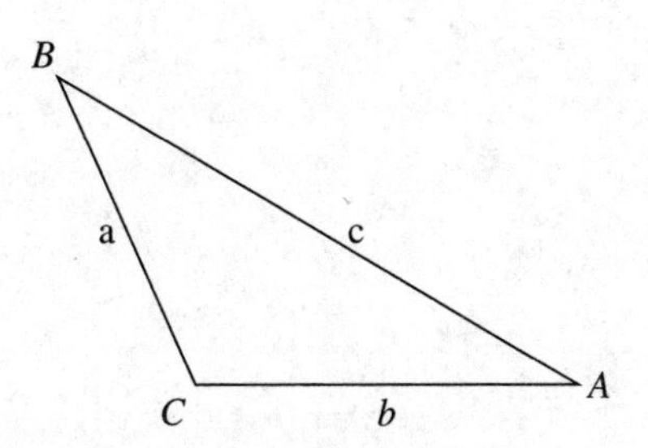

law of cosines: $c^2 = a^2 + b^2 - 2ab \cos C$

law of sines: $\frac{a}{\sin A} = \frac{b}{\sin B} = \frac{c}{\sin C}$

APPENDIX B

Summary of Important Formulas

$$d = vt$$

$$h = \frac{1}{2}at^2$$

$$\overline{v} = \frac{x_2 - x_1}{t_2 - t_1} = \frac{\Delta x}{\Delta t}$$

$$v = v_o + at$$

$$\overline{v} = \frac{v + v_o}{2}$$

$$v^2 = x_o^2 + 2a(x - x_o)$$

$$x = v_o t + \frac{1}{2}at^2$$

$$\overline{a} = \frac{v_2 - v_1}{t_2 - t_1} = \frac{\Delta v}{\Delta t}$$

$$\text{slope} = \frac{\Delta x}{\Delta t}$$

$$a = \frac{F}{m}$$

$$G = mg$$

$$F_{\text{static}} = \mu_s N$$

$$F_{\text{kinetic}} = \mu_k N$$

$$W = F \times I \cos\theta$$

$$\text{GPE} = wh = mgh$$

$$F_{\text{spring}} = kx$$

$$PE_{\text{spring}} = \tfrac{1}{2}kx^2$$

$$KE = \frac{1}{2}mv^2$$

$$W = \Delta KE$$

$$E = mc^2$$

$$P = \frac{\Delta W}{\Delta t}$$

$$P = \frac{\Delta E}{\Delta t}$$

$$P = \frac{F}{A}$$

$$P = hdg$$

$$F = hdgA$$

$$P_{\text{total}} = P_{\text{atmosphere}} + P_{\text{fluid}}$$

$$\text{IMA} = \frac{F_2}{F_1} = \frac{A_2}{A_1}$$

$$\text{efficiency} = \frac{\text{work output}}{\text{work input}} = \frac{\text{MA}}{\text{IMA}}$$

$$\text{density} = \frac{\text{mass}}{\text{volume}}$$

$$\text{sp. gr} = \frac{\text{density of substance}}{\text{density of water}}$$

$$p = mv$$

$$F\Delta t = m\Delta v = mv_f - mv_i$$

$$x_{\text{cm}} = \frac{m_1x_1 + m_2x_2}{m_1 + m_2}$$

$$a_c = v^2/r$$

$$F_c = ma = mv^2/r$$

$$v = 2\pi r/T$$

$$T = mv^2/r$$

$$s = r\theta$$

$$L = mvr$$

$$F_g = \frac{Gm_1m_2}{r^2}$$

$$Q = mc\Delta T$$

$$Q = mL$$

$$\Delta L = \alpha L_o \Delta T$$

$$\frac{P_1V_1}{T_1} = \frac{P_2V_2}{T_2}$$

$$\frac{V_1}{V_2} = \frac{T_1}{T_2}$$

$$P_1V_1 = P_2V_2$$

$$\Delta U = Q - W$$

$$\Delta S = Q/T$$

$$F_E = \frac{kq_1q_2}{r^2}$$

$$E = \frac{F}{q}$$

$$\Delta V = \frac{\text{work}}{q}$$

$$R = \frac{V}{I}$$

$$C = \frac{Q}{U}$$

$$F_m = qvB$$

$$F = ILB$$

$$P = IV = I^2R = V^2/R$$

$$v = f\lambda$$

$$\frac{1}{d_o} + \frac{1}{d_i} = \frac{1}{f}$$

$$M = \frac{h_i}{h_o} = -\frac{d_i}{d_o}$$

$$n = \frac{c}{v_{\text{medium}}}$$

$$n_1 \sin\theta_1 = n_2 \sin\theta_2$$

$$E = \frac{hc}{\lambda}$$

$$E = hf$$

$$E = hf = \text{KE}_{\text{max}} + \varphi$$

$$L = L_o\sqrt{1 - \frac{v^2}{c^2}}$$

$$T = \frac{T_o}{\sqrt{1 - \frac{v^2}{c^2}}}$$

$$m = \frac{m_o}{\sqrt{1 - \frac{v^2}{c^2}}}$$

APPENDIX C

Values of Trigonometric Functions

Note: When entering tables with an angle larger than 45°, select such angle from the right-hand side and obtain values in the column corresponding to the function at the *bottom* of the page.

Angle	Sin	Cos	Tan	Cot	Sec	Cosec	
0°	.0000	1.0000	.0000	∞	1.000	00	90°
1	.0174	.9998	.0175	57.29	1.000	57.30	89
2	.0349	.9994	.0349	28.64	1.001	28.65	88
3	.0523	.9986	.0524	19.08	1.001	19.11	87
4	.0698	.9976	.0699	14.30	1.002	14.34	86
5	.0872	.9962	.0875	11.43	1.004	11.47	85
6	.1045	.9945	.1051	9.514	1.006	9.567	84
7	.1219	.9925	.1228	8.144	1.008	8.206	83
8	.1392	.9903	.1405	7.115	1.010	7.185	82
9	.1564	.9877	.1584	6.314	1.012	6.392	81
10	.1736	.9848	.1763	5.671	1.015	5.759	80
11	.1908	.9816	.1944	5.145	1.019	5.241	79
12	.2079	.9781	.2126	4.705	1.022	4.810	78
13	.2250	.9744	.2309	4.331	1.026	4.445	77
14	.2419	.9703	.2493	4.011	1.031	4.134	76
15	.2588	.9659	.2679	3.732	1.035	3.864	75
16	.2756	.9613	.2867	3.487	1.040	3.628	74
17	.2924	.9563	.3057	3.271	1.046	3.420	73
18	.3090	.9511	.3249	3.078	1.051	3.236	72
19	.3256	.9455	.3443	2.904	1.058	3.072	71
20	.3420	.9397	.3640	2.747	1.064	2.924	70
21	.3584	.9336	.3839	2.605	1.071	2.790	69
22	.3746	.9272	.4040	2.475	1.079	2.669	68
23	.3907	.9205	.4245	2.356	1.086	2.559	67
24	.4067	.9135	.4452	2.246	1.095	2.459	66
25	.4226	.9063	.4663	2.145	1.103	2.366	65
26	.4384	.8988	.4877	2.050	1.113	2.281	64
27	.4540	.8910	.5095	1.963	1.122	2.203	63
28	.4695	.8829	.5317	1.881	1.133	2.130	62
29	.4848	.8746	.5543	1.804	1.143	2.063	61
30	.5000	.8660	.5774	1.732	1.155	2.000	60
31	.5150	.8572	.6009	1.664	1.167	1.942	59
32	.5299	.8480	.6249	1.600	1.179	1.887	58
33	.5446	.8387	.6494	1.540	1.192	1.836	57
34	.5592	.8290	.6745	1.483	1.206	1.788	56
35	.5736	.8192	.7002	1.428	1.221	1.743	55
36	.5878	.8090	.7265	1.376	1.236	1.701	54
37	.6018	.7986	.7536	1.327	1.252	1.662	53
38	.6157	.7880	.7813	1.280	1.269	1.624	52
39	.6293	.7771	.8098	1.235	1.287	1.589	51
40	.6428	.7660	.8391	1.192	1.305	1.556	50
41	.6561	.7547	.8693	1.150	1.325	1.524	49
42	.6691	.7431	.9004	1.111	1.346	1.494	48
43	.6820	.7314	.9325	1.072	1.367	1.466	47
44	.6947	.7193	.9657	1.036	1.390	1.440	46
45	.7071	.7071	1.0000	1.000	1.414	1.414	45
	Cos	Sin	Cot	Tan	Cosec	Sec	Angle

APPENDIX D

International Atomic Masses

ATOMIC MASS	NAME OF CHEMICAL ELEMENT	SYMBOL	ATOMIC NUMBER
1.0079	Hydrogen	H	1
4.0026	Helium	He	2
6.941	Lithium	Li	3
9.0122	Beryllium	Be	4
10.811	Boron	B	5
12.0107	Carbon	C	6
14.0067	Nitrogen	N	7
15.9994	Oxygen	O	8
18.9984	Fluorine	F	9
20.1797	Neon	Ne	10
22.9897	Sodium	Na	11
24.305	Magnesium	Mg	12
26.9815	Aluminum	Al	13
28.0855	Silicon	Si	14
30.9738	Phosphorus	P	15
32.065	Sulfur	S	16
35.453	Chlorine	Cl	17
39.0983	Potassium	K	19
39.948	Argon	Ar	18
40.078	Calcium	Ca	20
44.9559	Scandium	Sc	21
47.867	Titanium	Ti	22
50.9415	Vanadium	V	23
51.9961	Chromium	Cr	24
54.938	Manganese	Mn	25
55.845	Iron	Fe	26
58.6934	Nickel	Ni	28

(*continued*)

ATOMIC MASS	NAME OF CHEMICAL ELEMENT	SYMBOL	ATOMIC NUMBER
58.9332	Cobalt	Co	27
63.546	Copper	Cu	29
65.39	Zinc	Zn	30
69.723	Gallium	Ga	31
72.64	Germanium	Ge	32
74.9216	Arsenic	As	33
78.96	Selenium	Se	34
79.904	Bromine	Br	35
83.8	Krypton	Kr	36
85.4678	Rubidium	Rb	37
87.62	Strontium	Sr	38
88.9059	Yttrium	Y	39
91.224	Zirconium	Zr	40
92.9064	Niobium	Nb	41
95.94	Molybdenum	Mo	42
98	Technetium	Tc	43
101.07	Ruthenium	Ru	44
102.9055	Rhodium	Rh	45
106.42	Palladium	Pd	46
107.8682	Silver	Ag	47
112.411	Cadmium	Cd	48
114.818	Indium	In	49
118.71	Tin	Sn	50
121.76	Antimony	Sb	51
126.9045	Iodine	I	53
127.6	Tellurium	Te	52
131.293	Xenon	Xe	54
132.9055	Cesium	Cs	55
137.327	Barium	Ba	56
138.9055	Lanthanum	La	57
140.116	Cerium	Ce	58
140.9077	Praseodymium	Pr	59
144.24	Neodymium	Nd	60
145	Promethium	Pm	61
150.36	Samarium	Sm	62
151.964	Europium	Eu	63
157.25	Gadolinium	Gd	64
158.9253	Terbium	Tb	65
162.5	Dysprosium	Dy	66
164.9303	Holmium	Ho	67
167.259	Erbium	Er	68
168.9342	Thulium	Tm	69
173.04	Ytterbium	Yb	70
174.967	Lutetium	Lu	71
178.49	Hafnium	Hf	72
180.9479	Tantalum	Ta	73
183.84	Tungsten	W	74
186.207	Rhenium	Re	75
190.23	Osmium	Os	76
192.217	Iridium	Ir	77

ATOMIC MASS	NAME OF CHEMICAL ELEMENT	SYMBOL	ATOMIC NUMBER
195.078	Platinum	Pt	78
196.9665	Gold	Au	79
200.59	Mercury	Hg	80
204.3833	Thallium	Tl	81
207.2	Lead	Pb	82
208.9804	Bismuth	Bi	83
209	Polonium	Po	84
210	Astatine	At	85
222	Radon	Rn	86
223	Francium	Fr	87
226	Radium	Ra	88
227	Actinium	Ac	89
231.0359	Protactinium	Pa	91
232.0381	Thorium	Th	90
237	Neptunium	Np	93
238.0289	Uranium	U	92
243	Americium	Am	95
244	Plutonium	Pu	94
247	Curium	Cm	96
247	Berkelium	Bk	97
251	Californium	Cf	98
252	Einsteinium	Es	99
257	Fermium	Fm	100
258	Mendelevium	Md	101
259	Nobelium	No	102
261	Rutherfordium	Rf	104
262	Lawrencium	Lr	103
262	Dubnium	Db	105
264	Bohrium	Bh	107
266	Seaborgium	Sg	106
268	Meitnerium	Mt	109
272	Roentgenium	Rg	111
277	Hassium	Hs	108
	Darmstadtium	Ds	110
	Ununbium	Uub	112
	Ununtrium	Uut	113
	Ununquadium	Uuq	114
	Ununpentium	Uup	115
	Ununhexium	Uuh	116
	Ununseptium	Uus	117
	Ununoctium	Uuo	118

APPENDIX E

Score Sheet

Number of questions correct: ____________

Less: 0.25 × number of questions wrong: ____________

(Remember that omitted questions are not counted as wrong.)

Raw score: ____________

RAW SCORE	SCALED SCORE	RAW SCORE	SCALED SCORE	RAW SCORE	SCALED SCORE	RAW SCORE	SCALED SCORE	RAW SCORE	SCALED SCORE
75	800	52	740	29	600	6	470	–17	300
74	800	51	730	28	590	5	460	–18	290
73	800	50	730	27	590	4	460	–19	290
72	800	49	730	26	580	3	460		
71	800	48	720	25	580	2	450		
70	800	47	720	24	570	1	450		
69	800	46	710	23	560	0	440		
68	800	45	710	22	560	–1	400		
67	800	44	700	21	550	–2	390		
66	800	43	690	20	540	–3	390		
65	800	42	680	19	540	–4	380		
64	800	41	680	18	530	–5	380		
63	790	40	670	17	520	–6	370		
62	790	39	670	16	510	–7	370		
61	780	38	660	15	510	–8	360		
60	780	37	650	14	500	–9	360		
59	770	36	650	13	490	–10	350		

Glossary

absolute zero — the lowest possible temperature at which all molecular motion would cease; 0 K

acceleration — a change in velocity divided by the time required for the change to occur

alpha decay — the spontaneous emission of an alpha particle, which is a helium nucleus, by certain radioactive substances

ammeter — a device used to measure current

amplitude — the maximum displacement of the particles of a medium

angular acceleration — the rate of change in angular velocity over time

angular displacement — a vector measure of the rotation of an object about an axis

angular momentum — the product of the mass, velocity, and radius of motion

angular velocity — the rate of change of angular displacement with respect to time

Archimedes' Principle — the rule stating that an object immersed in a fluid will experience a buoyant force equal to the weight of the fluid it displaces

atomic number — the number of protons in the nucleus of an atom

average speed — the total distance an object travels divided by the time during which it traveled that distance

average velocity — the total change in displacement divided by the time during which the displacement occurred

balanced forces — a combination of forces on an object that result in a net force of zero

beat — the interference caused by two sets of sound waves with only slightly different frequencies

Bernoulli's Principle	the rule stating that as the speed of a moving fluid increases, the pressure exerted within the fluid decreases
beta decay	the spontaneous emission of an electron or a positron by certain radioactive substances
binding energy	the energy required to break apart an atomic nucleus; the difference in energy between the nucleons when they are separate and when they are bound together
Boyles' Law	the law that states that volume is inversely related to pressure if temperature is held constant
buoyant force	the upward force exerted by a fluid on an object placed in it
capacitor	an electrical device used to store charge
center of mass	the point at which all the mass of an object can be considered to be located
centripetal acceleration	the rate of change in velocity of a mass moving uniformly in a circle at constant speed; directed radially inward toward the center of the circular path
centripetal force	the force directed inward along the radius of an object's circular path
Charles' Law	the law that states that volume is directly proportional to temperature if pressure is held constant
circuit	a closed path of conducting materials through which electricity can flow
coefficient of friction	the ratio of the friction force between two surfaces and the normal force between those surfaces
concave lens	a converging lens; a transparent piece of glass or plastic that is thicker in the middle than at the ends
concave mirror	a converging mirror; a reflective surface that curves away from an object
conduction	the transfer of heat between materials that are in contact with one another
constructive interference	the phenomenon that occurs when two waves combine in such a way that the amplitude of the resulting wave is greater than either of the two individual waves
convection	the transfer of heat by the movement of a heated substance, such as by currents in a fluid
convex lens	a diverging lens; a transparent piece of glass or plastic that is thicker at the ends than in the middle

convex mirror — a diverging mirror; a reflective surface that bulges toward the object

Coulomb's Law — the law that states that the magnitude of the force between charged particles is proportional to the product of the two charges and varies inversely as the square of the distance between them

critical angle — the angle of incidence at which the angle of refraction is 90° relative to the normal

cross product — the combining of vectors in such a way that the result is a vector

current — the amount of charge moving through a conductor per second

density — the mass of a substance divided by its volume

destructive interference — the phenomenon that occurs when two waves combine in such a way that the amplitude of the resulting wave is less than either of the two individual waves

diffraction — the bending of waves around an obstacle or through an opening

direct relationship — a correlation in which two variables increase or decrease simultaneously

displacement — a change in position of an object, which is a vector quantity

distance — the length between two points, which is a scalar quantity

domain — a large group of atoms with net spins that align

Doppler effect — a relative change in frequency due to motion of a sound of source or its observer

dot product — the combining of vectors in such a way that the result is a scalar

elastic potential energy — the stored potential energy resulting when an object is deformed or distorted, such as a compressed spring

electric field intensity — the magnitude of the electric field

electric field — the force exerted on a charged particle by a charged object in the region around the object

electric potential — the amount of work per unit charge required to move a charge from infinity to another point in an electric field

electromagnet — a solenoid with an iron core, which forms a magnet that can be controlled

electromagnetic induction — the process through which an electric current can be induced by a changing magnetic field

electromagnetic spectrum the arrangement of electromagnetic waves in order of wavelength

electromagnetic wave a traveling disturbance produced by vibrating charges; travels as a series of vibrating electric and magnetic fields at right angles to one another

electromagnetism the relationship between electricity and magnetism

electroscope an instrument used to determine the presence of small electric charges

energy the ability to do work or cause change to a system

entropy a measure of the disorder of a system

force a push or a pull

free-body diagram a picture used to compare the direction and magnitude of the forces exerted on an object

frequency the number of revolutions or waves per unit time

friction the resistive force that opposes the motion of an object as a result of the contact between two surfaces

gamma decay the spontaneous emission of high-energy photons by certain radioactive substances

generator a device that transforms mechanical energy into electrical energy

gravitational constant the constant G, which has a value of 6.67×10^{-11} $N \cdot m^2/kg^2$, and is used in the equation to calculate the gravitational force

gravitational force the force of attraction between any pair of objects as a result of their masses

gravitational potential energy the energy an object has because of its position in a gravitational field

half-life the time it takes for half of the radioactive nuclei in a sample to decay

heat energy the transfer of thermal energy from a warmer substance to a cooler substance

heat engine a device that converts heat to mechanical energy by doing work

heat of fusion the amount of energy required to change a unit mass of a substance from a solid to a liquid at the melting point

heat of vaporization the amount of energy required to change a unit mass of a substance from a liquid to a gas at the boiling point

impulse the product of the average force exerted on a mass and the time interval over which the force is exerted; vector quantity

index of refraction	the ratio of the speed of light in a vacuum with the speed of light in a specific medium
indirect relationship	a correlation in which the dependent variable changes opposite to the independent variable
instantaneous velocity	the velocity of an object at a particular instant
internal energy	the total amount of energy of the particles and includes potential energy in addition to kinetic energy
isotopes	atoms of the same element with different numbers of neutrons
joule	the unit of energy equal to one newton-meter
Joule's Law of Heating	the law that states that the heat produced is directly proportional to the square of the current, the resistance, and the time
kinetic energy	the energy an object has as a result of its motion
kinetic friction	the resistive force that opposes the movement of an object already in motion
kinetic theory	the description of matter as being made up of small particles that are in constant motion
law of charges	the law that states that unlike charges attract one another and like charges repel one another
law of conservation of energy	the law that states that the total energy of a system remains constant; no new energy is created and no energy is destroyed
length contraction	the phenomenon in which an observer at rest relative to a moving object traveling at relativistic speeds would observe the length of the object to be shorter than it would be at rest relative to the observer
longitudinal wave	a wave in which the particles of the medium vibrate parallel to the direction of motion of the wave
mass defect	the difference between the mass of the unbound nucleons and the mass of the bound nucleons
mechanical energy	the total of the potential and kinetic energy of a system
mechanical wave	a traveling disturbance that requires a medium through which to travel
momentum	the product of the mass and velocity of a moving object; vector quantity
motor	a device that transforms electrical energy into mechanical energy
natural frequency	the frequency at which a system that has been disturbed will vibrate once there are no disturbing forces

normal force	the reaction force exerted on an object by a surface in contact with the object
nuclear decay	the process through which unstable nuclei release particles or energy in order to gain stability
nuclear fission	a nuclear reaction in which one large nucleus splits to form smaller nuclei
nuclear fusion	a nuclear reaction in which two smaller nuclei combine to form a larger one
Ohm's Law	the law that states that resistance is directly proportional to voltage and indirectly proportional to current
parallel circuit	a circuit through which current can follow multiple paths
Pascal's Principle	the rule stating that any external pressure applied to a confined static fluid is distributed uniformly throughout the fluid
photoelectric effect	the phenomenon in which electrons along the surface of a metal are emitted when electromagnetic energy with a certain minimum frequency is incident on the metal
photon	a discrete bundle, or quantum, of electromagnetic energy
pitch	the characteristic of sound that describes how high or low it is perceived; determined by the frequency of the sound wave
potential energy	the energy an object has as a result of its position or condition
power	the rate at which work is done or energy is used
pressure	the force per unit area
radiation	the transfer of heat by electromagnetic waves, or particles given off by radioactive substances
radioactivity	the spontaneous emission of radiation
real image	an image formed by a mirror or lens that exists where rays meet, and which therefore can be focused on a screen
reflection	the process through which light bounces off a surface
refraction	the bending of a light ray when it passes at an angle from one medium to another
resistance	the opposition to the flow of charges offered by a material
resonance	the tendency of a system to vibrate at a greater amplitude at some frequencies than others

resonant frequency a natural vibrating frequency of an object

right-hand rule a method of finding the direction of a magnetic field around a current-carrying wire or the force acting on a wire or charge in magnetic field

scalar quantity a quantity that is described by magnitude without regard to direction; examples include mass and temperature

scientific notation a shorthand notation for writing very large or small numbers using a coefficient that is greater or equal to 1 and less than 10 multiplied by base 10 to an exponent

series circuit a circuit through which current can flow through each element without branching

significant figures digits in a number that are known with some degree of certainty

Snell's Law of Refraction the law that states that the ratio of the sine of the angle of incidence to the sine of the angle of refraction is equal to the relative index of refraction for the two mediums

solenoid a series of closely spaced coils of wire

special theory of relativity the theory proposed by Einstein that suggests that the speed of light is constant in all reference frames, despite any relative motion between an observer and the light source, and the laws of physics are the same in all inertial reference frames

specific heat the amount of heat required to raise the temperature of one mass unit by one degree Celsius

standing wave a stationary wave pattern formed when two sets of waves with equal wavelength and amplitude interfere

static friction the resistive force that opposes the start of motion between two surfaces in contact

strong force a short-range force that holds the nucleons of an atom together despite the like charges of the protons

temperature the average kinetic energy of the molecules in a sample of matter

thermal energy the total kinetic energy of all the particles in a sample of matter

threshold frequency the minimum frequency of the incident electromagnetic energy on a metal for the photoelectric effect to be observed

time dilation	the phenomenon in which an observer at rest relative to a moving object traveling at relativistic speeds would observe the length of time to be longer than it would be at rest relative to the observer
torque	the tendency of a force to cause rotation about an axis, which is measured as the product of the force and the length of the lever arm
total internal reflection	the process in which light incident on the boundary between two mediums, passing from the medium with the higher index of refraction, is reflected back into the original medium
transverse wave	a wave in which the particles of the medium vibrate perpendicular to the direction of motion of the wave
unbalanced forces	a combination of forces on an object that do not result in a net force of zero
uniform circular motion	motion around a circle at a constant speed
uniformly accelerated motion	motion in which an object accelerates at a constant rate
vector quantity	a quantity that is described by both magnitude and direction; examples include force and acceleration
virtual image	an image formed by a mirror or lens that exists where rays appear to meet, but do not, and therefore cannot be focused on a screen
voltage	potential difference, or the amount of electric potential
voltmeter	a device used to measure voltage
wave	a disturbance that carries energy from one location to another
wavelength	the distance between similar points on consecutive waves
weight	the product of mass and the acceleration due to gravity at a particular location
work	the product of the force exerted on an object and the distance the object moves as a result of that force
work-energy theorem	the idea that the work done on a system or by a system equals the change in the energy of the system